Springer Series on Bio- and Neurosystems

Volume 10

The Springer Series on Bio- and Neurosystems publishes fundamental principles and state-of-the-art research at the intersection of biology, neuroscience, information processing and the engineering sciences. The series covers general informatics methods and techniques, together with their use to answer biological or medical questions. Of interest are both basics and new developments on traditional methods such as machine learning, artificial neural networks, statistical methods, nonlinear dynamics, information processing methods, and image and signal processing. New findings in biology and neuroscience obtained through informatics and engineering methods, topics in systems biology, medicine, neuroscience and ecology, as well as engineering applications such as robotic rehabilitation, health information technologies, and many more, are also examined. The main target group includes informaticians and engineers interested in biology, neuroscience and medicine, as well as biologists and neuroscientists using computational and engineering tools. Volumes published in the series include monographs, edited volumes, and selected conference proceedings. Books purposely devoted to supporting education at the graduate and post-graduate levels in bio- and neuroinformatics, computational biology and neuroscience, systems biology, systems neuroscience and other related areas are of particular interest.

The books of the series are submitted for indexing to Web of Science.

More information about this series at http://www.springer.com/series/15821

Jose L. Contreras-Vidal •
Dario Robleto • Jesus G. Cruz-Garza •
José M. Azorín • Chang S. Nam
Editors

Mobile Brain–Body Imaging and the Neuroscience of Art, Innovation and Creativity

Springer

Editors
Jose L. Contreras-Vidal
NSF IUCRC BRAIN Center
University of Houston
Houston, TX, USA

Jesus G. Cruz-Garza
Electrical and Computer Engineering Department
University of Houston
Houston, TX, USA

Chang S. Nam
Department of Industrial and Systems Engineering
North Carolina State University
Raleigh, NC, USA

Dario Robleto
Cullen College of Engineering
University of Houston
Houston, TX, USA

José M. Azorín
Departamento de Ingeniería de Sistemas y Automática
Universidad Miguel Hernández de Elche
Elche, Alicante, Spain

ISSN 2520-8535 ISSN 2520-8543 (electronic)
Springer Series on Bio- and Neurosystems
ISBN 978-3-030-24325-8 ISBN 978-3-030-24326-5 (eBook)
https://doi.org/10.1007/978-3-030-24326-5

This Springer imprint is published by the registered company Springer Nature Switzerland AG
The registered company address is: Gewerbestrasse 11, 6330 Cham, Switzerland

Series Editor's Foreword

Studying brain activities when people create art or perceive art is a fascinating area that opens new directions in bio- and neuro systems research and related technologies.

Perhaps the citation referenced in this book: "The uncertain nature of art has its advantages. It leads to constant experiment and questioning" (Harold Rosenberg, 1972) best describes the aim and the achievement of the book.

These are questions that research in bio- and neuro systems need to answer in the future, such as:

- How does the brain work differently when people create paintings or perceive pictures?
- How does the brain work when people create or perceive music?
- What is the healing power of art and how it can be utilised for mental health problems?
- Can people create paintings and music through their brain signals only using brain–computer interfaces (BCI)?
- What is the role of neurotechnology in art and the impact of art on new neurotechnology?
- How can we understand the dynamic interaction between biological molecules, like antibodies, brain activities and creativity?
- How people synchronise their brain activities when communicate between each other and what is the advantage and the disadvantage of that?

These and many other questions are addressed in this book from different perspectives, such as personal experience, scientific experiments, visual presentation, commentaries, open questions, speculations for the future and that makes the book an interesting reading setting new challenges to science, art, technology and the society.

Nikola Kasabov
Series Editor
Professor of Neural Networks and Knowledge Engineering
Knowledge Engineering and Discovery Research Institute
Auckland University of Technology
New Zealand

Foreword

The limits of my language are the limits of my world.
—Ludwig Wittgenstein

Based on the lives we live, we all speak a multitude of languages. The variances in our professional languages can seem as simple as differences in accents or as complicated as learning a new alphabet.

Exploring the intersections between art and neuroscience can seem to require fluency in two dramatically different languages. Although the languages may differ, deep similarities exist between the disciplines. Both incorporate experimentation, investigation of humanity's biggest questions, and perpetual work to understand more fully our world and the lives of those who inhabit it. Recognizing these similarities, despite any linguistic differences, those who learn the language of the other and work in the shared space between the arts and neuroscience can more fully examine and celebrate such elusive topics as creativity, understanding, memory, and motion.

Working with art museum visitors, I frequently witness the wonders of artistic experience. Through our teen artist program, I observe students visually rendering and emotionally processing their lives through the medium of paint, sculpture, or photography. During a bilingual toddler story time, I watched children form language and create meaning through the shapes and colors depicted in the art around them. Leading a tour for visitors with dementia, I witnessed participants who respond to the artwork with dramatically increased verbal engagement and who are able to form and access memories in ways not possible outside of the museum.

Through these observations and countless others, I am a researcher. I research the ability of the arts to inform and reflect the lives of visitors. Yet, my research focuses on observation and lacks the element of explanation. As described by Juliet King, Associate Professor in the Art Therapy Department at The George Washington University and Adjunct Associate Professor in the Department of Neurology at the Indiana University School of Medicine, building a more complete understanding will require the observations of educators, artists, and art therapists *alongside* the research and observations of colleagues in the neurosciences.[1] This shared work will determine how it is that art plays a significant role in making meaning and in recovering memory. In the space between art and science, current and future practitioners and researchers will determine how artistic practices influence and are influenced by neurology.

A frequent obstacle to collaborations between art and neuroscience has been the too often disconnected spaces in which the disciplines work. As institutions are striving to break down isolated work in academics, university museums can provide one public space for scientific experimentation. In their 2018 talk "Museum as Laboratory", artist Dario Robleto and Professor Jose L. Contreras-Vidal, Ph.D., Director of the Noninvasive Brain–Machine Interface

[1]King, Juliet. "A Revitalized Synthesis: Art Therapy, Neuroscience and Mobile Brain–Body Imaging." 49th Annual Conference, American Art Therapy Association, November 1, 2018, Miami, FL. Conference Presentation.

Systems Lab at the University of Houston, spoke about their collaborative experiment at The Menil Collection in Houston in 2016 and the benefits to using Mobile Brain–Body Imaging technology to study artistic experiences in public spaces.[2] For scientists, working in a museum, rather than a traditional laboratory setting, provides real-life experiences to monitor and evaluate. For the museum, public experiments can expand the educational role of the institution and create opportunities for visitors to learn about current brain imaging technologies, as well as consider—and possibly observe—this intersection between their art experience and their neurology. At a university art museum, these benefits multiply as students at all levels of learning—undergraduate through post-doctoral—engage in the process.

The realities of this combined work are as complicated as they are critical. Recognizing the logistical barriers to interdisciplinary work and the immense rewards that collaborative projects can offer, the *2016* and *2017 International Conferences on Mobile Brain–Body Imaging and the Neuroscience of Art, Innovation, and Creativity* created opportunities for experts in both fields to convene. Thought-leaders and practitioners exploring connections between art, neuroscience, engineering, media, industry, education, and medicine assembled to share research and knowledge, as well as to identify challenges and opportunities of their work. Most importantly, the community that gathered developed shared plans for future experimentation and exploration that supports cooperative efforts between disciplines.

This book represents dedicated work and the enthusiastic spirit of these convenings. Like the conferences, this text celebrates a multitude of backgrounds and expertise, giving equal significance to the scientific theory and evidence represented, as well as holding critical space for the artistic experience and representation. The impact of this work on educational settings is spotlighted and is one of several case studies on how this work directly impacts individuals and communities.

Most significantly, this wide-ranging and deeply collaborative text encourages all readers to learn from these critical partnerships, to speak multiple languages, and to join the conversation.

Jessica Kay Ruhle
Director of Education and Public Programs
Nasher Museum of Art at Duke University
Durham, United States

[2]Robleto, Dario and Jose L. Contreras-Vidal. "Museum as Laboratory". Nasher Museum of Art at Duke University, January 31, 2018, Durham, NC. Public Lecture.

Acknowledgements

This book is based upon work supported by the National Science Foundation (NSF) under Grants No. IIS-1631608 and IIS-1745835 (Division of Information and Intelligent Systems) to Prof. Jose L. Contreras-Vidal. Specifically, the editors would like to acknowledge the support of the NSF for the Doctoral Consortia at the *2016* and *2017 International Conferences on Mobile Brain–Body Imaging and the Neuroscience of Art, Innovation and Creativity* held in Cancun (July 24–27, 2016) and Valencia (September 10–13, 2017), and the preparation of this book. Additional support by the NSF under Grants No. IIS-1421948 and BCS-1551688 to Dr. Chang S. Nam is also acknowledged. This material was also supported in part by Brain Pool program funded by the Ministry of Science and Information and Communications Technology (MSICT) through the National Research Foundation of Korea (award No. 2018H 1D3A2001409). Any opinions, findings, and conclusions or recommendations expressed in this material are those of the author(s) and do not necessarily reflect the views of the NSF or MSICT.

The participation of international faculty and students attending the *2016* and *2017 International Conference on Mobile Brain–Body Imaging and the Neuroscience of Art, Innovation and Creativity* (Valencia, Spain) was supported in part by the Office of Naval Research Global Award No. CSP - N62909-16-1-2065 and N62909-17-1-2191. Any opinions, findings, and conclusions or recommendations expressed in this material are those of the author's and do not necessarily reflect the views of the Office of Naval Research. Support is also acknowledged from the Conselleria de Educación, Investigación, Cultura y Deporte—Generalitat Valenciana, Spain (No. AORG/2017/123). Spanish researchers were supported in part by the Universidad Miguel Hernández de Elche, Spain. Special thanks to the Institut Valencià d'Art Modern (IVAM, Valencia, Spain).

Participation of art faculty, visual artists and performing artists, and students from the arts and humanities was made possible by funding from the Division of Research, the Cullen College of Engineering and the Kathrine G. McGovern College of the Arts at the University of Houston, and the Alliance for the Arts in Research Universities (a2ru). Additional support was received from the NSF Industry-University Cooperative Research Center for Building Reliable Advances and Innovations in Neurotechnology (IUCRC BRAIN) at the University of Houston.

The editors recognize the support of the Brain–Computer Interface Hackathon (including the BR41N.IO—The BCI Designers Hackathon) held at the 2017 Valencia Meeting by the IEEE Brain Initiative, the IEEE Systems, Man, and Cybernetics Society, g.Tec (GmbH, Austria), and Brain Products (GmbH, Gilching, Munich, Germany). Additional support from the following companies is gladly acknowledged: Technaid (Arganda del Rey, Madrid, Spain), Emotiv (San Francisco, USA), Muse (Ontario, Canada), and OpenBCI (Brooklyn, NY, USA).

Finally, the Editors would like to acknowledge the students from the Laboratories for Brain–Machine Interface Systems at the University of Houston (USA) and the Universidad Miguel Hernández de Elche (Spain) for their logistic support, as well as to all participants of both conferences.

Jose L. Contreras-Vidal
Dario Robleto
Jesus G. Cruz-Garza
José M. Azorín
Chang S. Nam

Contents

Part V How Disruptive Neurotechnologies Are Changing Science, Arts, and Innovation

Part VI How Can the Arts and Neuroscience Describe and Promote the Processes of Learning and Creativity in K-12 and Higher Education?

Part VII The Future of Neuroaesthetics

Editors and Contributors

About the Editors

Jose L. Contreras-Vidal Ph.D., IEEE Fellow, is Cullen Distinguished Professor of Electrical and Computer Engineering, and Director of the NSF Industry-University Collaborative Research Center on Building Reliable Advances and Innovations in Neurotechnologies (IUCRC BRAIN) at the University of Houston. He has pioneered the development of non-invasive brain–machine interfaces to control robots, exoskeletons, virtual avatars, and prosthetic limbs to restore motor function in individuals with spinal cord injury, stroke, and limb amputation. He has also pioneered the concept of "The Museum and the Classroom as Laboratories" to understand the brain in action and in context of free-behaving individuals, with applications to medical devices, art therapy, aesthetics, neuroscience, and regulatory science. His work has been supported by the NIH, NSF, VA, DARPA, ONR, Foundations, donors, and industry. Dr. Contreras-Vidal's research has been highlighted by The Economist, Nature, Science, Science News, Der Spiegel, NSF, Wall Street Journal, SFN, O&P, Scientific American, NPR's Science Friday, and Neurology Today among others. His career development in biomedical engineering has been highlighted in the magazine Science.
Webpage: http://www.ece.uh.edu/faculty/contreras-vidal
Webpage: https://www.houstonmethodist.org/faculty/joseluis-contreras-vidal/
IUCRC BRAIN: http://brain.egr.uh.edu/

Dario Robleto is a transdisciplinary artist based in Houston, Texas. He currently serves as Artist-in-Residence in Neuroaesthetics at the University of Houston's Cullen College of Engineering and as Artist-at-Large at Northwestern University's McCormick School of Engineering. In 2016, he was appointed as the Texas State Artist Laureate.

Jesus G. Cruz-Garza is a Ph.D. candidate in Electrical Engineering at the University of Houston and Graduate Fellow on Translational Research with Houston Methodist Research Institute and University of Houston. He graduated from BS Engineering Physics at Tecnologico de Monterrey, Mexico.

José M. Azorín is the Director of the Brain–Machine Interface Systems Lab and Full Professor at Miguel Hernández University of Elche (UMH), Spain. He holds a Ph.D. in Robotics. He has been a visiting professor at the University of Houston (USA) and at Imperial College London (United Kingdom).

Chang S. Nam is a professor in the Department of Industrial and Systems Engineering at North Carolina State University. His research interests include brain–computer interfaces and cognitive neuroscience. He is the editor-in-chief of the journal "Brain–Computer Interfaces" and the main editor of the book "Brain–Computer Interfaces Handbook: Technological and Theoretical Advances" (CRC Press, 2018).

Contributors

Hassan Aleem Interdisciplinary Program in Neuroscience, Georgetown University, Washington D.C, USA

Luis Antonio Mercado Cerda Universidad Autónoma de Nuevo León, FIME, San Nicolás de Los Garza, NL, Mexico

José M. Azorín Brain–Machine Interface Systems Lab, Miguel Hernández University of Elche, Elche, Alicante, Spain

Laurie Baefsky College of Arts & Media, University of Colorado Denver, Denver, CO, USA

Janet Biggs Independent Artist, Brooklyn, NY, USA

Anthony Brandt Shepherd School of Music, Rice University, Houston, TX, USA

Emilie V. Brotherhood Dementia Research Centre, UCL Queen Square Institute of Neurology, University College London, London, UK;
Created Out of Mind, The Hub, Wellcome Collection, London, UK

Paul M. Camic Created Out of Mind, The Hub, Wellcome Collection, London, UK;
Salomons Centre for Applied Psychology, Canterbury Christ Church University, Tunbridge Wells, Kent, UK

Aaron Colverson Department of Ethnomusicology, University of Florida, Gainesville, FL, USA

Jose L. Contreras-Vidal IUCRC BRAIN, University of Houston, Houston, TX, USA

Ivan Correa-Herran Department of Neuroscience, Georgetown University, Washington D.C, USA;
Facultad de Artes, Universidad Nacional de Colombia, Bogotá, Colombia

Sebastian J. Crutch Dementia Research Centre, UCL Queen Square Institute of Neurology, University College London, London, UK;
Created Out of Mind, The Hub, Wellcome Collection, London, UK

Jesus G. Cruz-Garza IUCRC BRAIN, University of Houston, Houston, TX, USA

Brian Day Created Out of Mind, The Hub, Wellcome Collection, London, UK;
Whole-Body Sensorimotor Lab, Department of Clinical and Motor Neurosciences, UCL Queen Square Institute of Neurology, University College London, London, UK

María José Delgadillo Department of Hispanic Studies, University of Houston, Houston, USA;
IUCRC BRAIN, University of Houston, Houston, USA

Arne Dietrich Department of Psychology, American University of Beirut, Beirut, Lebanon

Jason L. Eriksen College of Pharmacy, University of Houston, Houston, TX, USA

Shane Esola Department of Mechanical Engineering and Mechanics, Drexel University, Philadelphia, PA, USA

Jo Ann Fleischhauer Cullen College of Engineering, University of Houston, Houston, TX, USA

Jorge A. Gaxiola-Tirado CINVESTAV, Monterrey's Unit, Apodaca, NL, Mexico

Lukas Gehrke Biological Psychology and Neuroergonomics, Technische Universität Berlin, Berlin, Germany

Stephen Grossberg Departments of Mathematics & Statistics, Psychological & Brain Sciences, and Biomedical Engineering, Center for Adaptive Systems, Boston University, Boston, MA, USA

Norberto M. Grzywacz Interdisciplinary Program in Neuroscience, Georgetown University, Washington D.C, USA;
Department of Neuroscience, Georgetown University, Washington D.C, USA;
Department of Physics, Georgetown University, Washington D.C, USA;
Graduate School of Arts and Sciences, Georgetown University, Washington D.C, USA

Christoph Guger g.tec Medical Engineering GmbH, Graz, Austria

Mariale M. Hardiman John Hopkins University School of Education, Baltimore, MD, USA

Charles R. Harrison Created Out of Mind, The Hub, Wellcome Collection, London, UK

Lauren Hayes Arts, Media and Engineering, Arizona State University, Tempe, AZ, USA

Guillermo Herrera-Arcos Tecnologico de Monterrey, School of Engineering and Science, Monterrey, Mexico

Noah Hutton Independent Scholar, New York, USA

Eduardo Iáñez Brain–Machine Interface Systems Lab, Miguel Hernández University of Elche, Elche, Alicante, Spain

Girija Kaimal Creative Arts Therapies Department, Drexel University, Philadelphia, PA, USA

Anastasiya E. Kopteva IUCRC BRAIN, University of Houston, Houston, TX, USA;
Department of Theater and Dance, University of Houston, Houston, TX, USA

Adam Lopez School of Biological Sciences, University of Missouri-Kansas City, Kansas City, KS, USA

Silvia Moreno Brain–Machine Interface Systems Lab, Miguel Hernandez University of Elche, Elche, Alicante, Spain

Chang S. Nam Department of Industrial and Systems Engineering, North Carolina State University, Raleigh, NC, USA

Mario Ortiz Brain–Machine Interface Systems Lab, Miguel Hernández University of Elche, Elche, Alicante, Spain

Daniel Pimentel Media Effects and Technology Lab, University of Florida, Gainsville, FL, USA

Maria Pombo Department of Psychology, Georgetown University, Washington D.C, USA

Dario Robleto Artist in Residence in Neuroaesthetics, IUCRC BRAIN, University of Houston, Houston, TX, USA

Stephanie M. Scott Department of Media Communications, Colorado State University, Fort Collins, CO, USA

Jill Sonke Center for Arts in Medicine, University of Florida, Gainesville, FL, USA

Justin Tanner School of Biological and Health Systems Engineering, Arizona State University, Tempe, AZ, USA

Tess Torregrosa Chemical Engineering Department, Northeastern University, Boston, MA, USA

Kenneth Veldman Vlamm Inc., Groningen, The Netherlands

Melissa S. Walker National Intrepid Center of Excellence, Walter Reed National Military Medical Center, Bethesda, MD, USA;
Creative Forces: NEA Military Healing Arts Network, National Endowment for the Arts, Washington, D.C., USA

Anthony J. Woods Wellcome, London, UK

Andreas Wulff-Abramsson Department of Architecture, Design and Media Technology, Aalborg University Copenhagen, Copenhagen, Denmark

Jokūbas Žiburkus Department of Biology and Biochemistry, University of Houston, Houston, TX, USA

Abbreviations

a2RU	Alliance for the Arts in Research Universities
ADHD	Attention-deficit/hyperactivity disorder
aFILM	Anchored Filling-In Lightness Model
AHLTA	Armed Forces Health Longitudinal Technology Application
AI	Artificial Intelligence
AIR	Artist-in-residence
AoW	Art of War
AR	Augmented Reality
ART	Adaptive Resonance Theory
ATR-N	Art Therapy Relational Neuroscience
AUT	Alternative Uses Test
B2AI	Brain to Art interfaces
BCI	Brain–Computer Interface
BHLAW	Blurred highest luminance as white
BOA conference	Brain on Art conference, the short name used for the 2016 and 2017 International Conferences on Mobile Brain–Body Imaging and the Neuroscience of Art, Innovation and Creativity held in Cancun (July 24–27, 2016) and Valencia (September 10–13, 2017)
BRAIN Initiative	Brain Research through Advancing Innovative Neurotechnologies Initiative
CAD	Computer-aided design
CARE	Choices, Agenda, Resources, and Emotions
CCPT	Child Centered Play Therapy
CREATE	Creative Embodiment, Relational Resonating, Expressive Communicating, Adaptive Responding, Transformative Integrating, and Empathizing and Compassion
DARPA	Defense Advanced Research Projects Agency, USA
DASER	D.C. Art Science Evening Rendezvous
DMI	Digital musical instrument
DMN	Default Mode Network
DoD	Department of Defense, USA
DSP	Digital signal processes
EEG	Electroencephalography/Electroencephalogram
EMG	Electromyography
ERP	Event-Related Potential
ERSP	Event-related spectral perturbations
ESSA	Elementary and Secondary Education Act, the Every Student Succeeds Act
ETC	Expressive therapies continuum
FAÇADE	Form-And-Color-And-DEpth
FFT	Fast Fourier Transform

fMRI	Functional magnetic resonance imaging
FTD	Frontotemporal dementia
GAD-7	Generalized Anxiety Disorder 7
HCI	Human-computer interaction
HD	High definition
IEEE Brain Initiative	Institute for Electrical and Electronics Engineers Brain Initiative
IOP	Intensive Outpatient Program
IRB	Institutional Review Board
ISO	Intrepid Spirit One
IT	Inferotemporal cortex
IUCRC BRAIN	NSF Industry-University Cooperative Research Center for Building Reliable Advances and Innovation in Neurotechnology
LASER	Leonardo Art Science Evening Rendezvous
LGN	Lateral geniculate nucleus
LH	Left hemisphere
MEG	Magnetoencephalography
MoBI	Mobile Brain–Body Imaging, also referred to as Mobile Brain/Body Imaging
mPFC	Medial prefrontal cortex
MRI	Magnetic resonance imaging
mTBI	Mild Traumatic brain injury
NAFKI	National Academies Keck Future Initiative
NEA	National Endowment for the Arts, USA
NICoE	National Intrepid Center of Excellence
NOAH	National Organization for Arts in Health, USA
NSF	National Science Foundation, USA
OEF	Operation Enduring Freedom
OFC	Orbitofrontal cortex
OIF	Operation Iraqi Freedom
PARIESA	Practice and Research in Enactive Sonic Art
PCA	Posterior Cortical Atrophy
PCL-M/C	PTSD checklist-military and civilian
PDC	Partial Directed Coherence
PET	Positron emission tomography
PFC	Prefrontal cortex
PH	Psychological health
PHQ-9	Patient Health Questionnaire-9
PPA	Primary progressive aphasia
PPC	Posterior parietal cortex
PPI	Patient public involvement
PSD	Power spectral density
PTSD	Post traumatic stress disorder
qEEG	Quantitative electroencephalography
REM	Rapid eye movement
RGB	Red, green, blue
RH	Right hemisphere
SAR	Synthetic aperture radar
SM	Service Member
STEAM	Science, technology, engineering, art, and mathematics
STEAMM	Science, technology, engineering, arts, math, and medicine
STEM	Sciente, technology, engineering, and mathematics
SVM	Support vector machine

TBI	Traumatic brain injury
TTCT	Torrance Test of Creative Thinking
UDP	User Datagram Protocol
VA	Veteran Affairs
VR	Virtual Reality
WPA	Works Progress Administration

Introduction: The Confluence of Art, Neuroscience, and Creativity Through Mobile Brain–Body Imaging

Jose L. Contreras-Vidal, Jesus G. Cruz-Garza, Dario Robleto, José M. Azorín, and Chang S. Nam

Creativity and the experience of aesthetic reflection are two of the most profound mysteries of the human brain, both enabling us to continually innovate through problem-solving and express complex emotions that help define what it means to be human. The burgeoning field of neuroaesthetics offers a unique possibility to work in a genuinely interdisciplinary way, revealing a multilayered understanding of art and the brain. This book emerges from the *International Conferences on Mobile Brain–Body Imaging (MoBI) and the Neuroscience of Art, Innovation and Creativity*, the so-called Brain on Art conferences, held in Cancun, Mexico (2016) and Valencia, Spain (2017), respectively, to explore these topics. This book represents an intertwining of disciplines that investigate not only their products—art and data—but also something more substantive and unique, as we argue for the vital importance of lasting collaboration and dialogue between our fields.

Recognizing the increasingly cross-disciplinary nature of many scientific, artistic, educational, and medical challenges of our time, the Brain on Art conferences aimed to identify the opportunities for collaboration between these respective fields. Such partnerships promote innovation and novel problem-solving by challenging disciplines to think outside their area. Many topics were explored by both scientists and artists such as an overview of the field of neuroaesthetics; the advancements of MoBI technology in studying creativity in action and in context; neuroeducation; ongoing efforts to understand the brain through reverse engineering; engineering personalized creative art therapies; and a call for the value in artists and scientists working to engage the public's interest and involvement in cutting edge neuroscience. Additionally, various interactive programs at the nexus of the arts and sciences were designed to demonstrate the possibilities of these cross-disciplinary collaborations. For example, in an ongoing collaboration that incorporates the tradition of artist-designed games and "actions", an experimental design model for brain imaging and acquisition was performed. With the conference's emphasis on building the infrastructure to sustain long term, outside the box collaboration, a Doctoral/Postdoctoral Consortium Program was run in parallel with the single-track conference. This allowed trainees from the arts, science, and engineering fields to explore and develop their research interests in a workshop guided by a multidisciplinary panel of distinguished researchers, artists, and innovators. Moreover, the consortium provided the following:

- A curated setting where students and trainees could present their work in poster format and meet other students while engaging with established researchers from around the world.
- A rare opportunity for students to receive guidance and feedback on their current research from experts outside their field, promoting networking and career development.
- An opportunity to contribute to the conference goals through active participation and interaction with other students and researchers.
- Research collaboration and exploration at the nexus of the arts and science through the Brain–Computer Interface (BCI) Designers Hackathon.

The long-term goal of the Brain on Art conference series is to develop a Strategic Plan or Roadmap (refer to Part VII of this book for details) that (1) provides global leadership on collaboration between the creative arts, science, engineering, medicine and the humanities, (2) advances health

J. L. Contreras-Vidal (✉) · J. G. Cruz-Garza
IUCRC BRAIN, University of Houston, Houston, TX, USA
e-mail: jlcontreras-vidal@uh.edu

D. Robleto
Artist in Residence in Neuroaesthetics, IUCRC BRAIN, University of Houston, Houston, TX, USA

J. M. Azorín
Miguel Hernández University of Elche, Elche, Alicante, Spain

C. S. Nam
Department of Industrial and Systems Engineering, North Carolina State University, Raleigh, NC, USA

J. L. Contreras-Vidal et al. (eds.), *Mobile Brain–Body Imaging and the Neuroscience of Art, Innovation and Creativity*, Springer Series on Bio- and Neurosystems 10, https://doi.org/10.1007/978-3-030-24326-5_1

and education, (3) innovates engineering tools for the exploration of the brain in action and context, (4) transforms the design of living and working environments, and (5) accelerates innovation, global relations, and worldwide quality of life. To achieve this goal, the conference organizers convened thought leaders and innovators from academia, medicine, arts, education, and industry to discuss the myths, challenges, and opportunities of trans-disciplinary research at the intersection of these disciplines in the context of a 3-day meeting in a unique setting selected to foster discussion, interaction, and collaboration. The goal was to provoke day-long discussion across the following five critical areas leading to the development of a strategic plan for inclusion of the creative arts to foster creativity and innovation in science, engineering, and other fields:

1. How best to achieve an advanced understanding of human responses in health and disease to emotionally rich stimuli such as the creative arts, our physical environments, and our interactions with technology?
2. What is the best approach for uncovering basic neural mechanisms (e.g., reverse engineering the brain) underlying aesthetic and creative experiences?
3. What are the artistic, scientific, and engineering challenges that affect collaboration and innovation?
4. How best to study and promote creativity?
5. How to design new tools for understanding and promoting innovation, health, and wellness?

Mobile Brain–Body Imaging and the Neuroscience of Art, Innovation and Creativity is a trans-disciplinary, authoritative, collective, multimedia effort to critically uncover the challenges and opportunities for transformational and innovative research and performance at the nexus of art, science, and engineering.

Special features: This multimedia book is written for the academic scholar (e.g., undergraduate, graduate, and postdoctoral trainees), professionals from academia, government, industry, and clinical centers, independent researchers, artists, and the casual reader interested in understanding emergent innovations in neuroaesthetics and creativity. The book contains languages, design features (e.g., illustrations, diagrams, etc.), and multimedia content to develop a conversational bridge between the disciplines involved.

Main benefits: This book addresses a set of universal and timeless questions with a profound impact on the human condition, including: How do the creative arts and aesthetic experiences engage the brain and mind and promote innovation? How do arts–science collaborations employ aesthetics as a means of problem-solving and create meaning through aesthetic problem-solving? How do the creative arts and neuroscience advance understanding of individuality, social cognition, improve health and promote lifelong learning? How do neurotechnologies affect science and artistic expression and collaboration? How do the arts and citizen scientists in the public sphere innovate neuroscience studies, informal education, and outreach?

This book is divided into seven parts, dedicated respectively to the confluence of art, neuroscience, and creativity through contemporary mobile brain–body imaging (MoBI) technology. Each part includes an Introductory section that summarizes the individual contributions while providing context. Part I is dedicated to neuroaesthetics and creativity, and addresses a relevant and timely research question, "How do the creative arts engage the human mind and promote creativity and innovation across fields?" Chapters in Part I critically address historical (Chapter "Unknown and Solitary Seas: Angelo Mosso's Nineteenth-Century Discovery of Imaging Dreams Through the Cerebral Pulse " by Robleto), philosophical (Chapter "Art and Neuroscience: A State of the Union" by Hutton), neuroscientific (Chapter "Brain Mechanisms of Creativity: What We Know, What We Don't" by Dietrich), and cross-disciplinary resources (Chapter "Theme and Variations as a Window into the Creative Mind" by Brandt) for the study of creativity and aesthetics.

Part II gathers chapters dealing with challenges and knowledge that the arts and sciences create. Chapters in Part II provide the reader with three case studies of research and performance at the nexus of art and science. Biggs, Ericksen and Žiburkus (Chapter "Art-Science Collaborations: How to Break Boundaries Without Breaking Trust") tell us how to break typical art–science boundaries without breaking thrust; Hayes (Chapter "PAR-IESA: Practice and Research in Enactive Sonic Art") provides her experiences in the practice and research of enacting sonic art; while Cruz-Garza, Kopteva, Fleischhauer (Chapter "Into the Mind of an Artist: Convergent Research at the Nexus of Art, Science, and Technology"), and Contreras-Vidal describe the approach, challenges, and lessons learned from a long-term art–science collaboration seeking to uncover the mind of an artist.

Part III is dedicated to brain mechanisms of aesthetic perception. Chapters in Part III summarize a powerful computational model of how the human brain sees art and how artists make it (Chapter "How We See Art and How Artists Make It" by Grossberg), followed by a neuroscience study that asks the question "Is Beauty in the Eye of the Beholder or an Objective Truth?" (Chapter "Is Beauty in the Eye of the Beholder or an Objective Truth? A Neuroscientific Answer" by Aleem, Pombo, Correa-Herran, and Grzywacz).

Part IV presents the cognitive and medical applications of art-neuroscience research with an emphasis on physical and mental health. Chapters in Part IV address current approaches to creative art therapy for the treatment of traumatic brain injury and post-traumatic stress (Chapter "Outcomes of Art Therapy Treatment for Military Service Members with Traumatic Brain Injury and Posttraumatic Stress at the National Intrepid Center of Excellence" by Walker), visual self-expression for health and wellbeing (Chapter "Brain on Art Therapy-Understanding the Connections Between Facilitated Visual Self-expression, Health, and Well-Being " by Kaimal), and shaping perceptions of dementia through art and science (Chapter "Created Out of Mind: Shaping Perceptions of Dementia Through Art and Science" by Crutch, Harrison, Brotherhood, Camic, Day, and Woods).

Part V explores disruptive neurotechnologies, specifically brain–computer interfaces (BCIs), and how they can change science, arts, and innovation. Chapters in Part V represent a collection of seven sci-art projects by diverse teams of graduate students and postdoctoral trainees that participated in the BCI Hackathon at the 2017 Brain on Art Conference in Valencia, Spain. These innovative projects included health, neurofeedback, art-making, medical device development, and augmented reality applications.

Part VI describes the processes of learning and creativity the arts and neuroscience can represent and promote in the contexts of K-12 and higher education. Chapters in Part VI address the roles of the arts in promoting creativity and learning (Chapter "The Arts, Creativity, and Learning: From Research to Practice" by Hardiman), and the confluence of arts, technology, and wellbeing (Chapter "Intersectionality: The Confluence of Arts, Technology, and Wellbeing" by Baefsky and Sonke).

Finally, Part VII is dedicated to developing a Roadmap for the field of Neuroaesthetics and Creativity, including how to promote meaningful art–science communication and collaboration, support emerging convergent research directions linking art, science, engineering, medicine and the humanities, and innovate trans-disciplinary training and education (Chapter "Towards a Roadmap for Neuroaesthetics" by Contreras-Vidal, Robleto, and Cruz-Garza). This chapter also outlines programs and activities to scale-up the conversation, inclusivity, diversity, and vertical application of art–science research collaborations to address societal challenges.

Jose Luis Contreras-Vidal, Ph.D., IEEE Fellow, is Cullen Distinguished Professor of Electrical & Computer Engineering, and Director of the NSF Industry-University Collaborative Research Center on Building Reliable Advances and Innovations in Neurotechnologies (IUCRC BRAIN) at the University of Houston. He has pioneered the development of noninvasive brain–machine interfaces to control robots, exoskeletons, virtual avatars, and prosthetic limbs to restore motor function in individuals with spinal cord injury, stroke, and limb amputation. He has also pioneered the concept of "The Museum and the Classroom as Laboratories" to understand the brain in action and in context of free-behaving individuals, with applications to medical devices, art therapy, aesthetics, neuroscience, and regulatory science. His work has been supported by the NIH, NSF, VA, DARPA, ONR, Foundations, donors, and industry. Dr. Contreras-Vidal's research has been highlighted by The Economist, Nature, Science, Science News, Der Spiegel, NSF, Wall Street Journal, SFN, O&P, Scientific American, NPR's Science Friday, and Neurology Today among others. His career development in biomedical engineering has been highlighted in the magazine Science.

Webpage: http://www.ece.uh.edu/faculty/contreras-vidal

Webpage: https://www.houstonmethodist.org/faculty/joseluis-contreras-vidal/

IUCRC BRAIN: http://brain.egr.uh.edu/

Jesus G. Cruz-Garza is a Ph.D. candidate in Electrical Engineering at the University of Houston and Graduate Fellow on Translational Research with Houston Methodist Research Institute and University of Houston. He graduated from BS Engineering Physics at Tecnologico de Monterrey, Mexico.

Dario Robleto is a transdisciplinary artist based in Houston, Texas. He currently serves as Artist-in-Residence in Neuroaesthetics at the University of Houston's Cullen College of Engineering and as Artist-at-Large at Northwestern University's McCormick School of Engineering. In 2016, he was appointed as the Texas State Artist Laureate.

José M. Azorín is the Director of the Brain–Machine Interface Systems Lab and Full Professor at Miguel Hernández University of Elche (UMH), Spain. He holds a Ph.D. in Robotics. He has been a visiting professor at the University of Houston (USA) and at Imperial College London (United Kingdom).

Chang S. Nam is a professor in the Department of Industrial and Systems Engineering at North Carolina State University. His research interests include brain–computer interfaces and cognitive neuroscience. He is the editor-in-chief of the journal "Brain–Computer Interfaces" and the main editor of the book "Brain–Computer Interfaces Handbook: Technological and Theoretical Advances" (CRC Press, 2018).

Part I

Neuroaesthetics and Creativity: How Do the Creative Arts Engage the Human Mind and Promote Creativity and Innovation Across Fields

Introduction

Jose L. Contreras-Vidal[1] and John D. Talasek[2]

[1]IUCRC BRAIN, University of Houston, Houston, TX, USA
[2]Cultural Programs of the National Academy of Science, Washington D.C., USA

> *The future of the world depends on us being better collaborators.*
>
> Fabrizio Hochschild, Assistant Secretary General of the United Nations

As one walks into the general assembly room of the United Nation's building in New York City, one sees a mural by Brazilian artist Cândido Portinari entitled War and Peace (1952–1956). It depicts a range of suffering from the result of war and conflict. No weapons are depicted in the large-scale painting—only their impact on lives. As one leaves the General Assembly space, one is confronted with the partner painting reflecting peace. Strategically placed these two paintings are as Ban Ki-moon stated, Portinari's call to action, "the terrible toll of war and the universal dream for peace". Entering into the space where discussions for the betterment of the global welfare are of concern, delegates are reminded of the gravity and importance of what is before them. The message of hope as they leave is in theory the intended and ultimate outcome. The paintings provide a platform to focus a constellation of ideas around a major goal (peace) and the thoughtful context is intended to encourage those ideas into a reality.

As Portinari's War and Peace exemplifies, artists have long known something about the workings of the human mind and our senses. Art can function as an interface to alter perception and form a platform for ideas. Art, as it functions at the United Nations, mindfully alters the space such that those who enter are invited to consider a collective idea or goal with the possibility of focusing or altering a perception. The idea of examining audience response as a way of understanding the impact of art practice, from the individual to the community level, is the basis of a still fairly new field of neuroaesthetics and will be explored further in this Introduction and Part I of this book, as it was a thread that ran throughout the *2017 International Conference on Mobile Brain–Body Imaging and the Neuroscience of Art, Innovation and Creativity*—usually referred to as the Brain on Art Conference—gathering in Valencia, Spain. In addition to neuroaesthetics and creativity, conference attendees pondered the impact of collaboration and dialogue at the nexus of art, science, engineering, education, and medicine to consider the range of possibilities of mindful engagement and the potential societal impacts of these collaborations. The benefit of such wide engagement of epistemologies challenges perceptions and allows us to look at problems anew. What impact could this have on understanding of our disciplines and the construction of knowledge? What impact could there be on technological advancements as well as the betterment of society and of our individual lives? Perhaps this was our own call to action similar to that of Portinari's, with no less urgency—to reach across disciplines in order to be informed, challenged, and inspired for the betterment of society.

There are antidotes throughout the history of ideas suggesting seeds of inspiration germinated from conversations from across disciplines. Between 1765 and 1813, The Lunar Society of Birmingham was an informal gathering of prominent thought leaders that included industrialists, natural philosophers, artists, and intellectuals. The name Lunar Society came out of the habit the group had of meeting during the full moon which pragmatically helped make their journey's home in the late evening easier given there was no street lighting. Although the formal list of members was never recorded and is often disputed, common participants included Erasmus Darwin, Richard Lovell Edgeworth, James Watt, and Josiah Wedgwood. Perhaps an example of cross-disciplinary exchanges that is more germane to a conference on art and brain research is that of the intellectual exchanges in Vienna at the turn of the twentieth century.

Examined extensively in his book *The Age of Insight: The Quest to Understand the Unconscious in Art, Mind, and Brain, from Vienna 1900 to the Present* (2012), Eric Kandel explained the impact that such conversations had on both art and cognitive science. In the early 1900s, artists and scientists in Vienna met in salons where the exchange of ideas led to breakthroughs in psychology, brain science, literature, and art. Kandel draws striking lines of influence between thought leaders including Freud, Schnitzler, Klimt, Kokoschka, and Schiele that exemplify the potential of cross-pollination between intellectual structures of thought.

In the spring of 2018, the National Academies of Sciences, Engineering, and Medicine's Board on Higher Education and Workforce released a consensus report on integrating art and humanities with STEM disciplines in higher education. Entitled *Branches of the Same Tree*, the report synthesized evidence that suggests the area between disciplines is fertile ground for exploration. Such integrative learning has the potential for producing creative problem solvers, critical thinkers, and better collaborators within diverse groups—characteristics that are actively sought after by employers in the twenty-first century. An appendix of the report lists well over 200 examples of integration in higher education that already exist. This same appetite in institutions of learning (both formal and informal) partially explains the explosion of other salon-style conversations that are convening on an international scale. For example, *Leonardo: The International Society of Art Science and Technology* has fostered a network of over 32 salons around the world including the United States, Canada, Great Britain, Europe, Tehran, and Brazil. *LASER (Leonardo Art Science Evening Rendezvous)* has become an active network that allows local communities to connect on a global scale. The salons are often grass-roots efforts to foster a desire within communities for this type of dialogue on an ongoing basis. Intellectual communities are hungry for these exchanges and the creation of spaces where they can be fostered.

Ideas Come from the Unexpected

All the more important to foster conversations between our established silos. This is not intended to break down the silos but rather to build connections between them that allow for advancements through creativity and discovery. Why is this important? In the fall of 2017, Fabrizio Hochschild, Assistant Secretary General (ASG) of the United Nations spoke at the New World Frontier Forum in Cambridge, Mass. He spoke of common global threats that included security, climate change, and food and water accessibility among others. The audience of this gathering was a unique blend of thought leaders from a broad range of backgrounds. The very last sentence he spoke was that, "The future of the world depends on us being better collaborators." This was a call to action that was the verbal equivalent of Portinari's. The ASG, in addressing a diverse group of people from science, engineering, medicine, art, humanities, dance, and so on, was suggesting that it was going to take all of the tools in our toolbox—all areas of human inquiry and accumulated knowledge—to address the needs of the planet and imagine—and build—the future that is possible.

In the same vein, the *Brain on Art* Conference in Valencia and In this part, book asked the question: How do the creative arts engage the human mind and promote creativity and innovation across the fields? In the next three chapters, the interplay between art, neuroscience, psychology, philosophy, and engineering is brought to light from four points of views and from three disciplinary perspectives. First, Dario Robleto, an American transdisciplinary artist, researcher, writer, teacher, and "citizen-scientist" writes a passionate account of what it was likely the first attempt, albeit indirectly, to measure the brain's blood flood to examine the "human dream". Robleto astutely recognizes in the work of the nineteenth century Italian physiologist, scientist, and inventor Angelo Mosso, the power of observation, perseverance, and insight as he recorded essentially the brain's pulsating movements emerging during the sleep of Giovanni —a young boy who has suffered a fractured skull resulting in a large opening that served as a window to peek into the inner workings of his brain.

Noah Hutton, a filmmaker and writer, engages the mind while discussing the pitfalls of "treating art as an exotic stimulus and the brain as a universalized end-domain for us to plant our flag of understanding". Hutton goes further to conclude that a laudable goal for neuroaesthetics should be "the reduction of subjectivities to their constituent parts so that they can then be put back together again in more sublime ways". The neuroscience reader will recognize Hutton's proposal to be vaguely reminiscent of the neuroanatomical brain architecture in which the occipital, parietal, and temporal lobes of the brain are thought to decompose the (multimodal) sensory world into their basic components (such as color and edges in the case of visual images), just to be put back together in the frontal lobe where contextual information such as memory and emotions can "personalize" such complex aesthetic stimuli leading to individualistic aesthetic experiences and judgment, making in fact every one of us an "artist" that creates our own aesthetic perception and judgment of the world.

Arne Dietrich, a psychology professor from Lebanon and a scholar on creativity and consciousness, predicts dire consequences for the neuroaesthetics field if the *status quo* remains unchanged. Dietrich uncovers theoretical and paradigmatic inconsistencies in past approaches to the study of creativity and emphasizes the fact that we still know very little about the neural basis of creativity. But Dietrich provides us with five suggestions that could help correct this knowledge gap. He starts by suggesting that highly likely possibility that creativity is fully embedded and distributed in the brain. He then proposes considering at least three types of creativity (deliberate, spontaneous, and flow), which may differentially use complementary networks for implicit and explicit information processing in the brain. These parallel "creative" pathways may benefit of (learned) predictive brain representations of the world and evolutionary neural computations in support of sightedness and creativity.

We would like to conclude this Introduction to Part I by emphasizing that this book aims not only to define and propose ways to uncover how creativity works but also discuss the importance of why we should care about studying creativity. We hope the exciting and timely chapters in this book clarify the importance of creativity on knowledge production, technological advancement, health, and social-economic wellbeing in the twenty-first century.

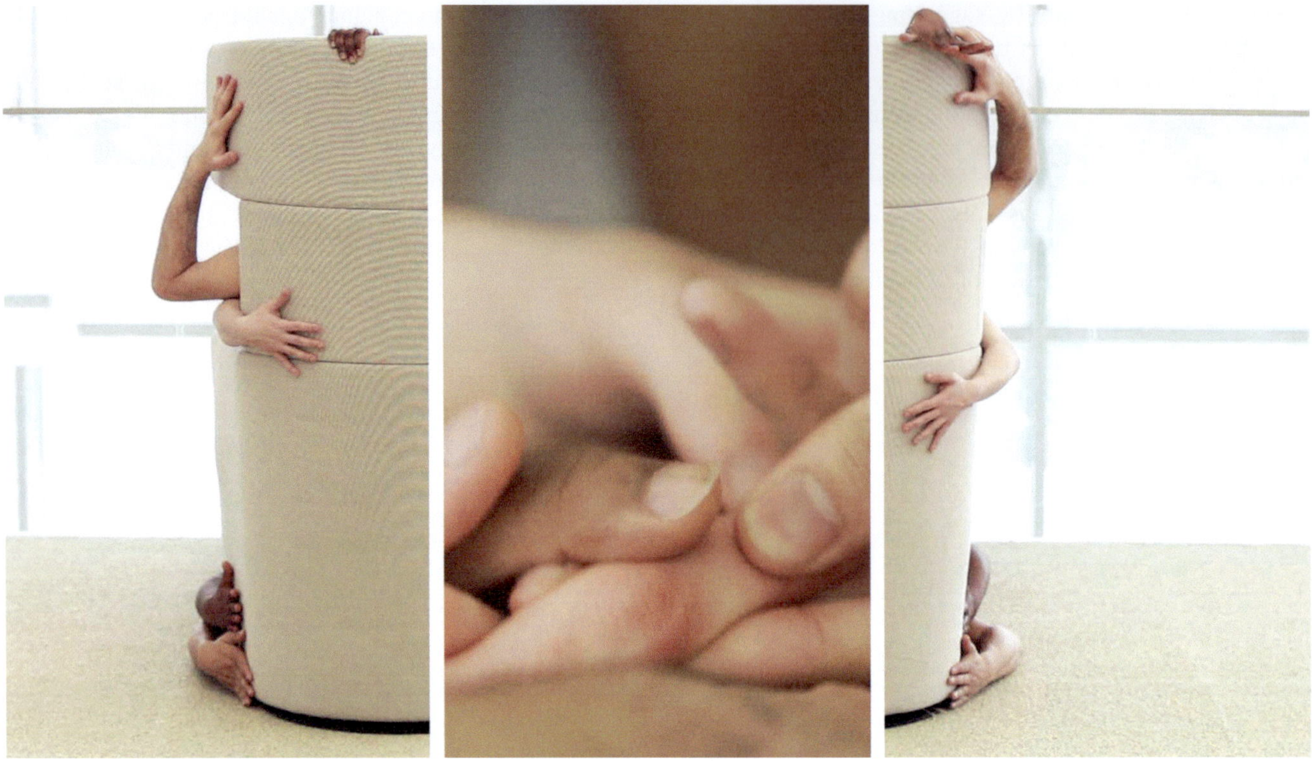

Miriam Simun ALLOW YOUR FINGERTIPS THEIR BRAINS (Exercise #16) Inkjet print on cotton paper, 10" × 5.5" 2018 Training Transhumanism (I WANT TO BECOME A CEPHALOPOD) psycho-physical training regimen for evolving the future of the human, based on the model of the cephalopod. The regimen seeks to develop within the human new sensitivities and capacities for a world marked by ever-increasing ecological and technological change Training Transhumanism (I WANT TO BECOME A CEPHALOPOD) was developed by Miriam Simun while a researcher at the MIT Media Lab, in collaboration with choreographer luciana achugar

Unknown and Solitary Seas: Angelo Mosso's Nineteenth-Century Discovery of Imaging Dreams Through the Cerebral Pulse

Dario Robleto

On a quiet, brisk evening in 1877 in Turin, Italy, the snow from a passing winter still on the ground, something extraordinary was occurring in a side room down a long, darkened dormitory hall of an insane asylum: a human dream was traced in smoke. This startling scientific feat was accomplished by a true innovator and artist/scientist hybrid, the Italian physiologist Angelo Mosso (1846–1910). As with many of his experiments into the recesses of the brain, the line between data and poetry was provocatively blurred. If this moment is remembered today at all though, it is not necessarily recognized as an important marker in the still young field of neuroaesthetics. As an artist fascinated by the collaborative possibilities between the arts and neuroscience, I reflect on this day quite a bit as it taps into so many things I feel bond those fields: pushing the threshold of the sensitivity of observation; driving the innovative use of materials in making the invisible visible; and creating new questions, images and models that probe and provoke our never-ending desire to investigate the nature of consciousness and creativity.

As I curiously set foot into this world of creativity and the brain—even becoming a test subject and collaborating on neuroscientific studies with viewers interacting with my artwork—I am open and ready to contribute to and absorb whatever new layers of meaning modern neuroscience can bring to my understanding of the creative life I have committed to [2, 4]. But as is often the case in my work, historical curiosity sets in and my mind turns to those who have come before. I start to ponder: How is the real-time recording of the blood flow and electricity in my brain even possible? When, where, and who first attempted to materially record the long-assumed immateriality of thoughts, emotions, memories, creativity, and dreams? Was this a problem best suited for physiology or metaphysics?

D. Robleto (✉)
Artist in Residence in Neuroaesthetics, IUCRC BRAIN,
University of Houston, Houston, TX, USA
e-mail: robletodario@gmail.com

We live in an incredible era of images and their making. There are over eight billion videos viewed a day on Facebook, the Hubble telescope inches closer to seeing the first starlight ever to flicker on in the universe, and we have clear images of individual neurons firing in our brains. Two of the technologies we use today to image the brain, the functional magnetic resonance imaging (fMRI) and the scalp electroencephalogram (EEG), are so commonly in use that, even for the public, the idea that we can image a living human brain in thought, experiencing emotion or battling disease, is no longer so remarkable. If anything, and especially from the more vocal criticisms within the humanities, we are in a full brain-imaging fatigue-stage, with numerous popular science articles claiming such things as love, addiction, faith, aesthetics, and other complex mental states of being are somehow "explained" through mapping blood flow through the brain. This contemporary criticism is fair, further illuminating long-held explanatory divides between scientific measurement and one's subjective experience in the construction of meaning, especially when that meaning arises from art and aesthetic experience [1, 8]. However, before we could ever have such debates, I would instead like to focus on the remarkable beginnings of the first attempts to physiologically image the interior of our living brains. In the grand arc of scientific history, this capability is very recent, and it is worth pausing and reflecting on it, from both an artistic and scientific point of view, as one of our greatest achievements in making the invisible visible. Even if today we have lost some of our awe when it comes to peering into the human brain, when we revisit this cold night in the asylum in the late 1870s, when Angelo Mosso essentially invented the modern concept of real-time brain imaging, we can be reminded of what a radical leap into the unknown it was, and the palpable sense of joy, curiosity, and even melancholy that such tests produced (Fig. 1).

Angelo Mosso was a brilliant scientist and inventor with far-reaching interests, writing books on everything from the pulse, the brain, emotions, fatigue, and archeology. As a

J. L. Contreras-Vidal et al. (eds.), *Mobile Brain–Body Imaging and the Neuroscience of Art, Innovation and Creativity*,
Springer Series on Bio- and Neurosystems 10, https://doi.org/10.1007/978-3-030-24326-5_2

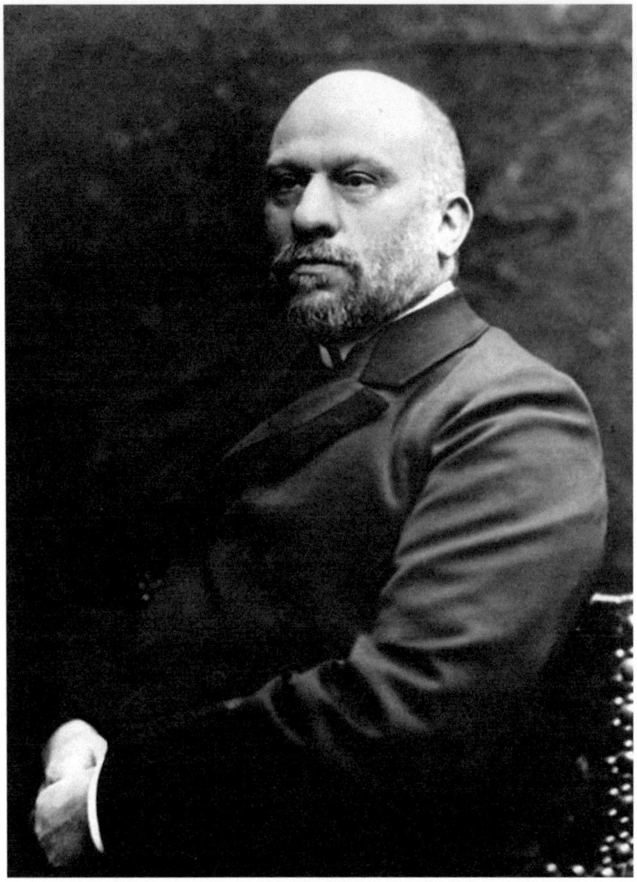

Fig. 1 Professor Angelo Mosso (1846–1910)

physiologist though, he continually turned his interest to the problem of blood flow and mental states. Today it may be common knowledge that mental activity requires blood flow to the brain, but up until the late nineteenth century this was unproven. At the time, because of the difficulty of accessing and probing a living human brain, it was still unclear how physically to study such a phenomenon. Mosso's search was no less than a quest to prove the materiality of consciousness. With all the difficulty and debate such a quest raises even today, we can properly imagine his dilemma over a century ago.

Mosso's confidence in the subject was partly built on recent technological strides made on that other organ of great mystery and scientific debate—the heart. The term he would coin in his studies on the brain—the "cerebral pulse," or the heartbeat in the brain—point to this lineage. Only a few decades before, other great scientists such as Karl von Vierordt and, especially, the French physiologist Étienne Jules Marey had built the recording machines (the kymograph, sphygmograph, and cardiograph), and the conceptual methodology (the graphic method), that allowed for the permanent visual inscription of interior physiological processes in the living body [5, 9].

Frustrated with the limitations of the human senses, language, and memory to record and archive fast moving, imperceptible and internal biological phenomena, these scientists, through the graphic method, revolutionized the field of medical imaging by translating these phenomena outside the body through another medium. Marey was especially suspicious of the role of language in scientific communication stating in his landmark 1878 publication, La Méthode graphique, "Let us reserve the insinuations of eloquence and the flowers of language for other needs; let us trace the curves of phenomena that we want to know and compare them" [5, pp. iii–vi]. For Marey, language was a system of communication devised long before the objectives of science, and he did not trust it was sufficient for expressing and transmitting this interior narrative of life because of its fluctuations and possibilities for misunderstanding.

These devices all worked under a similar method, which was to absorb the energy of bodily movement (a pulsing artery, inhalation, electrical discharges, internal sound waves), through an air- or water-filled membrane or spring that would then make an attached stylus pulse in unison. The stylus would then trace out the white curvilinear forms (pulse waves, flatlines, etc.) on a piece of blackened paper, which was attached to a rotating cylindrical drum. To allow the delicate stylus head to render these vibrations, a frictionless and exquisitely sensitive medium was needed. These scientists turned to the material that humans used to record the first images of themselves within the time of caves: the powdery residue of soot from a flame. That this astonishingly fragile medium was selected for practical reasons makes it no less poetic that the first heartbeats and pulses ever scientifically recorded were traced in the vestiges of candle flames that burned and were extinguished almost 150 years ago.

Like Marey, Mosso believed that movement conditioned all life. From chemical interactions within cells, the electricity propelling muscle contraction or kinesthetic movements like walking or inhalation, to move was to live. For Mosso, consciousness must also have a corresponding relationship to movement, and the graphic method was the best system science had in place to uncover it. But to only understand the graphic method as advancement in scientific measurement would be to miss the more complex ambitions of the effort. Historically, the heart, and eventually the brain, were the two most contested sights in the body in ancient debates about the physical location of one's identity, emotions, intellect, and even the immortal soul. Across time and cultures the heart, for example, was considered the literal conduit for the soul between the material and immaterial realms and therefore widely regarded as unknowable and off limits to scientific investigation. Even if physiologists could overcome the seemingly insurmountable technological hurdles of accessing the interior living body, for the sciences to probe, touch, measure, operate or even look upon a living human heart or brain was a taboo of the highest order.

If some scientists of this era were willing to venture past these taboos, there were still remnants of unresolved entanglements between the mystical-religious and scientific. We can glimpse this in the language of their ambitions. Mosso, as Marey did before him, spoke of a universal "natural language of life" hidden just past our sensory capabilities which awaited decipherment. Like a hieroglyph holding the potential of ancient wisdom, each crest or trough in the waveform was a potential letter in this invisible grammar. Mosso hoped "to wrest from Life its secret," which would be revealed by continually refining inscription devices sensitive enough to peer into this ephemeral movement of life. Very literally, Mosso wanted to "see how the brain writes when it guides the pen itself" [7, p. 77].

One of his significant adaptations to these graphical devices was the invention of the plethysmograph. This apparatus used a water-filled glass cylinder fully encased around the arm or foot, allowing for the pulsations in the limb to expand and contract the water pressure, which was controlling the movement of the recording stylus across the soot-covered paper. The device was so successful that it led the esteemed psychologist William James, who was at the time working to build the principles of human psychology on physiological foundations, to echo in his 1884 paper, "What is an emotion?", the scientific hopes of finally revealing the hidden mysteries of the interior living body:

> The researches of Mosso with the plethysmograph have shown that not only the heart, but the entire circulatory system, forms a sort of sounding-board, which every change of our consciousness, however slight, may make reverberate. Hardly a sensation comes to us without sending waves of alternate constriction and dilation down the arteries of our arms. [3, pp. 191–92]

With the success of this device, Mosso's great leap in thinking was to ask if the blood flow to the surface of the brain also acted as a sort of sounding-board to the changes in our emotional and intellectual states. However, like the problems of working on a living, beating heart, there was no way to non-invasively access and record the living, thinking brain in real time. Mosso was left with the unfortunate task of searching for patients in hospital wards who had, usually through a terrible accident, a significant enough head injury that part of their skull was removed, exposing sections of the brain that were only covered by a thin layer of skin.

To this end, he used an adapted version of Marey's cardiograph—a device designed to record heart sounds directly from the chest wall. This approach made practical sense as he was essentially trying to record pulsating movement from the brain, like the heartbeat produced through the chest wall. Mosso faced unique problems with his patients regarding how to secure a recording device into the crevices of

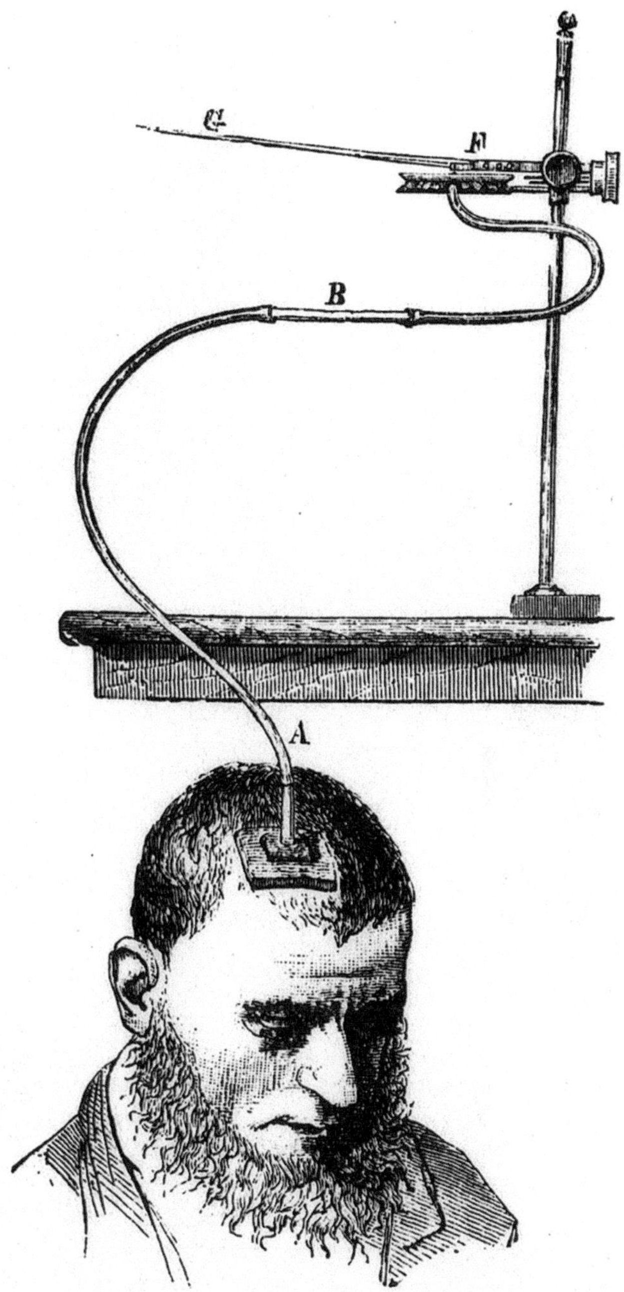

Fig. 2 Example of Mosso's device for recording the cerebral pulse

damaged skulls. He solved this problem by perfectly fitting a molded plate of gutta-percha (a natural latex) into the skull opening, maintaining a slight air gap between the plate and the brain surface, which produced an airtight fit. The plate was equipped with a glass tube at its center so that when the pulsating brain forced air out, the pressure change was transferred to a recording arm inscribing the waves into the soot-covered paper (Fig. 2).

With these brain "autographs," as he sometimes referred to them, Mosso laid the foundation for a vital scientific field of today: brain imaging as it relates to blood flow. From cave walls to canvases, in the context of the millennia-long artistic exploration of how much we can reveal about ourselves through the simple act of drawing a line on a surface, he had also invented an entirely new scientific visual language of self-reflection: hidden aspects of our psyche were now made manifest in the ripples of a wave. Mosso was deeply interested in the physiology of emotional states that seemed beyond our conscious control and conducted many groundbreaking experiments into phenomena such as pallor, blushing and trembling, even writing the first book-length study devoted to the human emotion of fear in 1891, which was translated into English in 1896. Along with these pioneering studies, he would also go on to record the brain under several scenarios—solving math problems, inducing emotions, inhaling ammonia or other drugs, fasting, epileptic fits, and even forcing patients to pass out. But as groundbreaking as these recordings were, Mosso went further. He did not only want to establish that there was a physiological connection between general psychological states and circulation, but he also wanted to know if the specificity of emotions or the exact conditions present when consciousness arises were recordable. Was there undiscovered meaning bubbling in the shape of the curves? In his landmark publication Fear he clarified his investigation:

> The serious aspect of the question is, that physiologists would like to catalog many qualities which we have always considered as the most noble of our character, the most sublime feelings of human nature, amongst the automatic movements and more material instincts in the lower story of the brain. [7, p. 77]

From the current ongoing efforts to define consciousness to the investigation of the neurobiological underpinnings of emotions or aesthetic experience, Mosso's original quest still resonates with us today in ways he could not have imagined.

As he refined the sensitivity of his machines, Mosso turned his attention to a little scientifically explored and long thought inaccessible experience: waking from sleep. From ancient folklore to Greek philosophers, the mysterious condition of sleep has confounded many great thinkers. Why, for example, did one lose their self-awareness, agency, and consciousness while sleeping? In other words, where did "you" go when sleeping and how were you restored with full memory and continuity of self each time you awoke? Mosso hoped that the unconscious mind would reveal some new relationship between matter and thought, and the threshold between sleeping and waking states would potentially be the crucial moment to measure the cerebral blood flow as the material conditions for consciousness were regained.

In 1877, Mosso continued his investigations into sleep states with experiments performed on an 11-year-old boy, Giovanni Thron, who had been living in an insane asylum. The young boy, when he was only 18 months old, had taken a terrible fall from a terrace that fractured his skull, causing a major concussion to the brain. He would soon develop epileptic fits and signs of insanity, causing his family to commit him to the asylum for the rest of his life. The damage to young Giovanni's brain was so severe that it halted his intellectual development, forever locking him in a mental state before his fall. Mosso would become quite fond of Giovanni, remarking on his beauty, smile, and sweet nature, not unlike a large baby. But it was the profound tragedy of Giovanni's stunted life that most impacted Mosso's time with him. Although the boy was now mostly mute, a single verbal-intellectual remnant, now turned into a plea for the unobtainable, persisted for his short life—he would repeat, "I want to go to school" constantly throughout the day. Though Giovanni's state was tragic, Mosso recognized the rare opportunity before him and with great compassion began a series of recordings that would capture recesses of the mind even he had not predicted [6, 7].

When Mosso visited him in the insane asylum, he had a large opening in his skull above the right eye, the fracture having never closed. Because of the delicacy of the recording device and absolute stillness that was required, Mosso would wait with care and patience night after night for the often-agitated Giovanni to fall into deep sleep.

As he first set out to explore these unknown recesses of the brain, as seemingly remote as the still unknown true depth of the bottom of the sea, the difficulty and macabre melancholy of the moment was never lost on Mosso. By the light of a small lamp, he quietly passed by the darkened corners and rooms filled with asylum patients unmoored from their minds. With some patients naked in the freezing winter night, sitting up or bound to their beds, he would plead with them to remain still and quiet as he worked on Giovanni, only to be met with shrieks and vacant eyes. Even under ideal conditions, he was already pushing the edge of capturing delicate and precise measurements. Under these conditions, many nights he left alone and discouraged, wondering if his experiments would succeed.

But one evening, after a severe epileptic attack had exhausted the boy, Mosso had the stillness he needed to probe this sleeping/waking boundary. Although a thin layer of skin had since grown over the gap in Thron's fractured skull, the pulsating brain was still clearly seen and felt under the surface. Crafted perfectly to fit into Giovanni's wound, Mosso would apply the gutta-percha plate to the opening and was able to record the boy for a few hours while he slept. Like a scientific lullaby, Mosso called out the young boy's name, "Giovanni," in between minute-long periods of silence. As Mosso's consciousness-detecting machine dutifully unspooled its waveforms, it registered that the young boy's unconscious brain perceived the calls to him, signified by the increased blood flow and elevation in the line, whether he understood the meaning of them or not [6, p. 78].

But one night in his ongoing exploration into the mysterious conditions of sleep something unexpected happened that would offer Mosso an insight he had not anticipated:

> It was one of the most interesting sights to observe in the stillness of night, by the light of a little lamp, what was going on in his brain, when there was no external cause to disturb this mysterious life of sleep... then came stronger blood-waves which flooded the convolutions, raising the height of the pulsations, which were automatically marked by the apparatus applied to the brain. We scarcely dared breathe. The one who was observing the instruments communicated with the other, who was watching over the patient, by pressing his hand. Looks full of interrogation and wonder would meet, and exclamations had to be forcibly repressed. [7, p. 73]

In his barely controllable excitement, Mosso was pondering a profound question: Had the team just recorded what was once thought beyond the reach of accessibility—the first physiological evidence of a human dream?

It is worth reflecting for a moment that even in the context of his era, as scientific tools of observation were undergoing a technological revolution, Mosso's work stands out. By the late nineteenth century, the microscope, telescope, and photography had revealed once unimaginable realms of the material world, redefining expectations on what scientists and the public had access to visualize and record. The invisible was literally made visible through such images as the first photographs of the sun's surface, lightning, magnetic fields, or the teeming microbial life in a droplet of water. The goal of Mosso's research, to detect the material conditions of consciousness as it was restored from a state of sleep, was equally stretching the threshold of poetic, philosophical, and scientific notions of sensitivity. If correct, and the team had recorded the seemingly more distant phenomena of the materiality of dreams—a boundary line no device had ever traversed—this wasn't only a startling demonstration of the viability of Mosso's approach but a conceptual expansion of the assumed scientific limits of observation (Fig. 3).

But the difference between physiologically recording a possible dream-state and deciphering its personal meaning was vast. Fortunately, Mosso's particular brilliance and openness allowed him to ponder these moments with young Giovanni in poetic ways no less important than the physiology, even going as far as stating that he did not need to "conceal the artist side of their investigations from the fear of desecrating science"—a courage across disciplines that is to this day no easy task for a scientist to embrace. Unsure yet of how to interpret the meaning of these unexpected disturbances, or what he was so lyrically referring to as "undulations," the always reflective Mosso gives some moving possibilities of what could be bubbling in this fragile, young mind:

> Did the face of his mother and the recollections of his early childhood grow bright in his memory, lighting up the darkness of his intelligence and making his brain pulsate with excitement? Or was it perhaps only a morbid phenomena, like the jerky movements of a broken wheel, or the index of a machine out of order, swinging idly to and fro? Or was it an unconscious agitation of matter, like the ebb and flow of an unknown and solitary sea? [7, pp. 73–74]

Of course, dreams, love, fear, aesthetic experience and the myriad emotional experiences that define our humanity are not sufficiently "explained" through the single lens of blood flow to the brain. Over a century after Mosso's breakthrough we are still struggling to define emotions and consciousness. It is interesting that even with all the promised hope of the graphic method and automatic inscription devices, with their potential for universal scientific clarity, bypassing the need for verbal or written language, it is still poetic language Mosso turns to when confronted with the inscrutable mystery of another's dream.

Watching Giovanni suffer through the seizures, sleeplessness, confusion, and agitation obviously left a deep impact on Mosso. "Of all the experiments I have ever performed with human subjects, these have cost me the greatest effort and have left the most profound impression," he remarked [6, p. 77]. The tragic irony of this breakthrough could not have been lost on Mosso: the first glimpse into a depth of our inner selves long thought impenetrable, our dreams as "unknown and solitary seas," was communicated to us through a broken mind that could no longer know itself. Giovanni would pass away soon after this test from acute anemia, but these few minutes of etchings in soot of his still-living and dreaming brain would immortalize him in a way the etchings on a gravestone never could.

In our insatiable curiosity to explore and refine our ability to peer inside the brain, it is important to pause a moment to remember all those scientists and test subjects who first

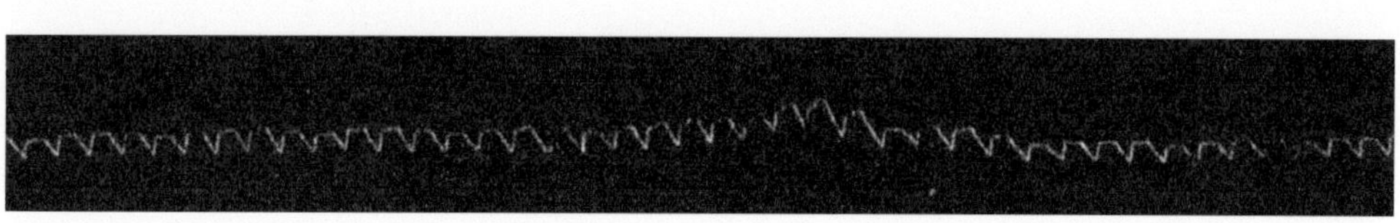

Fig. 3 Cerebral pulse tracing of Giovanni Thron while dreaming

established the paths we continue to tread. Brain imaging has, of course, revolutionized brain research, diagnostics, treatments, therapies and machine-brain interfaces, while investigations into dreaming, daydreaming, meditation, intuition, the subconscious, and the default mode network, to name a few, are some of the most exciting fields of research within the neuroscience of creativity today. But history offers us a necessary act of humility when we reflect on the fact that close to 150 years after Mosso's experiments we remain largely mystified by the waking, let alone the dream states of our creative minds.

Perhaps on a still-dark cave wall, the earliest attempt by a human to give form to their night apparitions remains. Certainly our poets and priests, through inspiration or revelation, have been struggling for millennia to translate and find meaning in their unconscious visions. Mosso's work rivaled these previous attempts with poetic and material fragility, and with profound physical and philosophical implications. We must remember he and young Giovanni's offering to this ancient quest: the first dream recorded and preserved through the smoke rings of a candle flame.

References

1. Ball, P.: Neuroaesthetics is killing your soul: Can brain scans ever tell us why we like art? (2013). Retrieved from http://nature.com/news/neuroaesthetics-is-killing-your-soul-1.12640
2. Cruz-Garza, J.G., Brantley, J.A., Nakagome, S., Kontson, K., Robleto, D., Contreras-Vidal, J.L.: Mobile EEG Recordings in an Art Museum Setting. IEEE Dataport (2017). https://doi.org/10.21227/h2tm00
3. James, W.: What is an emotion? Mind **9**, 188–205 (1884)
4. Kontson, K.L., Megjhani, M., Brantley, J.A., Cruz-Garza, J.G., Nakagome, S., Robleto, D., White, M., Civillico, E., Contreras-Vidal, J.L.: Your Brain on Art: Emergent Cortical Dynamics During Aesthetic Experiences. Front. Hum. Neurosci. **9**, 626 (2015). https://doi.org/10.3389/fnhum.2015.00626. Erratum in: Front. Hum. Neurosci. **9**, 684 (2015). PubMed PMID: 26635579; PubMed Central PMCID: PMC4649259
5. Marey, É.: La Méthode graphique dans le sciences expérimentales et principalement en physiologie et en médecine. G. Mason (1878)
6. Mosso, A.: In: Raichle, M.E., Shepherd, G.M. (eds.) Angelo Mosso's Circulation of Blood in the Human Brain (Trans.: Fabbri, C.N.). Oxford University Press ([1880] 2014)
7. Mosso, A.: Fear (Trans.: Lough, E., Kiesow, F.). Longmans, Green, and Co. ([1891] 1896)
8. Noë, A.: Art and the Limits of Neuroscience (2011). Retrieved from https://opinionator.blogs.nytimes.com/2011/12/04/art-and-the-limits-of-neuroscience/
9. Vierordt, K.: Die Lehre vom Arterienpuls in gesunden und kranken Zuständen: gegründet auf eine neue Methode der bildlichen Darstellung des menschlichen Pulses. Druck und Verlag von Friedrich Vieweg und Sohn (1855)

Dario Robleto is a transdisciplinary artist based in Houston, Texas. He currently serves as Artist-in-Residence in Neuroaesthetics at the University of Houston's Cullen College of Engineering and as Artist-at-Large at Northwestern University's McCormick School of Engineering. In 2016, he was appointed as the Texas State Artist Laureate.

Art and Neuroscience: A State of the Union

Noah Hutton

To see something—to do what one might call "viewing" art—is also to express oneself; it is an act that exists on a continuum with the same faculties required for the imagining and making of things. To engage with art is to bring equal parts active expectation and passive sensory collection. One affects the outcome of an entanglement with art simply by the act of bringing one's own body to the task, an apparatus chock full of the personal interior and intimately linked to the political exterior. In quantum physics, there is the colloquial understanding of *indeterminacy*, the strange reality that things at the quantum scale seem to behave in ways unlike the easily measurable matter around us. But within the field, there is significant theoretical divergence about the precise causal mechanisms at play when a quantum phenomenon is measured. Neils Bohr's theory about the act of measurement could be helpful to us in our journey in this chapter. For in Bohr's model, it is not the experimenter's willful gaze itself that affects the results of the experiment (this was Heisenberg's *uncertainty principle*); rather, it is the specific physical properties of the experimental apparatus itself, set to record the quantum phenomena, that governs the nature and possibility of the experimental results [1]. Heisenberg and Bohr represent a significant fork in the quantum road: in the former theory, the willful human gaze is enough to achieve causative force upon the object of the experiment; in the latter, the willful gaze is not enough—it must be extended to include the apparatus of the engagement, treated as an equally significant mediator of experience, a thing that spells out the possible results through the arrangements of its physical structure. So too in aesthetic encounters: we may bring ourselves willfully, but the act of engagement relies upon the distributed apparati of culture, social context, a curated viewing context, personal memory, and what you ate for breakfast. In this sense, we might say that a Bohrian framing of aesthetic engagement reminds us that we meet the world halfway, and that we ought to interrogate the apparati of our engagements with equal fervor as we regard the pristine aesthetic objects themselves.

Let us first examine the aims and means of *neuroaesthetics*, which takes seriously the fact that we all have brains, and tries to sidestep the forces of biographical context and sociopolitical particularity by identifying first principles of seeing, evaluating, and creating art, hoping we will be able to plant our interdisciplinary flag somewhere in the mushy folds of a universal human brain.

1 The Universal Aesthetic Object

Whereas some traditions of art history have perhaps overemphasized the explanatory role of biography in shaping art (complete with, for example, a wall label that packages a neat story of Van Gogh's time in Arles beside a vibrant canvas), the discipline of neuroaesthetics—the scientific probing of how the brain views, evaluates, and creates art—has run in the opposite direction, leaving behind the psychology and biographical sketch of the creator in favor of parallel objectifications of the art object as a scientific stimulus and the brain object as a shared universal, harkening a return in both cases to a version of the Kantian ideal of beauty and universal meaning. In the face of chaos, plasticity, and the general indeterminacy of life and art, could this attempt to pull out universals from the muck of chaotic differences be an overcorrection? If certain strains of the humanities fall into the trap of overvaluing biographical details as explanatory signposts, do we find an opposite but equally blinding trap in the standardized corridors of neuroaesthetics?

Though it was German philosopher Alexander Baumgarten who coined the term "aesthetics" in 1750, it was with Immanuel Kant's 1790 treatise *Critique of Judgement* that the focus on beauty as a universal property of aesthetic objects was born. Other philosophers, like Leibniz, shifted

N. Hutton (✉)
Independent Scholar, New York, USA
e-mail: noahhutton@gmail.com

J. L. Contreras-Vidal et al. (eds.), *Mobile Brain–Body Imaging and the Neuroscience of Art, Innovation and Creativity*, Springer Series on Bio- and Neurosystems 10, https://doi.org/10.1007/978-3-030-24326-5_3

the focus to the observer, but were equally interested in aesthetic universals. Leibniz treated aesthetic objects in relation to the emotional valence in their holder or perceiver, specifically interested in their relationship to pleasure:

> *Pleasure* is the feeling of a perfection or an excellence, whether in ourselves or in something else. For the perfection of other beings is also agreeable, such as understanding, courage, and especially beauty in another human being, or in an animal or even in a lifeless creation, a painting or a work of craftsmanship, as well [7, p. 697].

There has been much refutation and complication of this kind of universal aesthetic idealism in the ensuing decades, the most potent of which decenters the program of universal cultural norms from a decidedly Eurocentric point-of-view. Yet the notion of beauty and pleasure as measurable aesthetic features across all brains persists in the field of neuroscience, egged on by the supposed universalism of the neural structures underpinning aesthetic engagement. In this sense, though the last century of art theory and criticism and the gradual development of the neuroscience of aesthetic experience have destabilized the notion of the lone artistic genius, neuroaesthetics has perhaps overcorrected into its reductionist roots, gradually walking itself toward the attractive promise of genericized universalism, and in the process risking an erasure of cultural specificity and personal context. In its search to "crack the code" of how aesthetic meaning is shaped in "the brain", the desire is to place all humans in the same brain-bucket, but as of yet the question remains of how big of a bucket will be needed to fit the totality of aesthetic engagement into its scientific program.

Indeed, those who turn to neuroscience for an explanation of beauty display an optimism in their interdisciplinary pursuits that is anchored by two core beliefs. One is the general, unshakable logic of materialism: everything we experience, think, or do is tethered to our nervous system, and thus illuminating that system naturally informs the study of things we see and create, like art [6]. The second looks to evolutionary science to inject the weight of history into the pursuit, for if evolution has shaped our brains, its teachings might help explain why we all do what we do, even if that doing seems to take varying eventual forms, a "cheesecake" issue of cultural specificity that belies deep evolutionary similarities [2]. These beliefs are buoyed by the early promise of discoveries in neuroscience, which provided dazzling, albeit preliminary, accounts of how parts of visual cortex organize and process information, and how certain instincts for attraction and revulsion to visual information may indeed be hard-wired. Thus the optimism in the ability of neuroscience to explain the lofty questions bound up in seeing and making art is considered by the purveyors of neuroaesthetics to be a continuation of where the field is heading, rather than a blind leap with no hard evidence, as certain outspoken critics of neuroaesthetics would attest [8].

So for those eager to set forth answers now, the question looms large: are the returns from the present-day interdisciplinary dialogue of neuroaesthetics original and of use? And if so, in returning to our Bohrian model of engagement, are brains themselves not the ultimate apparati? For in their biases and their predictions, their squeezing of perception through narrow bands of sensory input, and their vast oceans of memory containing all the context of a person's life, what else—beyond the detailed account of this neurobiological instrument—must we seek out?

2 A Fateful Encounter

Ideally, interdisciplinarity is the act of two fields entwining to produce something new. This newness ought to happen in the space opened up by the encounter, inaccessible by either field left to its own devices. I first studied art history and neuroscience separately as an undergraduate and failed to figure out a way to relate those two interests to one another. I became interested in neuroaesthetics precisely because I saw it as an interdisciplinary field by definition: it had two things smashed together in its name, and it needed both to exist: the breathtaking science of the brain encountering the world of culture and aesthetics. Perhaps neuroscience could offer something that the humanities could not; perhaps the humanities could offer science something it desperately needed as it began probing the seat of subjective experience. I began blogging about the field, interviewing neuroaesthetics researchers, and though their insights and research are illuminating and worth longer discussions, I want to turn now to a description of one fateful encounter at one specific interdisciplinary event. I do this because I believe that too often in the annals of theoretical essays and critical nonfiction the actual real-world friction that occurs when disciplines rub up against one another is lost, replaced by long-winded arguments (as I've likely been doing so far, here). But for me, personally, it was only when I ran into some of the fiercest critics of my own interests that I began to widen my perspective on aesthetic engagement, so I offer this account as a snapshot of interdisciplinarity, a moment in time that reshaped my thinking.

We were at the 2013 Venice Biennale[1] as part of a symposium organized by *The Association for Neuroaesthetics* to respond to the work of the performance artist Tino Seghal, who had been making sensational waves in the art world not only for the nature of his work but for the nature of his post-contractual art transactions, which famously avoid written documentation and insist on unconventional definitions.

[1]More information and video recordings available at http://aon.neurobureau.org/venice-symposium-2013/.

In the Giardini, the main gallery space in the Biennale park, a group of us—philosophers, neuroscientists, art historians, and me—came upon Seghal's esoteric piece (which would go on to win the top prize at the Biennale) involving several performers who would rotate into sitting and laying positions in the center of a large hall throughout the day, voicing slowed-down versions of pop songs and other vaguely familiar incantations. Standing next to me in our group was Olaf Blanke, who investigates the mysterious fluidity of body perception, and Vittorio Gallese, one of the co-discoverers of mirror neurons and the leader of a subfield of cognitive science now known as "embodied simulation" [3].

Standing across the room was the philosopher Alva Noë, Berkeley professor and author of *Out of Our Heads*, a manifesto of "embodied cognition" (not to be confused with Gallese's theory of *embodied simulation*, the process of empathically simulating the actions of others inside one's own motor system without necessarily acting externally—Noë's *embodied cognition,* on the other hand, argues for extending our concept of the boundaries of the human mind out of the brain and into the external environment, literally). Earlier that year, Noë had written an opinion piece for the *New York Times* entitled "Art and the Limits of Neuroscience" in which he railed against any neuroscientific approach to art and aesthetics, and even to understanding consciousness. I wrote what now appears to me to be a somewhat bitter paragraph-by-paragraph response to Noë's article, and though I still would contest the overreach of his statements, there is something that continues to ring true about one of the baseline critiques he offered in his piece:

> What is striking about neuroaesthetics is not so much the fact that it has failed to produce interesting or surprising results about art, but rather the fact that no one — not the scientists, and not the artists and art historians — seem to have minded, or even noticed. What stands in the way of success in this new field is, first, the fact that neuroscience has yet to frame anything like an adequate biological or "naturalistic" account of human experience — of thought, perception, or consciousness [8].

If someone asked me to quickly describe what neuroscience has produced that is of interest to art—what the true bumper crop of neuroaesthetics has been—I might unfurl a laundry list of findings, mostly from visual neuroscience, and plenty from Gallese's explorations into the empathic motor system. I would hope to convince you based on the sheer quantity of experiments that in one way or another neuroscience has offered something worthwhile about how we understand the creative, perceptual, or evaluative process. You might notice that this list would be made up of small and finite experimental examples, many tethered to the coarse explanatory weight of neuroimaging.

With the list spooled out, you might wonder whether there's an overarching theoretical framework that could tie all of this together, could connect the dots between the silos of the research community. If I were trying to summon an overarching theoretical framework in a book, as many have, I might present pieces of visual art, music, dance, or films along the way that would each dovetail with discussions of scientific studies on related aspects of perception, emotion, or memory, as I tell you how I think the brain works and why we make art and why these pieces of art I presented are how they are and why many people consider them to be great.

But what I would still be missing—and indeed what much of neuroscience seems to still be missing—is that overarching theory, what Noë calls a "naturalistic account of human experience." What can neuroscience really add to art theory, practice, and criticism that is of clear and present use—and vice versa? Do we need a neural theory of consciousness before any overarching theory of neuroaesthetics can bear full weight? And how do we avoid the omnipresent trap—prevalent in books on art and neuroscience as well as in sexy public-facing discussions between artists and scientists—of ascribing neuroscience onto art, where the former is treated as the ultimate Truth and the latter as the exotic, intuitive Other?

The next day, we sat around a table in front of a small audience gathered in the Peggy Guggenheim Library in Venice and discussed Seghal's work. Art historians described what the work reminded them of; neuroscientists described how the work might be experienced (a routine that can dance perilously close to suggestions of how it might be *explained*) by means of certain regions, connections, and processes of the brain.

When it was my turn to present, I first introduced the distinction between *descriptive* neuroaesthetics (science that correlates activity in brain regions to features in artworks which seem to depend on the functions of those regions), and *experimental* neuroaesthetics (a more mature line of work, where experiments are devised to study the perceptual process itself, rather than matching things up with the art after the fact).

But when it came to speaking specifically about Seghal's work, I fell into the same old trap of descriptive neuroaesthetics, of talking broadly about the brain and letting the specificity of the artwork slip away, just as Noë had warned it would. I wanted to respond to what Noë had said in his opening remarks, that talking about how "art activates us" is a mistake, and that art should be thought of as "providing us an activity to activate the work of art."[2] To me his insistence on keeping the conversation outside the brain and never bringing in a thread of cognitive science into a symposium organized by the *Association for Neuroaesthetics* seemed

[2] I rely here on an unpublished transcript of the event provided to me by the *Association for Neuroaesthetics*.

particularly stubborn. So in my response to Seghal's piece I described how the knowledge of two seemingly opposing cognitive processes—top-down processing and bottom-up sensory perception—are themselves locked in an ongoing piece of interior performance art. Because of how long it took me to sink into the rhythms and vocabularies of Seghal's piece in the Giardini, the relationship between this interior dance of top-down and bottom-up—the expectations based on experiences of previous work mixing with the actual sensory information arriving in the moment—seemed appropriate to discuss at a neuroaesthetics symposium. But in doing so I slid down that perilous cliff of *explanation*, letting the art recede into a mirage of a neuroscientific catch-all.

As soon as I finished giving these opening remarks, my misstep was brought to the foreground by the art historian Sigrid Weigel, who immediately challenged my comments. "When you talked about top-down and bottom-up, not only the metaphor irritates me, but also the question of how one can bring neuroscience into art history and the other way around," she said. "When seeing and reflecting on Seghal's work, I would say, this—this is not enough." Weigel's issue was with the dominance of visual neuroscience, which she rightly sees as too often taking precedence over motor, auditory, or more complex emotional systems when infusing neuroscience into discussions of art.

After I added some assurances that my comments were not meant to explain anything, but rather to "add a layer that could enrich and expand the discussion as opposed to explaining or limiting," a full-fledged turf battle broke out. Art historian Michael Diers asked why I am so interested in art: "Is it to ennoble your neuroscience?"

Vittorio Gallese interjected on my behalf, responding to Diers:

> Why are you so puzzled? Let's look at the past and progress will ensue. When Warburg was in Florence he was heavily reading Charles Darwin, and I don't think he read Darwin to ennoble the history of art or the other way around. People are curious. So why should we prevent ourselves from an additional perspective just because we cross boundaries? Are we afraid of losing our specificity? I don't see why so many people are puzzled, afraid, angry, confronting themselves with these topics from people from other fields.

The art historians claimed that neuroscience always skews discussions of art toward the visual; the neuroscientists protested. Alva Noë returned to his entirely valid stump speech about art disappearing from neuroaesthetic discussions, that it "is never actually made the focus of attention, why? Because what we end up looking at is something as a stimulus, but of course everything is a stimulus, there is no human experience without the brain, there is also no human experience without the body and a situated animal interacting dynamically with the environment."

I left the symposium scratching my head: if the point was to find new approaches to art through the infusing of neuroscience (hence the *Association for Neuroaesthetics*), where were the new ideas?

I have come to agree with Noë that art disappears from many neuroaesthetic papers, books, and public discussions [8]. In these cases, the art is treated as a mere stimulus, a rocket booster that can be discarded on the way to X, where X is inevitably a brain-based answer. But while Noë does point out the pitfalls of this rocket-booster approach, he does not integrate cognitive science in any meaningful way into his discussions, and thus I don't believe his approach offers a new way of approaching art. Noë's insistence on shifting the discussion out of the head and into the environment, thereby neglecting neuroscience altogether, may clear away the shaky causal foundation of early neuroaesthetics but eventually ends up feeling just as devoid of new ideas as that which he seeks to destabilize. His coldness toward neuroscience is just another way to draw battle lines in the dialogue between the humanities and the sciences, an all-too-easy territorialism that promotes more turf battles than it opens new questions.

My hunch is that there is a false appraisal of neuroscience that dead-ends interdisciplinary presentations, including my own schpiel in Venice. In such situations, neuroscience is mistakenly (by scientists and philosophers alike) treated as an end-domain: a place we arrive at for an answer and in turn receive quixotic scientific visuals of the brain. From *Neuromania* to Noë, backlash in this context makes sense: the current answers to weighty questions about art and existence are weak placeholders that gain steam from the nebulous authority of anything *brain*, but in the end the paucity of the current understanding of the brain betrays any hopes at an appropriately complex view of cause and effect. It is in these shortcomings that the dead ends of current dialogues are sensed and the regressive backlash against future attempts sown. Whether it flows from genuine excitement over early indications from neuroscience research itself or comes in reaction to backlash from the humanities, the overhyping of neuro-truth as an end-domain has led those of us actively interested in interdisciplinary dialogues to the precipice of our own disappointment: the sinking feeling that neuroscience might not be able to land us on that moon where we'd hoped to one day plant our flags and write a universal guide to aesthetic engagement. Then we arrive at an event like the one in Venice to try to find new connections between the arts and sciences, but all too often interdisciplinarity resembles the now-withered concept of bipartisan political compromise, where, like a bill that is stripped of its most potent actions so that it can receive a majority vote, the attempt to bridge a divide ends up leaving behind the most virtuous elements of each field in the pursuit of a valorized middle ground. It's that strange feeling in the room after an interdisciplinary exchange, when it seems the artist and scientist have talked at and through one another, but not really *with* one another.

So in hopes to surmount the twin challenges of false end-domains and false niceties, in recent interdisciplinary exchanges I've been testing out a new approach to engaging with the neuroscience of aesthetics, one that acknowledges the strong gravitational pull of a neural end-domain but offers a hand-picked analogy from the space race era to replace the classic image of a flag-planting triumph. I argue that when the humanities, social sciences, or any other discipline engages with—or is engaged by—the neurosciences, the metaphor we ought to keep in mind is that of Apollo 13, for it was in that near-disaster that the human agents were able to transform their intended end-domain from the ominous site of an inevitable crash-landing to the engine for their slingshot back to Earth, and thus a source of renewed momentum.

In the same way, we might imagine aesthetic engagements from the perspective of the arts and art history that swing close to neuroscience for its new ideas, tuning into the undeniably attractive force of its material lessons about the seat of human subjectivity, but remaining acutely aware that they may never offer end-all answers to our individual questions about art and the imagination, let alone scale up to universals to touch all of aesthetic experience. The Apollo 13 approach is ultimately more curious about how the gravitational pull of neuroscience can help us get back to the personal and the political; how its transformational knowledge can re-activate and re-engage us as active aestheticians.

3 Toward Future Engagements

A new trend in neuroaesthetics suggests a way in which aesthetic engagement, infused with explanatory momentum from brain science, can ask new questions of the personal and the political. In an ironic twist, this insight arrives by means of a line of research that concerns the *un-engaged* brain, at rest, and its lessons as to how meaning is formed during artistic engagements.

The Default Mode Network (DMN) is a distributed network of brain regions whose activity seems to reappear in the valleys between the peaks of outward-focused attention, when you're not necessarily *doing* anything. The spike in research interest about the DMN marks a fundamental paradigm shift in neuroscience, one that goes against the traditional modus operandi of brain scanning, wherein a researcher measures the effect on the brain of active engagement with a certain stimulus, usually coming from the external world. Indeed, "finding a network of brain areas frequently seen to decrease its activity during attention-demanding tasks was both surprising and challenging," notes Marcus Raichle of Washington University, "because initially it was unclear how to characterize their activity in a passive or resting condition." [9, p. 416]. In approaching the DMN, Raichle's work has pointed toward the need to reorient our binary notions of *active* versus *inactive*, for with the DMN we find the omnipresent "baseline" brain, the parts that brain imaging studies always seek to cancel out so that the true point of "activation" can be seen. It turns out that the full apparatus of aesthetic engagement involves not just our active, willful gaze, but the "resting" brain itself— the stars in the sky, ever-present behind the bright beams of the day.

The DMN consumes most of the energy metabolized by the brain as a whole. It's an omnipresent, baseline state, but it is most active during the in-between moments when you're staring up at the ceiling, riding on the train, reflecting at the end of a long day—moments which seem to be tethered in study after study to activity in regions such as the angular gyrus, the posterior cingulate cortex, and the medial prefrontal cortex, which are regions that have been implicated in autobiographical thinking, and in the relation of the self to other people, events, and planning for the future.

Though these are massive areas of the brain to be tossing around in any kind of ultimate explanatory way, it hasn't stopped some researchers from beginning to probe how the DMN may be involved in aesthetic engagement. Neuroscientist Ed Vessel devised an appraisal system for viewers to rate a wide range of artworks—from abstraction to portraiture and landscapes—while lying in an fMRI scanner [10]. The participants were shown the artwork for a brief interval, then given four seconds to submit a rating on a scale of 1–4 of how powerful, pleasing, and profound they found the image. Vessel's key finding is that for ratings of 1–3, the DMN showed fairly low activity, with subtle, linear increases as evaluations improved. But for the top rating of 4, there was a dramatic, step-like jump in activity, as if the DMN fully "came online" for the highest aesthetic appraisals. At these moments, the sensory areas involved in viewing the art stayed online as well—a rare co-activation of two networks that usually exhibit toggling behavior, depending on whether you're focusing attention outward or at "resting state", looking inward. Vessel described this step-like activation of the DMN during "4" ratings as a "signature" of peak aesthetic response, and argued that this activity supports the notion that the DMN is about self-referential processing, as in, "*I* love this painting."

We might rephrase the conclusion of this line of research as suggesting that "self-relevance is an integral part of aesthetic experience." For someone coming at this from the humanities, this big takeaway still ends up sounding like a self-evident, intuitive truth known to the arts and art historical practices for eons. The results may speak volumes for the neuroscience of the DMN and its relationship to aesthetic appraisal, and they may make significant progress from past studies, which is how much of science works; for art theory, though, these statements can continue to sound like counting to ten.

What if we paused for a moment to consider an interpretative use of the DMN that was not just about bringing personal taste to the act of engagement, but also social and political taste? What is the Default Mode Network of a larger unit than the individual; say, contemporary capitalism? And in asking such a question, how can we better orient ourselves with the aesthetic preferences of the world around us?

Columbia's Zuckerman Institute recently named Jeff Koons as its first artist-in-residence, centering that work above other possibilities (say, a local Harlem artist). Koons' work, like other mega-successful contemporary artists, is made for and sold to a global financial elite. What is the nature of a society governed by those whose peak aesthetic experience brings a DMN-associated brain network online that sees a piece of themselves in Jeff Koons' ironically disengaged, meta-upon-meta, wealth-signifying balloon-poodles?

For if we buy into the adage often tossed around in the arts that "the personal is political," then in our moments of active engagement with aesthetic objects we must begin to treat not just artworks but also the full human apparatus—the brain, and all its contexts, active and passive—as a site of politics. In this way, aesthetics becomes a critical weapon, and neuroaesthetics the site of a forthcoming battle.

For in the gap between the sweet-nothings of neuroscience and the hallowed hallways of art history and criticism, brain-platitudes—like Koons-platitudes—would have us believe they're not intimately linked to the technocratic and ethically-fraught world that continues to reproduce itself around them. In a world where more than half of the U.S. BRAIN Initiative was funded by DARPA, where tech companies edge ever-closer to creating their own brain-modeled, deeply-learned algorithms to maximize revenue, what is the full range of what twenty-first century neuroscience will pursue and enable? And how can we paint a more complete picture of its aims, complicate its platitudes, so that we can engage with the world with a more complete awareness of the apparati in which we engage?

The neuroscientist would undoubtedly stand behind the ethical shield of disease prevention and treatment—an entirely valid stance. But as reductionistic methods probe deeper into the subjectivity of the human mind, and as pharmaceutical corporations pump their prices and pathologizations, this Valid and Good stance is no longer enough. This is what Yuval Noah Harari alludes to in *Homo Deus*, noting that "No clear line separates healing from upgrading. Medicine almost always begins by saving people from falling below the norm, but the same tools and know how can then be used to surpass the norm" [4, p. 51].

By omitting any such contextual discussion, valorizations of neuroscience thereby ignore the ethical and political embeddedness of the field, and thus risk distancing the general public from the crucial interrogations ahead of us. Over and over again we instead receive unfettered utopianism, as in Eric Kandel's closing lines to *Reductionism in Art and Brain Science*: "the new science of mind seems on the verge of bringing about a dialogue between brain science and art that could open up new dimensions in intellectual and cultural history" [5, p. 189]. I wish I was as optimistic, but the world I see around me suggests a more ethically complex picture of some of the places brain science may be employed, like DARPA gobbling up those BRAIN insights to help drone operators stop sweating their remote deeds and start forgetting them faster. One might protest that a chapter on aesthetic engagement is the last place to mount such a political critique. I would argue that the exact opposite is true. As neuroscience increasingly encroaches on the domain of human subjectivity—our sacred imaginative apparatus itself—the discussion of aesthetic engagement, of what happens in the brain when we view or create art, becomes merely a proxy conversation for the real change ahead: the reduction of subjectivities to their constituent parts so that they can then be put back together again in more sublime ways, if you can afford it.

References

1. Barad, K.M.: Meeting the Universe Halfway: Quantum Physics and the Entanglement of Matter and Meaning. Duke University Press, Durham (2007)
2. Changizi, M.: Harnessed: How Language and Music Mimicked Nature and Transformed Ape to Man. BenBella Books, Dallas (2011)
3. Gallese, V.: The empathic body in experimental aesthetics—embodied simulation and art. Empathy 181–199 (2017)

4. Harari, Y.N.: Homo Deus: A Brief History of Tomorrow. McClelland & Stewart, S.l. (2015)
5. Kandel, E.: Reductionism in Art and Brain Science: Bridging the Two Cultures. Columbia University Press, S.l. (2016)
6. Livingstone, M., Hubel, D.: Vision and Art: The Biology of Seeing. Abrams, New York (2014)
7. Loemker, L. (Trans., ed.) Leibniz Philosophical Papers, p. 697. Chicago University Press, Chicago. (1956)
8. Noë, A.: Art and the Limits of Neuroscience (2011). Retrieved from https://opinionator.blogs.nytimes.com/2011/12/04/art-and-the-limits-of-neuroscience/
9. Raichle, M.E.: The brain's default mode network. Annu. Rev. Neurosci. **38**(1), 433–447 (2015)
10. Vessel, E.A., Starr, G.G., Rubin, N.: Art reaches within: aesthetic experience, the self and the default mode network. Front. Neurosci. **7** (2013)

Noah Hutton is a writer and filmmaker who has presented work at the Venice Biennale, Society for Neuroscience, Wellcome Collection, Rubin Museum of Art, and elsewhere. He directed the documentary films Crude Independence (SXSW 2009) and Deep Time (SXSW 2015). He studied art history and neuroscience at Wesleyan University.

Brain Mechanisms of Creativity: What We Know, What We Don't

Arne Dietrich

In response to "Hey Yogi, I think we are lost", Yogi Berra, former player and general manager of the New York Yankees once said: "Yeah, but we are making great time." It's hard to think of a more fitting depiction for the present-day, ill-conceived efforts to identify the mechanisms of creativity in the brain. Let me come right out and say it. It's phrenology.

Sure, studying creative thinking in the lab, under tightly controlled conditions, isn't the easiest way to make a living as a psychologist. Even for the wilderness of human thinking, creative ideas seem to be deliberately designed to defy empirical inquiry. They pop up as they please and, when they do, they hit you like the hammer of Thor. No wonder we have always mystified them—visits from the muse and light bulbs come to mind. Most brain scientists would rather try to nail jelly to the wall.

Yet, finding the cognitive and neural mechanisms of creativity is a topic that couldn't be any more central to our humanity. Suppose an advanced alien lifeform visits Earth to investigate if *Homo sapiens* is worth saving. Suppose further that they don't have a portable consciousness-detector, a small antenna-held gizmo they can conveniently hold to our heads to check for signs of inner musings. What would they identify as the defining characteristic of being human? Taking a quick look around and seeing what we have done with the place, they'd be hard pressed to put any other item on top of their list than our creativity and inventiveness. We are an intensely creative species and there isn't an element of the periodic table we haven't tinkered with to utterly transform the world we live in. All progress in the arts, sciences, and engineering originates from the capacity to change existing thinking patterns, break with the present, and create something new. Creativity, and its derivative products—the knowledge and artifacts that make up human culture—is the quintessence of our humanity.

A. Dietrich (✉)
Department of Psychology, American University of Beirut, Beirut, Lebanon
e-mail: ad12@aub.edu.lb

There is also a very practical goal here. By uncovering the nuts and bolts of how a three-pound pile of electrified biochemistry conceives of mathematical theorems, invents kitesurfing, creates beautiful art, discovers the laws of nature, thinks of space rockets, and designs buildings that look like sea shells, we might be able to enhance this process. It'd be an instant game-changer for any nation or company that gets an initial handle—not on how to invent new stuff—but on how to improve the invention machine itself that invents all the new stuff. From Silicon Valley to your local arts school, from the world's medical associations to the U.S. military, everyone would be all over any paradigm promising to get us closer to that prospect.

Given all this, one would think that the neurocognitive mechanisms of creativity are the subject of intense research efforts in the behavioral and brain sciences, with dozens of labs and massive funding involved. But this is not the case. Only about a handful of labs tackle this vexed problem empirically and funding is practically nonexistent. One has to asked why, given the paramount importance of the topic. For reasons we lift from the muddied waters here, the experimental study of creativity did not develop over the past 70 years like other areas of psychology—relentlessly forwards and upwards, in case you weren't paying attention. While there has been a veritable explosion of knowledge about the mind, creativity has become the most striking exception to this success story. It is hard to think of a mental phenomenon so central to the human condition that we understand so little. It is not too much to say that, at this point, we do not have a single sound mechanism to explain the extraordinary creative capacities of an Ludwig van Beethoven or Marie Curie. In other words, there is no mechanism, cognitive or neural, that we know for sure—with empirical evidence—that is responsible, enhances, or can explain in any way how creativity arises in the brain, let alone state why some individuals are more creative than others. This is remarkable, given the victory parade the neurosciences have been on for the past few decades.

J. L. Contreras-Vidal et al. (eds.), *Mobile Brain–Body Imaging and the Neuroscience of Art, Innovation and Creativity*, Springer Series on Bio- and Neurosystems 10, https://doi.org/10.1007/978-3-030-24326-5_4

1 How to Study Creativity in the Lab

So, what are cognitive neuroscientists to do if they want to catch creative thoughts as they occur? Obviously, they cannot simply take volunteers, shove them into the nearest brain scanner and tell them: now, please be creative! That is a given.

But that's exactly what they do. In this section, allow me to briefly explain the two paradigms that have probably generated well over 90% of anything you might have ever read about the brain mechanisms of creativity to see if you can go with them.

The first is divergent thinking, proposed by Joy Paul Guilford in his farewell address as president of the American Psychological Association in 1950. In it, he called for the study of creativity and backed his call to arms with a proposal on how to go about doing so, the concept of divergent thinking. The concept is defined as the ability to generate multiple solutions to open-ended questions. This conceptualization was quickly taken up by others because it represented the hope to bring a hitherto intractable problem into the folds of empirical science. Several standardized testing methods for "creativity" were subsequently developed. The most popular of these psychometric tools—to this very day—are Torrance's [12] Torrance Test of Creative Thinking, which is entirely based on divergent thinking and Mednick's [9] Remote Associates Test, which is based on the related construct of associative hierarchies. The one most often used in neuroscience studies is a short version of the infamous Alternative Uses Test, or AUT as it is known in the field. It asks participants to generate alternative uses for common objects such as a brick, safety pin, or automobile tire.

To see what it does, take one minute and write down all the alternative uses of what to do with a brick. Then score the test on three factors. First is ideational fluency, which is simply the total number of ideas you had. If you came up with 8 items your score is 8. Second is flexibility, which is the number of different types or categories of ideas. If all you could think of were uses as a weapon (hit an enemy, throw at a blocked door, etc.) your score would be a meager 1. Third is originality. This is often assessed with the so-called consensual method, in which naïve judges rate unusual answers with 1 point, unique answers with 2 points, and so on. Suppose you had one unusual and one unique item, your originality score would amount to 3. Sum up all points (8 + 1 + 3) and your overall creativity score would stand at 12. And now for the real issue: Do you think this captures your creativity? Do you think that a score of 12 as compared to, say, 9 would tell you anything about a person's creative abilities? How do you think bone-fide, creative giants like Einstein or a Shakespeare would have scored on the AUT?

Whenever I give the AUT to my students in class and ask them the same questions, they laugh. No way, they say. I have given the AUT to groups of artists and watch them grow incandescent with anger upon the realization that scientists reduce their most prized possession to this. In all honesty, can we really expect a test that asks you to imagine alternative uses of a safety pin to pick a Mozart from a certified public accountant?

The AUT is actually part of the larger Torrance Test of Creative Thinking (TTCT; [12]), which does have some decent, real-world predictive properties. But as the cognitive psychologist Mark Runco [11], one of the 'custodians' of the TTCT, points out, the full TTCT is an hours-long test and, most importantly, not a test of 'creativity' but a test of divergent thinking. Unfortunately for us, and for progress, neuroscientists have dealt with such disclaimers in the handiest possible way. They ignored them. In neuroscience studies, the mini AUT is used (for scanning purposes), and the results are routinely proclaimed *as discoveries about creativity*. And as soon as the media gets involved, all inhibition is lost, and the overselling of the findings deteriorates into outright pseudoscience.

The second neuroimaging paradigm uses music—jazz improv mostly. Here a melodic string is completed in one of two conditions, a set pattern from memory (control condition), and an improv string ('creativity' condition), with variables like length, or cadence all held constant. In case you have doubts about the ecological validity of this paradigm, especially in a sterile brain-scanning lab, allow me to tell you a little anecdote. One of the participants—a famous jazz musician—in perhaps the single most famous experiment of this type [8] took part once in a panel discussion at a conference and was asked if he considered what he did in the improv condition to be creative. Without a whiff of hesitation, he said: No! It is the closest emperor-has-no-clothes moment I have ever seen at a conference.

2 The Rocky Horror Pixel Show

Neuroimaging creativity sounded like a good idea 20 years ago, a low-hanging fruit if there ever was one. There were ready-to-go 'creativity tests' in existence and the university's brand new neuroimaging center was just next door. How can you fail? For—surprise, surprise—some brain region was indeed firing away with extra oomph and the topic itself; gee, does it get any sexier than that? Alas, a bit of level-headed thinking would have saved a lot of grant money and us from this pixelated Potemkin village.

It's not my intention here to mount a general critique of neuroimaging technology. Far from it. The great neuro show of recent decades has revealed a great deal about the human brain and how it functions. At the same time, such a frenzied and hyped atmosphere invites excesses, and creativity research is perhaps the best example of one. Even if we bracket test validity and the artificial lab conditions (see [1]), there are two additional problems that render divergent thinking theoretically incoherent for neuroscience.

First, divergent thinking is a false category formation [4]. It only takes a moment's reflection to see that we can also be creative with convergent thinking, a fact everyone in the field acknowledges. For all the uplifting stories, the Einsteins riding on beams of light, the Newtons watching falling apples (a myth likely originating from Voltaire) or the Archimedes displacing bathwater, creative ideas can just as easily be the result of laborious trial and error. What would we otherwise make of Edison's "empirical dragnet" method that yielded a total of 1093 patents; Watson and Crick's algorithmic approach to testing the stability of DNA base pairs; Bach's assembly-line tactic to composing hundreds of cantatas; or the imaginative ways in which NASA engineers solved the problems of the otherwise doomed Apollo 13 mission? Since the exact opposite—convergent thinking—can also produce creative ideas, the obvious question arises of what, exactly, is creative about divergent thinking? No one has been willing, or able, to explain this to me. If both, divergent and convergent thinking, can lead to both, creative and non-creative thinking, the concept of divergent thinking as a proxy for creative thinking makes no sense. It is incapable of identifying the processes that turn normal thinking (whatever that is) into creative thinking.

The false category problem also applies to all other conceptions of creativity that have been tried over the years, such as defocused attention, remote associations, flow, madness, lateral thinking, low arousal, daydreaming, REM sleep, right brains, mindfulness, unconscious thinking, prefrontal cortex, or the default mode network. Given their opposites also lead to creativity, they all fail to carve nature at the right joints.

Second, divergent thinking is, like creativity itself, a compound construct consisting of many different, separate, and distributed mental processes with no one having the slightest clue what they are and in what mix. Although this problem is also widely acknowledged, there is no effort underway to break divergent thinking down further so as to link it to the kinds of processes we use to operationalize all other psychological phenomena, such as working memory, cognitive control, perceptual processes, semantic memory, inhibition, or executive attention. Remember that neuroscientists hunt for underlying mechanisms; that is, individual mental processes that can actually be identified by neuroimaging. Divergent thinking, therefore, does nothing to address the initial problem, the fallacy of taking creativity as a monolithic entity. In fact, the concept is used today the way it was initially conceived by J. P. Guilford in 1950 [6]!

For neuroimaging, all three confounds—validity, false category formation, compound construct—combine to make defeat certain. Simply put, if you fail to isolate the subject matter of interest in your study, you cannot use neuroimaging to search for mechanisms. You just don't know what the brain image shows!

3 Neuroanatomically-Updated Phrenology

Phrenology is to neuroscience as astrology is to psychology, the quintessential pseudoscience of the discipline. Mention the P-word to a group of neuroscientists and you won't be able to finish your sentence so eager would they be to interrupt you and put ideological distance between themselves and phrenology. Trouble is, however, the basic error that fueled phrenology some 200 years ago is nearly impossible to shake and it keeps on popping up.

Franz Joseph Gall, the father of phrenology, associated a total of 27 regions with specific mental faculties, all without so much as a shred of evidence. The list contained such doozies as a center for mirthfulness, combativeness, marvelousness, secretiveness, and, my personal favorite, the organ of philoprogenitiveness—which he located, if you must know, just above the middle of the cerebellum. It's easy to laugh at this bumps-on-the-skull idea today but few people appreciate the enormous popularity phrenology enjoyed at the time. In Victorian Britain, it ranked with Darwin's theory of evolution.

For creativity, the best-known example is, of course, the right-brain theory. It emerged in the 70s and has proven to be a particularly vicious mutation of phrenology, probably because it was the first to metastasize to a global audience. You can still find a business seminar on how to think with both sides of your brain and an endless supply of books and magazines promising an easy step-by-step program on how to tap into your creative right-brain potential. You might as well ask someone to make better use of the thalamus.

The unlikely story of the right brain developed from split-brain patients who had their corpus callosum cut to manage life-threatening epilepsy. Subsequent research showed that there are indeed several cognitive functions that are lateralized—language most prominently. The generalization commonly extracted from this is that the left hemisphere (LH) is more skilled at analytical tasks, such as sequential reasoning, while the right hemisphere (RH) is more skilled at tasks requiring synthesis, such as seeing the whole of a picture (see [5]).

Anyone with two neurons to rub against one another can readily imagine what happened next. For, no matter how careful such a trend is formulated, generalizations that feed into our phrenological thinking habits inevitably set the stage for a broadside of flag-waving oversimplifications. And creativity was such an easy target, given our overly romantic view of it. Isn't regular old thinking conscious, analytical, and systematic, perfectly suited, in other words, to the logicality of LH? And isn't creative thinking intuitive, primal, holistic, and delightfully irrational, custom-made, as it were, to the mysterious ways of RH? Sadly, LH has since been the go-to brain half of methodical and unimaginative engineers, while RH has become the creative playground of inspired artists. Today, everyone recognizes the right-brain theory as a dud, of course, but its persistence in the general public underscores the need, in clear and vivid form, of how important it is to systematically demolish ideas gone bad.

Phrenological thinking is so pervasive and the opportunities for flummoxing so abundant that the field is full of such misbegotten proposals. The next candidate in this wild-goose chase was the prefrontal cortex, which owed its brief moment in the limelight due to its general position at the apex of human mental faculties. It didn't stay there long as it was quickly followed, in about the mid-2000s, by an idea that was also just too irresistible for the myopic theorizing that has come to characterize this field—the default mode network (DMN). Proposed by Raichle et al. [10], this network of brain areas shows heightened activity at 'rest' and is thought to support mindwandering, daydreaming, or moments of introspective thought. As was the case for the right-brain theory, this struck all the right chords in some quarters. Researchers jumped on it and, for about 10 years or so, the DMN was the bull to ride.

Sharp-eyed readers will have recognized this as simply another false category formation. There is of course no reason to think that the other large-scale brain network, the central-executive network (CEN), which controls executive functions and shows heightened activity when we focus our attention on a specific task, couldn't also generate creative thoughts. As this became clear, the phrenological explanation evolved yet again. The latest twist is that creativity is purported to be associated with enhanced connectivity and a back-and-forth interplay between these networks (see for example, [2]).

The all-important thing to remember here is this. All of these claims are extracted from—wait for it—the one-minute AUT! What's more, the findings are not packaged and sold as studies of divergent thinking. Aside from the occasional qualifying remark—as if this turns the water into wine—the papers routinely make grand proclamations about the brain mechanisms of creativity! Clearly, we need a sanity check.

I could go on and on. In fact, I think I will. Creativity is a topic where respectable people, even those of the highest scholarly standing, regularly rise to levels of speculation that can safely be called imprudent. Creative people are supposed to use more of their brain—somehow, for no one can tell you exactly how, let alone link this to creativity—use less brain more efficiently (which is, come to think of it, the opposite claim), have more dopamine receptors (or was it serotonin?), have more densely packed neurons, or more synaptic connections, or a thicker corpus callosum. Indeed, one can find claims in the literature for the whole funhouse of brain structures in the telencephalon—hippocampus, visual cortex, temporal lobe, and, why not, the basal ganglia. The next thing in tow, given the drift of things, is surely the mirror neuron system or neurogenesis in the hippocampus. And let's also not forget the legions of creativity coaches, leadership consultants, and motivational speakers who circle the globe in a tireless effort to meme-launder all of this and emit it to audiences suffering from uncritical idolatry. This shouldn't surprise us, I guess, given the time-tested ability of pseudoscience to expand in a vacuum.

Will these phrenological thinking habits about creativity go extinct? I expect not. It's too seductive. We might not even have seen yet the high water mark of this failed research program.

4 Where Must We Go from Here?

To sum this up in no uncertain terms, we know next to nothing about the brain mechanisms of creative thinking. What we do know, though, is that the paradigms that have been tried so far have failed. That is something. And it follows from this that if we do not change our ways and try new approaches and ideas, it is unlikely that we will know more about it in 20 or 50 years time. So, clearly, we need a restart, preferably one with ideas that are theoretically and conceptually sound. Here are 5 suggestions (Table 1).

First is the Vaudeville conception [3]. We might think of creativity as a single and cohesive entity in psychological terms, a personality trait—as in, Steve Jobs had it (notice the singular) and my grandfather did not—but creativity, as such, might not exist as a distinct and separate entity at the neural level. That is to say, at the level of the brain, creativity

Table 1 Five suggestions of concepts or ideas that are to replace the phrenological thinking currently in vogue and that should be part of any mechanistic explanation of creativity

Vaudeville conception
Evolutionary thinking
Prediction system
Dual architecture
Types

If we are to take creativity as the multidimensional and complex thing that it is, we have to, at the levels of mechanisms, really take it as the multidimensional and complex thing that it is

does not translate into a distinct neural signature that we can associate with the psychological understanding we have of creativity. Think philoprogenitiveness! But the whole rationale of neuroimaging studies rests on the assumption that creativity, or divergent thinking, is a discrete thing in the brain and that that thing is detectable by neuroimaging tools. The underlying tacit assumption here is that there is such a thing as 'normal' thinking to which a separate and extra something—the creative bit—is specifically added to make the sparkling difference. Few people would probably subscribe to this position once it is laid bare, but without it neuroimaging makes no sense. And it wouldn't matter if the 'creative bit' is a place or a network, the tacit assumption is the same. By way of comparison, it's easy to point out England on a map, but you can't find all people in the world who speak English that way. The Vaudeville conception counters this monolithic entity fallacy and takes serious the view that creativity is a highly complex and multidimensional phenomenon. Too different is what scientists, entrepreneurs, designers, or ballet dancers must do to be creative in their respective domains. Creativity, then, is fully embedded and distributed in the brain, or in a word, everywhere. Asking neuroscientists for the location of creative thinking is like asking them for the location of thinking.

Second is evolutionary theory. Broadly speaking, when we think creatively and break new ground we can be said to explore an unknown problem space. We try out several different options or solutions in this space (variation) and pursue one but not the others (selection). Mathematically, creative thinking can therefore be described as a set of evolutionary variation-selection algorithms with varying degrees of sightedness [7]; that is, our exploratory walks through unknown solution spaces aren't blind like in biological evolution but rather informed, to various degrees, by expertise and educated guesses. Despite the broad agreement that the basic grammar and logic of evolutionary thinking applies to human creativity, the two-step evolutionary rationale has been nearly universally ignored in setting up empirical protocols. It's hard to imagine useful neuroimaging data from studies that blend variation and selection, given that both likely engage different cognitive processes and different brain areas.

Third is the brain's prediction system. Theorists have been converging from quite different quarters on the idea of prediction as a central purpose of brain function. It's a new and powerful paradigm in the neurosciences. I have recently proposed that predictive representations might be the neural mechanism for the partial sightedness in human evolutionary algorithms [3]. This represents a proposal of a mechanism for a *specific component* of the creative process, which is much more readily subjected to empirical testing than the whole of creativity.

Fourth is the brain's dual architecture. It's well known that the brain has two anatomically and functionally distinct information-processing systems, one implicit and one explicit. And both can be creative. Needless to say, creative output for either the implicit or the explicit system would involve different mental processes and brain areas. This, too, has been nearly universally ignored in the neuroscientific study of creativity. But it has far-reaching implications for the cognitive and neural mechanisms of different kinds of creativity, given that both systems are anatomically and functionally distinct.

Fifth are types of creativity. Given the highly complex and multifaceted nature of creativity, an obvious way to make it more tractable is to parse it into several different subtypes. I have proposed to initially divide creativity into three distinct types, a deliberate mode, a spontaneous mode, and a flow mode [3]. To avoid the pitfalls of previous such attempts—false category formation and compound construct—the three creativity types are explicitly defined and delineated from one another based on established concepts in cognitive psychology and neuroscience. They are thus valid subtypes in the sense that they can be theoretically defended. Biting off smaller pieces of the larger pie also makes for a more targeted, and realistic, line of attack that will eventually lead to more meaningful data about how creativity happens in the brain. And it might also finally stop the bad habit of making grandiose statements about creativity per se. Since different types of creativity contain opposing cognitive and neural mechanisms—focused versus defocused attention or DMN versus implicit system, for instance—any global claim about creativity as a whole will almost certainly qualify as phrenology.

References

1. Abraham, A.: The Neuroscience of Creativity. Cambridge University Press (2018)
2. Beaty, R.E., Benedek, M., Silvia, P.J., Schacter, D.L.: Creative cognition and brain network dynamics. Trends Cogn. Sci. **20**, 87–95 (2016)
3. Dietrich, A.: How Creativity Happens in the Brain. Palgrave Macmillan, London (2015)
4. Dietrich, A., Kanso, R.: A review of EEG, ERP and neuroimaging studies of creativity and insight. Psychol. Bull. **136**, 822–848 (2010)
5. Gazzaniga, S.M., Ivry, R.B., Mangun, G.R.: Cognitive Neuroscience, 4th edn. W. W. Norton, New York (2014)
6. Guilford, J.P.: Creativity. Am. Psychol. **5**, 444–454 (1950)
7. Kronfeldner, M.E.: Darwinian "blind" hypothesis formation revisited. Synthese **175**, 193–218 (2010)
8. Limb, C., Braun, A.: Neural substrates of spontaneous musical performance: an fMRI study of jazz improvisation. PLoS ONE **3**, e1679 (2008)
9. Mednick, S.: The associative basis of the creative process. Psychol. Rev. **69**, 220–232 (1962)

10. Raichle, M.E., MacLeod, A.M., Snyder, A.Z., Powers, W.J., Gusnard, D.A., Shulman, G.L.: A default mode of brain function. Proc. Nat. Acad. Sci. **98**, 676–682 (2001)
11. Runco, M.: Creativity. Ann. Rev. Psychol. **55**, 657–687 (2004)
12. Torrance, E.P.: Torrance Test of Creative Thinking. Personal Press, Lexington (1974)

Arne Dietrich is Chair and Professor of Psychology at the American University of Beirut. He is the author of Introduction to Consciousness (2007) and How Creativity Happens in the Brain (2015), and numerous research articles on creativity, altered states of consciousness, and the psychological effects of exercise.

Theme and Variations as a Window into the Creative Mind

Anthony Brandt

1 Introduction

A central question in the neuro-imaging of creativity is: what sort of tasks should subjects perform? As outlined in a review by Dietrich and Kanso [3], experimental tasks generally fall into three categories: divergent thinking exercises, artistic improvisation, and tests designed to identify "Eureka moments."

The Alternative Uses Test is a staple of experiments for divergent thinking: subjects are asked to proliferate alternative uses for common objects such as a tire, paper clip, or brick. The subjects are then evaluated for fluency (the number of ideas they develop), flexibility (how diverse their solutions are), and novelty (how uncommon).

The second category involves monitoring artists such as jazz musicians and painters as they create. Typically, the artist is given an open-ended prompt and is encouraged to freely improvise. Although experimental conditions often impose necessary constraints—for instance, on physical movement—the goal is to observe real-world practice.

Finally, in "Eureka moment" studies, subjects are given a problem to solve. For instance, in Mednick's Remote Associates Test, subjects are given word triplets and asked to find the word that links all three. What word relates to all three: dust, cereal, fish? Officer, cash, larceny? The hope is to see what is happening in the brain at the precise moment when the right answer springs to mind.[1]

The imaging results so far have been mixed. Reviewing the literature in 2010, Dietrich and Kanso write:

> The most forthright conclusion that can be taken away from the present review is that not a single currently circulating notion on the possible neural mechanisms underlying creative thinking survives close scrutiny. Indeed, the data are so highly variegated that even weak trends are difficult to make out. (839)

One central critique made by Dietrich and Kanso is that while researchers are often trying to identify which brain regions go on- or off-line during creative tasks, that effort is fundamentally flawed. The preponderance of evidence points to creativity being widely distributed in the brain. Instead of trying to "map" creativity—which Dietrich and Kanso liken to a "phrenological" approach—it would be more productive to study the cross-talk between brain regions and how neural networks collaborate. The recent interest in functional connectivity is a promising step in this direction [1, 6].

But the experimental paradigms may also need to be revisited. For instance, divergent thinking testing "has received substantial criticism of its validity and practical value" [8, 24]. Yet imaging studies—even those analyzing functional connectivity—still often rely on these stalwart tests as the basis for observing the creative mind in action.

It may be fruitful to explore new creative tasks. To that end, I would like to suggest a long-standing creative challenge in Western classical music: the theme and variations. As a way of viewing creativity in action, this musical form has several advantages: it is a well-formed task that is easy to explain. It has ample precedent in history: composers across several centuries have taken on the challenge, ranging from J. S. Bach's thirty *Goldberg Variations* to Frederic Rzewski's *The People United Will Never Be Defeated,* an hour-long set of variations on a Chilean protest song. It is

Electronic Supplementary Material
The online version of this chapter (https://doi.org/10.1007/978-3-030-24326-5_5) contains supplementary material, which is available to authorized users.

A. Brandt (✉)
Shepherd School of Music, Rice University, Houston, TX, USA
e-mail: abrandt@rice.edu

[1]The answers are "bowl" and "petty".

J. L. Contreras-Vidal et al. (eds.), *Mobile Brain–Body Imaging and the Neuroscience of Art, Innovation and Creativity*, Springer Series on Bio- and Neurosystems 10, https://doi.org/10.1007/978-3-030-24326-5_5

also a task that is readily adaptable to other fields, such as visual art, dance, and design.

Most notably, variations form would also enable scientists to explore a frequently overlooked aspect of the creative process: the tension between novelty and familiarity [2, 5]. The human brain embraces surprise but also likes to maintain a foothold in routine. Because the sweet spot between novelty and familiarity is hard to gauge, we constantly experiment with different recipes, sometimes incorporating a greater portion of the familiar, at other times a higher dose of surprise. A variations set is a vivid demonstration of this strategy: it is a process of derivation and extrapolation, laying out a range of solutions from the ordinary to the unexpected.

Adapting the theme and variations for the lab would build on pioneering experiments in "spontaneous creativity" by Limb and Braun [4]. In that study, jazz musicians were examined in an fMRI as they performed a simple scale and melody, and then improvised on both. The goal was to observe the contrasts between merely reproducing a musical source and doing something creative with it. The variation form would take this one step further: the goal would be to observe what is happening in the brain as it creates *multiple* derivations. It would thus merge aspects of a divergent thinking task and artistic improvisation.

2 The Theme and Variations as a Creative Paradigm

How does a theme and variations work? In Western classical music, the form follows a fairly consistent trajectory. A piece will open with the presentation of a theme that may be either original or borrowed (for instance, Beethoven based a variations set on "God Save the Queen"). That source theme then cycles over and over again, each time in a new way. Generally, the proportions and harmonic progression of the theme are preserved: that is, the variations of a 32-bar theme will be 32 bars long; and the underlying harmonic progression will be maintained, perhaps with small alterations. But, even as those givens remain in place, the sequence of variations offers constant novelty, thanks to embellishments and new figurations. Generally, each variation will establish a distinctive character, which it then maintains; that makes the contrast from one to the next more marked. Most crucially, in the course of a variations set, each one will generally drift further and further away from the theme, becoming more distorted and complex. To that end, it is not uncommon for the final variation to "break the mold" of the theme, expanding the proportions and deviating more dramatically from the underlying harmony.

Like the Alternative Uses test, fluency, flexibility, and novelty are all relevant. The number of variations is generally viewed as an index of the composer's imagination: the more, the better. For instance, Brahms composed twenty-eight on a theme by Paganini, Beethoven thirty-two on a theme by Diabelli. Like the Alternative Uses task, the variation sets depend on diversity and contrast, with musical variables such as rhythm, melodic figuration, register, dynamics, articulation, and tempo all in play. And the most inventive composers generate highly original solutions: for instance, in his *Diabelli Variations*, Beethoven bases one variation on a "slow motion" fragment of the theme that wanders harmonically in unique ways. In Rachmaninoff's *Rhapsody on a Theme by Paganini,* the composer created his famous Variation XVIII by flipping Paganini's theme upside down and transforming it into an elegiac melody—a totally novel result.

However, there are also crucial differences between this form and classic tests of divergent thinking. In the Alternative Uses test, an object is generally given a new function—a brick is used as paperweight—but the object itself is not necessarily altered. In a theme and variations, creative work is done directly on the source: the variations are not *alternative uses,* they are "substitute" versions of the theme. This is an important distinction. On the one hand, the brain certainly makes inventive leaps by finding alternative uses for existing tools: for instance, the heat-seeking technology used in the Javelin anti-tank missile has been repurposed as a malaria detector. But one only has to observe the endless series of variations that human minds have composed on everything from fonts to sneakers, doors, suitcases, and toothbrushes to recognize that refashioning the familiar into new guises is a central human enterprise [2]. Thus, a theme and variations might be a complementary way of studying how the brain generates a diversity of options.

In addition, in an Alternative Uses test, the subject does not necessarily proceed in an orderly way from the most obvious to the most unusual output: ideas may come to mind in a scatter-shot fashion, making the most original thinking harder to track. While the exemplars in the literature do not necessarily follow a uniform trajectory, there is a noteworthy tendency for a set of variations to get more and more distorted. That makes musical sense: it would be anti-climactic to present the more obvious variations *after* the more imaginative ones; and the step-by-step distancing from the source enables the transformations to be comprehensible in a way that suddenly leaping to the most far-out ones would not. As a result, asking subjects to scale their variations from the *mild* to the *extreme* is a reasonable instruction, and could provide data that is easier to analyze.

A set of variations is likewise different from an unstructured creative prompt. In a free composition, the sources of inspiration may be hidden or difficult to unpack; but in works like Brahms's *Handel Variations* or Gershwin's *Variations on "I Got Rhythm,"* the source theme is explicit, making the DNA of the composer's inventiveness easier to decode. Second, in a variations set, you get *complementary* outputs, each one contributing something that the others do not. The form is thus an interesting way to study a mind that must continually monitor itself, asking *what have I already done?* as it contemplates its next moves.

Is creating a theme and variations suited for experiments? Classical composition typically doesn't happen in real time: writing and notating a set of variations might take days or weeks. That makes it hard to document in the lab. However, variation form is close to jazz improvisation in two ways: first, during a typical jazz performance, each band member takes a turn at improvising, with the goal of contributing a new musical "perspective" on the theme. Second, from one night to the next, the *same* band member is challenged to improvise afresh, creating something akin to a real-time series of variations over time. Given their gifts with spontaneous creation, it should not be too much of a stretch to ask jazz musicians to create an extemporaneous series of variations in the lab.

In an unstructured improvisation, the creative process may be idiosyncratic and hard to compare across disciplines. Thanks to its well-articulated structure, a theme and variations may offer a useful way to bridge disparate fields. For instance, in the visual arts, Pablo Picasso created variations on paintings such as Velasquez's *Las Meninas* and Manet's *Le déjeuner sur l'herbe*—a crucial way he developed his craft. Similarly, Jasper Johns' flag series visually revisits the American flag, from changing its color scheme to blurring it with charcoal. Theme and variations are also a well-established form in dance. Diversifying options is relevant to any number of design tasks: architects routinely draft dozens of prototypes for a new building; Apple has an entire department just devoted to proliferating options for the iPhone box. Indeed, it is the basis of brainstorming, which is based on the "assumption that the larger the number of ideas produced, the greater the probability of achieving an effective solution" [7, 24]. Although a clear analog does not exist in writing, authors frequently create variations on themes: for instance, Shakespeare's *Macbeth* has been set in feudal Japan (Kurosawa's *Throne of Blood*), the modern Middle East (in a production by Grzegorz Jarzyna), and an upscale restaurant (the BBC's *Shakespeare Retold*); it is easy to envision designing a prompt to spur writers to create alternative versions of short texts or simple plot-lines. A study currently underway at the University of Houston offers a precedent for using a shared protocol in multiple modalities: while outfitted with wearable EEG monitors, trios of artists have created spontaneous, collaborative creations in creative writing, the visual arts, music, and dance based on the Dadaist creative game "Exquisite Corpse."[2]

3 An Example from the Literature: The Variations Movement of Schubert's "Trout Quintet"

The fourth movement of Schubert's "Trout" Quintet[3] is a representative example of a composer creating variations that move successively farther from his source.

The movement opens with the main melody of Schubert's song *Die Forelle* ("The Trout"), played by the strings.

The *Forelle* theme, which is D-Major, is characterized by a contour that rises and falls every two bars, and a jaunty dotted rhythm (Fig. 1).

Five variations follow. In the first, the piano enters, playing the theme largely verbatim in a higher register, while the strings play an animated accompaniment. From time to time, the piano embellishes the melody with added figuration; but, with the melody exposed and adhered to so faithfully, this variation is only a small step away from the original (Fig. 2).

The second variation moves a step further away: the melody shifts to an inner voice—the viola—making it less obvious. On top of it, the violin plays an elaborate solo, filled with sweeping runs from its lowest to highest registers and added chromaticism—that is, notes outside the main scale (Fig. 3).

In variation three, the theme is further effaced, like a carving worn down by time. First of all, the melody moves to the low register, where it is harder to hear; in addition, the cello and double bass play it in even values, without its characteristic dotted rhythm. Meanwhile, the piano has a fast-moving, attention-grabbing solo filled with leaps, runs, and more chromaticism (Fig. 4).

The fourth variation is even more radical: it is in the minor mode; and it unexpectedly shifts keys mid-stream. Furthermore, the first three variations each maintain a single

[2]In the "Exquisite Corpse" game, one artist "passes the baton" to the next, producing a collective creation. Often the next artist in line is only allowed to see the tail end of whatever the previous artist has made.

[3]Recording credit for Online Resource Resources in this chapter:Franz Schubert: Quintet in A-Major, "Trout," Op. 114.The Budapest String Quartet, George Szell, piano, Georges E. Moleux, contrabass.Bridge Records 9062.Courtesy of Bridge Records.

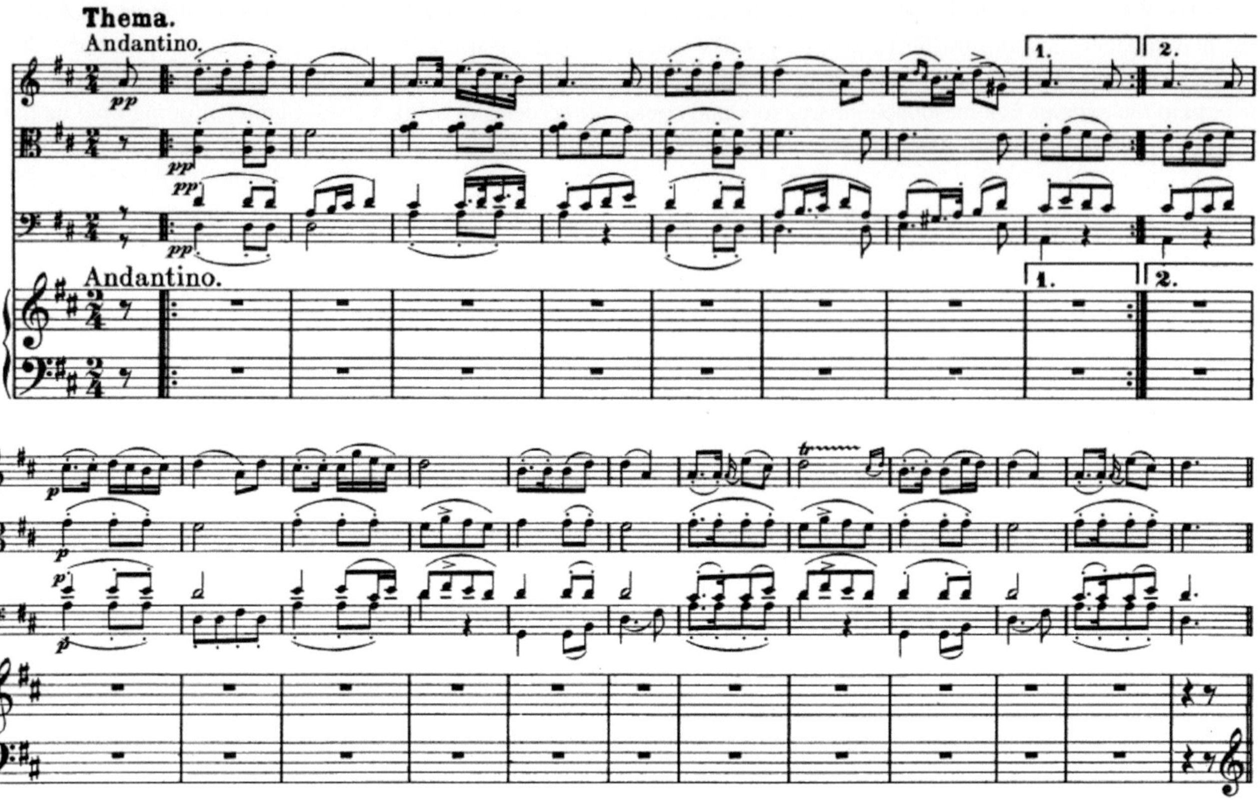

Fig. 1 The theme of the fourth movement of Franz Schubert: Quintet in A-Major, "Trout" (Online Resource 1)

character; but this variation oscillates in an almost bi-polar way between aggressive vigor and more plaintive tranquility. This music would not have come into being if it weren't for the *Forelle* theme: but now that connection is more remote and harder to recognize (Fig. 5).

Finally, the fifth variation introduces new distortions: it is in another key, and the harmony surprisingly mixes the major and minor modes. Most dramatically, all of the previous variations match the proportions of the original—but this one keeps going, its harmonic excursions eventually bringing it back to the home key. Like the fourth variation, this variation is a distant echo of the original. Yet even at his most extreme, Schubert never uproots the familiar entirely: thanks to its arching shape and dotted rhythm, the cello solo that runs throughout maintains an audible link to its source (Fig. 6).

As often happens, the movement closes with a "call-back" to the original theme, this time in a more literal arrangement of the song *Die Forelle.* Thus, Schubert's movement presents a paradigm of variation form: successive statements of the theme gradually retreat from their source but never let go of it entirely.

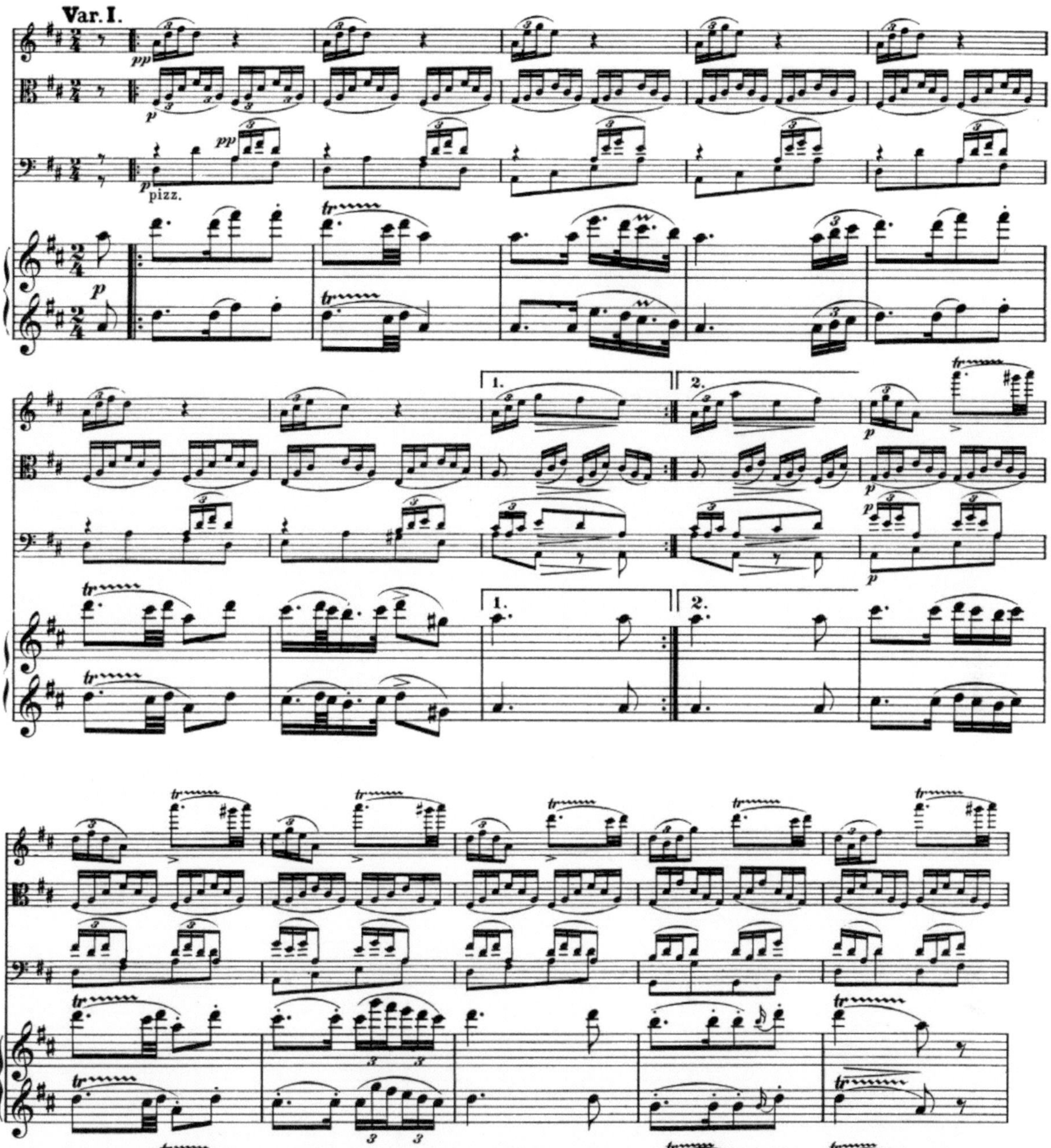

Fig. 2 Variation 1 (Online Resource 2)

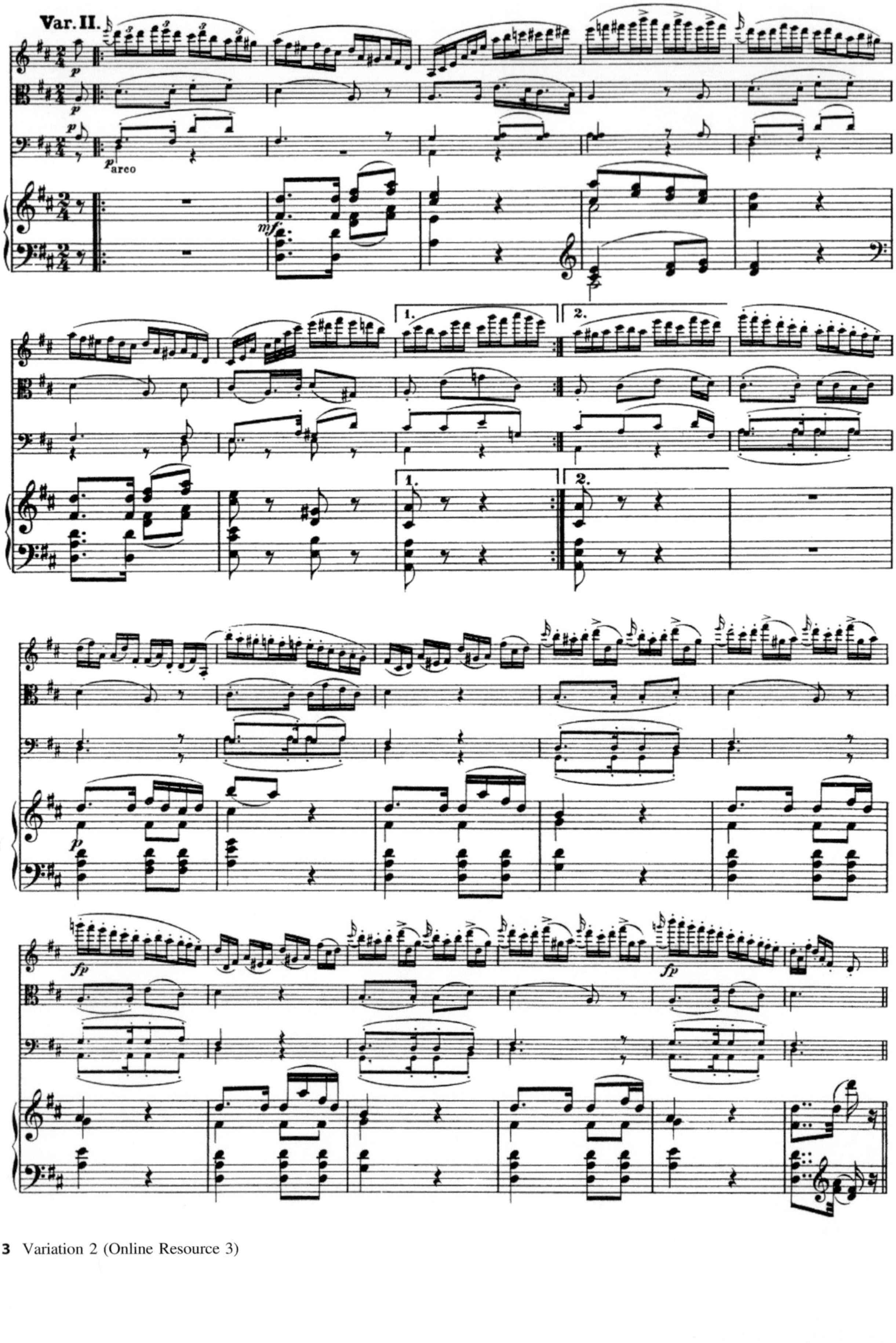

Fig. 3 Variation 2 (Online Resource 3)

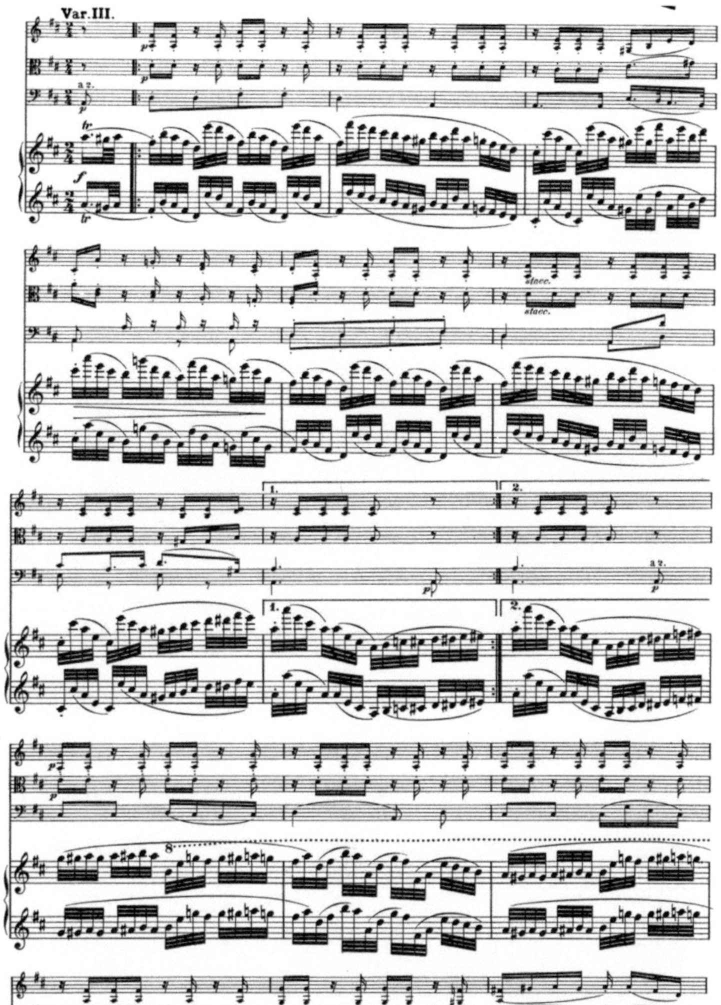

Fig. 4 Variation 3 (Online Resource 4)

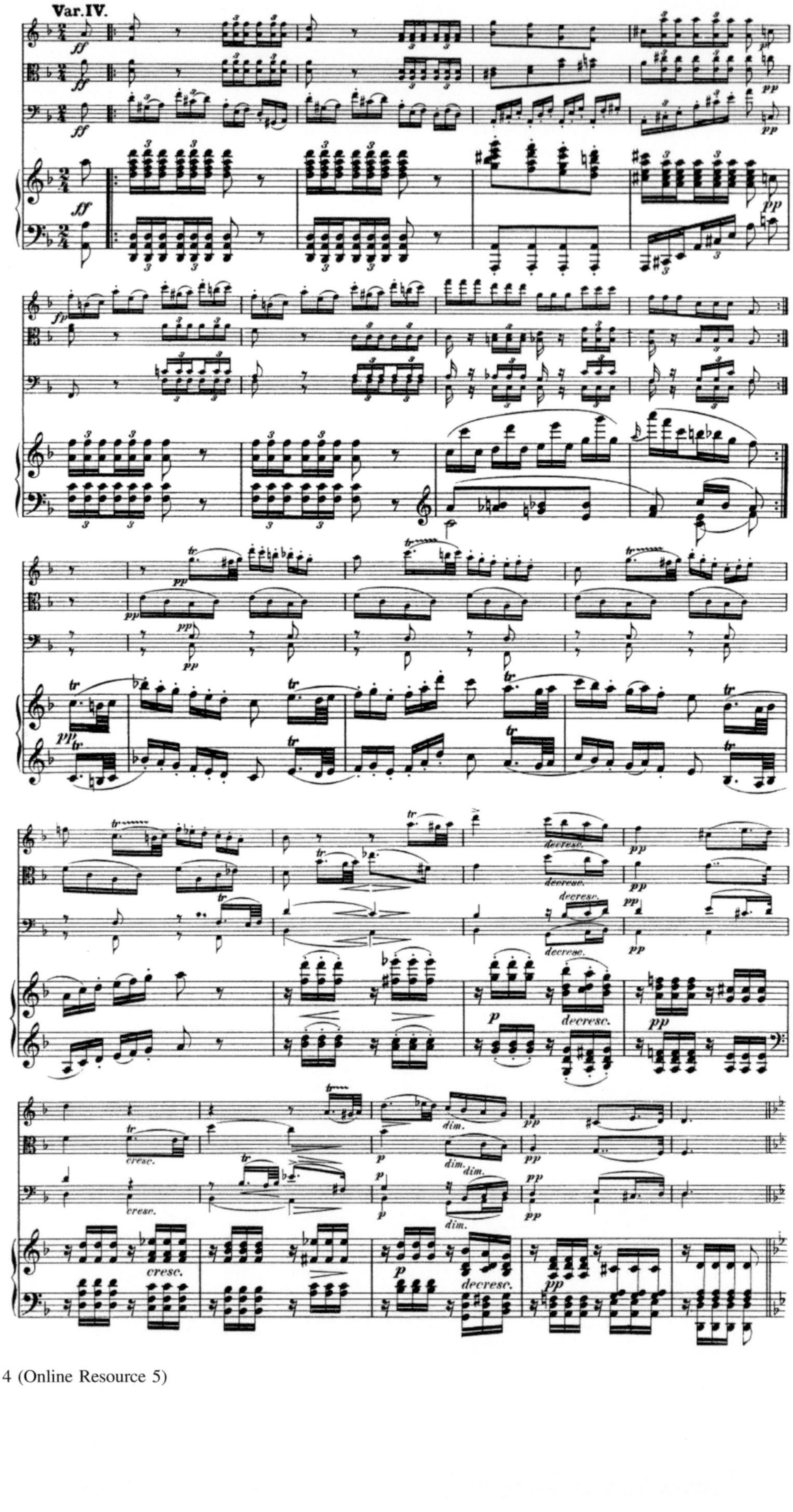

Fig. 5 Variation 4 (Online Resource 5)

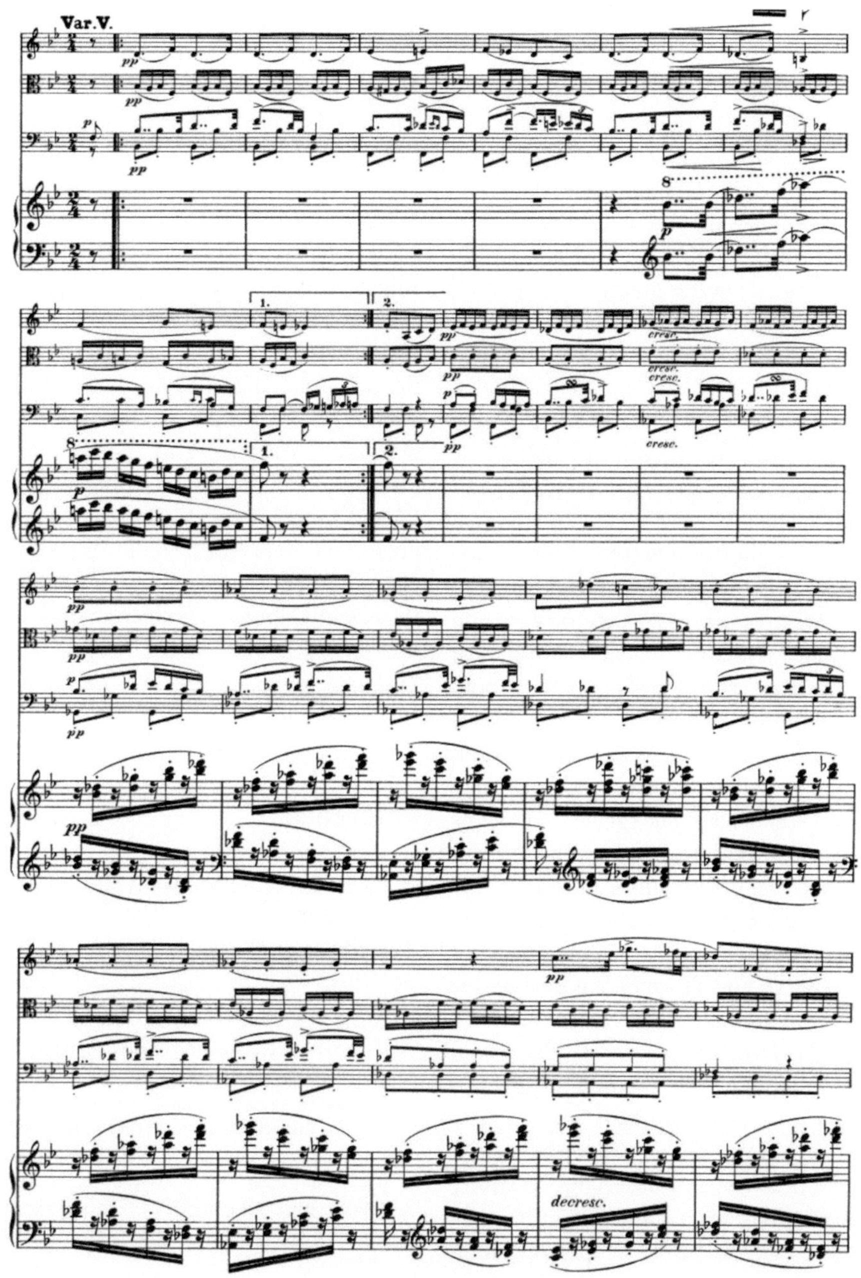

Fig. 6 Variation 5 (Online Resource 6)

4 Conclusion

What might imaging a musician composing a theme and variations tell us about the brain? In much of the creativity literature, there is a presumption that the brain primarily generates novel solutions through "oddball combinations." That is the basis for traditional divergent thinking tests, where there is a reward for the unlikeliest uses. Similarly, the Remote Associates test is based on the premise that people who are adept at finding the common ground between far-flung word triplets have the mental flexibility that promotes innovative thinking. It's why we're told to "Think outside the box." And it's why business consultants may challenge you to solve a problem in your company by pairing the problem with a random word (for instance, "sell more subscriptions" and "elephant."): the supposition is that arbitrary juxtapositions will stimulate novel thinking.

A musical theme and variations may be a window into another dimension of the creative process: as we venture into novel territory, we explore different distances from the familiar [2]. That's what leads car manufacturers to both upgrade current models and build concept cars, and fashion designers to make ready-to-wear clothing and more far-out haute couture. Furthermore, while creative thinking may certainly involve "random walks" and unexpected collisions of ideas, it is equally plausible that we often gradually drift away from precedent: what may sometimes appear to the conscious mind as a creative "leap" may actually consist an unconscious series of incremental steps. A theme and variations may be a way to observe how far-out ideas can evolve in the brain from an increasingly radical series of mutations.

As we look ahead to the future, it is important to avoid falling into the trap of phrenology that Dietrich and Kanso warn against: the imaging of creativity will require new techniques and strategies to observe neural collaboration. On top of that, it may require new ways of observing creativity in action. To that end, variations are a showcase of musical imagination, and have the potential to be deeply revealing about the creative mind.

Appendix: A Selection of Notable Theme and Variations

J. S. Bach: *Goldberg Variations*
Joseph Haydn: *Symphony No. 31, "Hornsignal,"* IV
Wolfgang Amadeus Mozart: *Piano Sonata in A-Major, K 331, I*
Quintet in A-Major for clarinet and string quartet, IV
Ludwig van Beethoven: *"Eroica" Variations* for piano
Symphony No. 3, "Eroica," IV
Diabelli Variations for piano
Felix Mendelssohn: *Variations serieuses* for piano
Franz Schubert: *Quartet no. 14 in d-minor, "Death and the Maiden," II*
Piano Quintet in A-Major, "Trout," IV
Johannes Brahms: *St. Anthony Variations* for orchestra
Handel Variations for orchestra
Variations on a Theme by Paganini, books 1 and 2
Sergei Rachmaninoff: *Rhapsody on a Theme by Paganini* for piano and orchestra
Benjamin Britten: *The Young Person's Guide to the Orchestra* (variations on a theme by Purcell)
Witold Lutoslawski: *Variations on a Theme by Paganini* for piano and orchestra
George Rochberg: *String Quartet no. 6*, III (variations on the Pachelbel Canon)
Frederic Rzewski: *The People United Will Never Be Defeated* for piano.

References

1. Beaty, R.E., Benedek, M., Wilkins, R.W., Jauk, E., Fink, A., Silvia, P.J., Neubauer, A.C.: Creativity and the default network: a functional connectivity analysis of the creative brain at rest. Neuropsychologia **64**, 92–98 (2014). https://doi.org/10.1016/j.neuropsychologia.2014.09.019
2. Brandt, A., Eagleman, D.: The Runaway Species: How Human Creativity Remakes the World. Catapult, New York (2017)
3. Dietrich, A., Kanso, R.: A review of EEG, ERP, and neuroimaging studies of creativity and insight. Psychol. Bull. **136**(5), 822–848 (2010). https://doi.org/10.1037/a0019749
4. Limb, C.J., Braun, A.R.: Neural substrates of spontaneous musical performance: an fMRI study of jazz improvisation. PLoS ONE **3**(2) (2008). https://doi.org/10.1371/journal.pone.0001679
5. Martindale, C.: The Clockwork Muse. Basic Books, New York (1990)
6. Takeuchi, H., Taki, Y., Hashizume, H., Sassa, Y., Nagase, T., Nouchi, R., Kawashima, R.: The association between resting functional connectivity and creativity. Cereb. Cortex **22**(12), 2921–2929 (2012). https://doi.org/10.1093/cercor/bhr371
7. Taylor, D.W., Berry, P.C., Block, C.H.: Does group participation when using brainstorming facilitate or inhibit creative thinking? Adm. Sci. Q. **3**(1), 23 (1958). https://doi.org/10.2307/2390603
8. Zeng, L., Proctor, R.W., Salvendy, G.: Can traditional divergent thinking tests be trusted in measuring and predicting real-world creativity? Creativity Res. J. **23**(1), 24–37 (2011). https://doi.org/10.1080/10400419.2011.545713

Anthony Brandt is a Professor of Composition at Rice University's Shepherd School of Music and Artistic Director of the new music ensemble Musiqa. His compositions include three chamber operas, as well as scores for orchestra, chamber ensembles, voice, theater, dance, art installations, and television. Dr. Brandt and neuroscientist David Eagleman's book The Runaway Species: How Creativity Remakes the World is being published in 13 countries. Dr. Brandt has also co-authored papers for the journals Frontiers and Brain Connectivity and a chapter in the upcoming Oxford Handbook of Music Psychology. He has organized three international conferences on "Exploring the Mind through Music" at Rice.

Part II

How Do the Arts and Sciences Challenge Each Other and Create New Knowledge Through Collaboration?

Introduction

Dario Robleto
Artist in Residence in Neuroaesthetics, IUCRC BRAIN, University of Houston, Houston, USA

There is an overarching question that drives this book and the conference it summarizes that is also highly relevant in the broader culture: What does true collaboration look like between disciplines and people from disparate fields? In an age of increasing specialization across all forms of knowledge production, doing the hard work of discovering common questions and developing the language and techniques for meaningful collaboration to occur would seem an unnecessary complication. But, increasingly, the complexity of our inquiries into nature through one domain is so expansive that they transcend their field of origin and require multiple disciplines to address them adequately. If we focus on the two fields that anchor this book—art and neuroscience, two disciplines that may not at first glance suggest commonalities—a question emerges that should inspire transdisciplinary curiosity: What is creativity? As an artist, and as I have learned from colleagues in the neuro and psychological sciences, it is humbling to realize how little we actually know about how this central behavior of the human condition arises from the brain.

Historically, the different methodological approaches between the arts and sciences have created a perception that each field cannot contribute to the other in meaningful ways beyond the purely referential. This disparity is not so surprising when we consider that the language of science—quantitative and predictive, tractable, and reproducible—does not seem to easily match up with the fluidity, subtlety, individuality, and context-specific states that we understand aesthetic and creative experience to encompass. Further, there is a deep philosophical tradition within the arts that art does not necessarily need to concern itself with the objective, factual, and verifiable standards much science is defined by. But it remains difficult to imagine how we will ever move closer to a deeper understanding of creativity without both a bottom-up scientific approach working in tandem with the top-down perspective from the artists themselves.

This is why the burgeoning field of neuroaesthetics presents exciting challenges to both fields. It is a rare opportunity when a scientific field's advancement largely depends on developing long-term and meaningful relationships and collaborations with artists and arts institutions. One of the challenges moving forward will be how the field stays rigorous to its scientific roots while remaining open to new ways of thinking within the arts. Similarly, the arts will need to remain open to the possibility that a neuroscientific understanding of the creative process can enrich their practice. Is there a consensus definition of creativity and aesthetics that applies to both the arts and neuroscience? What are the physical (neurological) underpinnings of creativity and aesthetics, and can they be recorded and quantified? Can the physical and anatomical understanding of creativity and the brain say anything revelatory about the lived, experiential relationship of creators and viewers to art?

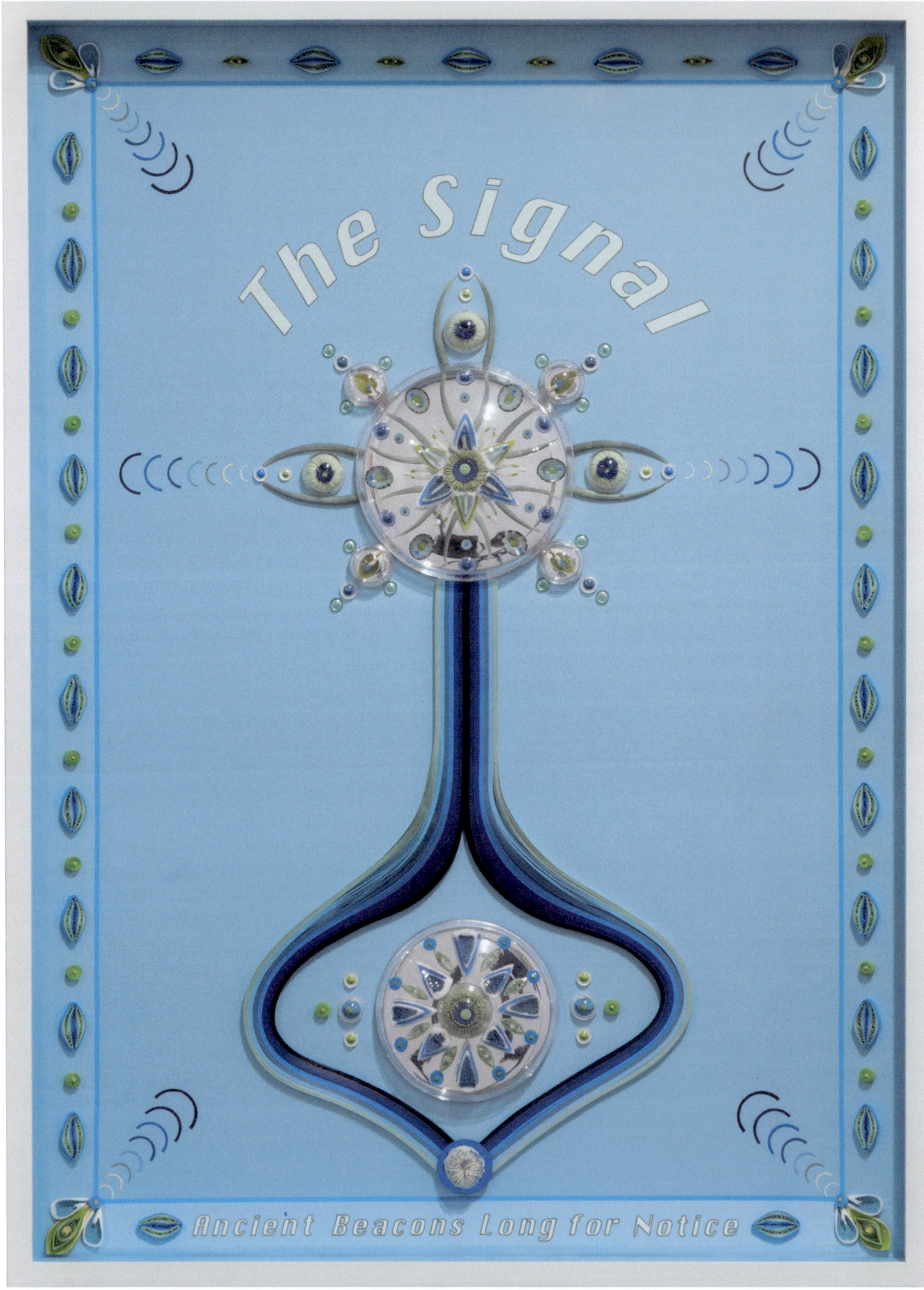

Dario Robleto
The Signal
Cut paper, various cut and polished seashells, green and white tusks, squilla claws, mushroom coral, colored powder pigments and beads, plastic domes, mirrored Plexiglas, foam core, glue, frame
34 × 24 × 3 inches
2018

Art-Science Collaborations: How to Break Boundaries Without Breaking Trust

Janet Biggs, Jokūbas Žiburkus, and Jason L. Eriksen

Nature is a temple in which living pillars
Sometimes give voice to confused words
Man passes there through forests of symbols
Which look at him with understanding eyes
Like prolonged echoes mingling in the distance
In a deep and tenebrous unity,
Vast as the dark of night and as the light of day,
Charles Baudelaire, Les Fleurs du Mal, 1857

1 Introduction

Janet Biggs (Fig. 1)—Is it possible to have a true collaboration between artists and scientists, and if so, is it beneficial to all parties involved? Or is the interaction between art and science exploitive and self-serving? Original source can be misrepresented, misunderstood or misused in striving to prove a scientific theory or make the best artwork possible. "Science" isn't just someone in a white lab coat holding a test tube and "art" isn't just Vincent van Gogh's Starry Night. Is there an achievable middle ground, a productive intersection between art and science, where individual integrity is maintained while improving the vision, execution, and impact of a project?

In 2014, I was invited to develop a new body of artwork for a solo exhibition at the University of Houston's Blaffer Art Museum. I had previously produced a video in 2009 that focused on Alzheimer's disease. As my personal experiences with the disease were increasing, with new family members being diagnosed, I was interested in creating a museum exhibition of video installations that traced both a personal narrative and the biological underpinnings of the disease.

At my request, the Blaffer Art Museum reached out to University of Houston faculty members who were actively engaged in research on Alzheimer's disease. Jason Eriksen, Ph.D., Associate Professor of Pharmacology and Jokūbas Žiburkus, Ph.D., Associate Professor, Department of Biology and Biochemistry, generously agreed to meet with me, open their research laboratories for filming and step into the ever-shifting terrain of a collaboration.

2 The Conversation

Janet Biggs—I envision a world where deep disciplinary knowledge, diverse community, and interdisciplinary communication drive research and problem-solving; a world where individual expertise is seamlessly integrated into collaborations, conversations, and everyday practices, but my personal experience leads me to believe that this world is a world yet to come, a possible future.

> The uncertain nature of art has its advantages. It leads to constant experiment and questioning. Rosenberg [1]

I have found that collaborations—two independent entities working together with equal weight to organize, create, or achieve something new—rarely exist. Occasionally, I've experienced collaborations that felt more like assisting or willingly working with a destructive interloper, whether acting inadvertently or consciously. This is not to say that this

J. Biggs (✉)
Independent Artist, Brooklyn, NY, USA
e-mail: janetbiggs1@gmail.com

J. Žiburkus
Department of Biology and Biochemistry, University of Houston, Houston, TX, USA

J. L. Eriksen
College of Pharmacy, University of Houston, Houston, TX, USA

J. L. Contreras-Vidal et al. (eds.), *Mobile Brain–Body Imaging and the Neuroscience of Art, Innovation and Creativity*, Springer Series on Bio- and Neurosystems 10, https://doi.org/10.1007/978-3-030-24326-5_6

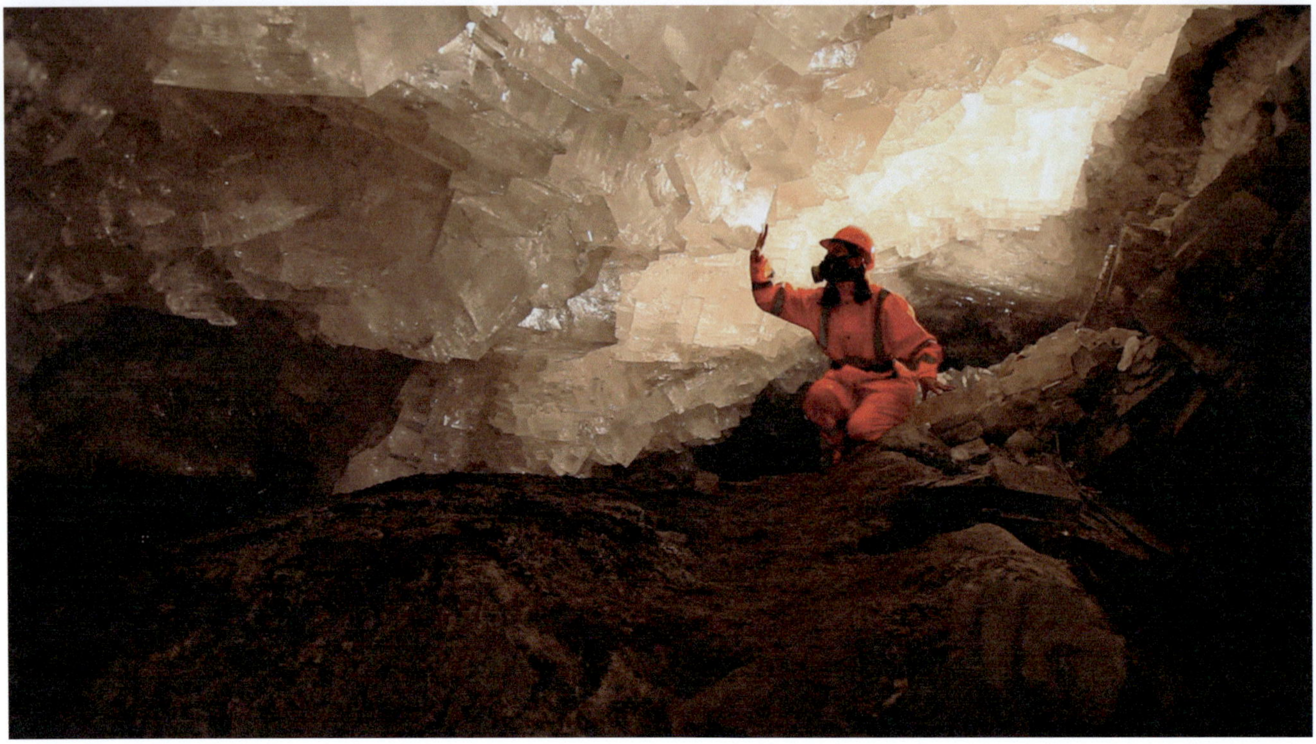

Fig. 1 Janet Biggs, Can't Find My Way Home (detail), 2015. Four-channel HD video installation with sound. Courtesy of the artist, Cristin Tierney Gallery (New York, NY), Analix Forever (Geneva, Switzerland) and CONNERSMITH (Washington, DC). Excerpt available at https://vimeo.com/130826328

kind of interaction can't be useful, successful and exceed expectations, but recognition of an individual's driving forces and initial expectations are essential for the future development of art and science collaborations.

My work focuses on individuals in extreme landscapes or situations, and often navigates the territory between art and science (Fig. 2). I have participated in, and led, art and science expeditions from the high-Arctic of Svalbard to inside Kawah Ijen, an active volcano in East Java. My work has taken me from areas of conflict in the Horn of Africa to Mars (as a crew member of a Mars Desert Research Station simulation mission). I have worked with neuroscientists, Arctic explorers, aerospace engineers, astrophysicists, Yemeni refugees, a gospel choir, and robots.

In 2009 and 2010, I participated in the Arctic Circle, an expeditionary residency program that brings together artists and scientists aboard a specially outfitted sailing vessel in the high-Arctic Archipelago of Svalbard. While artist-led, this program has given scientists working outside of established research avenues the opportunity to engage in real, physical, Arctic fieldwork instead of theoretical models.

Inversely, many science-led expeditions include someone from the arts in their proposal; an interdisciplinary team is more likely to receive funding. Both these models invite deeper, physical engagement and opportunities, even if not yet generating collaborative projects other than the shared goal of access.

Intersections and interactions can influence thought and behavior, producing unexpected new possibilities, inspiring new questions and conclusions, but occasionally they can present conflicts and misunderstandings. I once initiated an interaction with a laboratory as they entered the human-study phase of their research. I asked to be a participant in the study and film the process for an artwork. The lab generously agreed, giving me both information and unfettered access. Once back in my studio, I added in new images that made for a better artwork, but these additional images misrepresented and sensationalized the work being conducted by the lab. I upped the voltage, both literally and figuratively, by recreating and filming one aspect of the study in my studio, the creation of a negative memory by mild electric shock. Fearing that their future funding could be affected by my artwork's portrayal of dramatically heightened levels of electric shock, the laboratory requested I remove any specific reference to the study, the scientists, and the university conducting the research. This was a poignant lesson in the potential ramifications that can occur through misunderstanding and misrepresentation, and in the responsibility collaborating parties hold to maintain the integrity of each other's work.

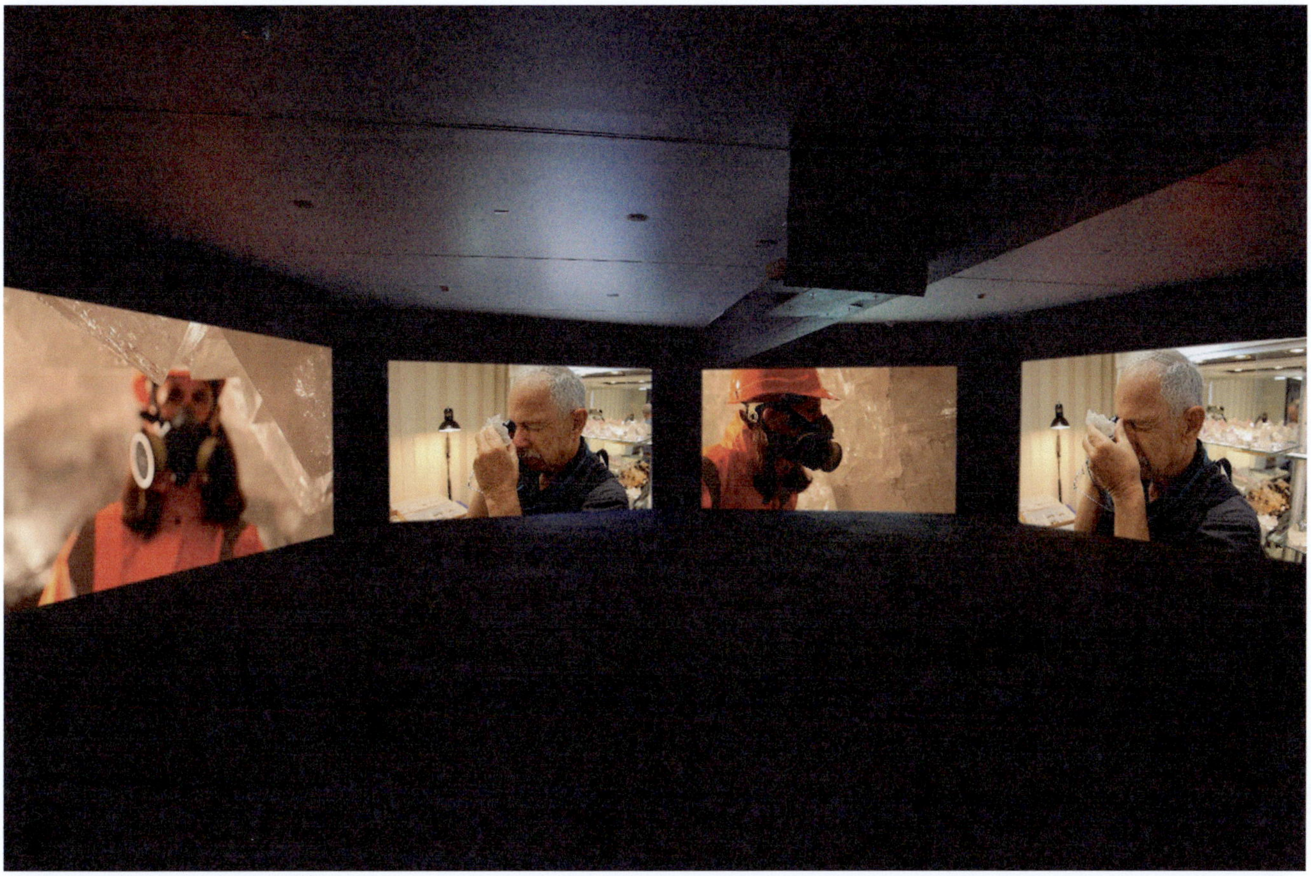

Fig. 2 Janet Biggs, *Can't Find My Way Home* (installation view, Blaffer Art Museum), 2015. Four-channel HD video installation with sound. Courtesy of the artist, Cristin Tierney Gallery (New York, NY), Analix Forever (Geneva, Switzerland) and CONNERSMITH (Washington, DC)

Jason Eriksen—Stories permeate science. As an example, the pioneering neuroscientist Charles S. Sherrington once offered this famous description of the cerebral cortex emerging from sleep:

> The great topmost sheet of the mass, that where hardly a light had twinkled or moved, becomes now a sparkling field of rhythmic flashing points with trains of traveling sparks hurrying hither and thither. The brain is waking and with it the mind is returning. It is as if the Milky Way entered upon some cosmic dance. Swiftly the head mass becomes an enchanted loom where millions of flashing shuttles weave a dissolving pattern, always a meaningful pattern though never an abiding one; a shifting harmony of subpatterns. [2]

Although we now know the brain is considerably more active during sleep than Sherrington's original description, the perception of the brain as an enchanted loom and similar poetical representations like these have shaped generations of scientists; these visions have had profoundly transformative impacts within scientific communities. Fundamentally, the process of scientific discovery begins with an overarching conceptual vision, based in part on fact and part on intuitive understanding. This vision leads eventually to the development of hypotheses and testable conceptual models that can be used to empirically establish the truthfulness of these hypotheses. While the testable model is the basis of scientific research, these tools, for all their assumed precision, remain analogues of real-world events, as they only capture some aspects of the thing in question. All models have hidden limitations that are eventually revealed over time.

As a scientist, I am most interested in questions related to the aging brain, with a focus on the pathogenic, destructive events found in neurodegenerative disorders. Alzheimer's disease is the leading cause of dementia, characterized initially by short-term memory loss, progression to serious changes in personality, and finally to complete incapacitation and loss of self in the late stages of disease [3]. A curious finding of the disease is that many disease-associated changes in the brain begin decades before an awareness of the disease or clinical signs are present. The Alzheimer's disease field is rich with hundreds of empirically developed and tested models created to study various aspects of the disease, and over the last three decades, we have learned a great deal about the changes that occur in the brains of patients afflicted with disease (Fig. 3). Despite the successes, we've also become increasingly aware of the limitations of the models that are used to study the disease. The stories that

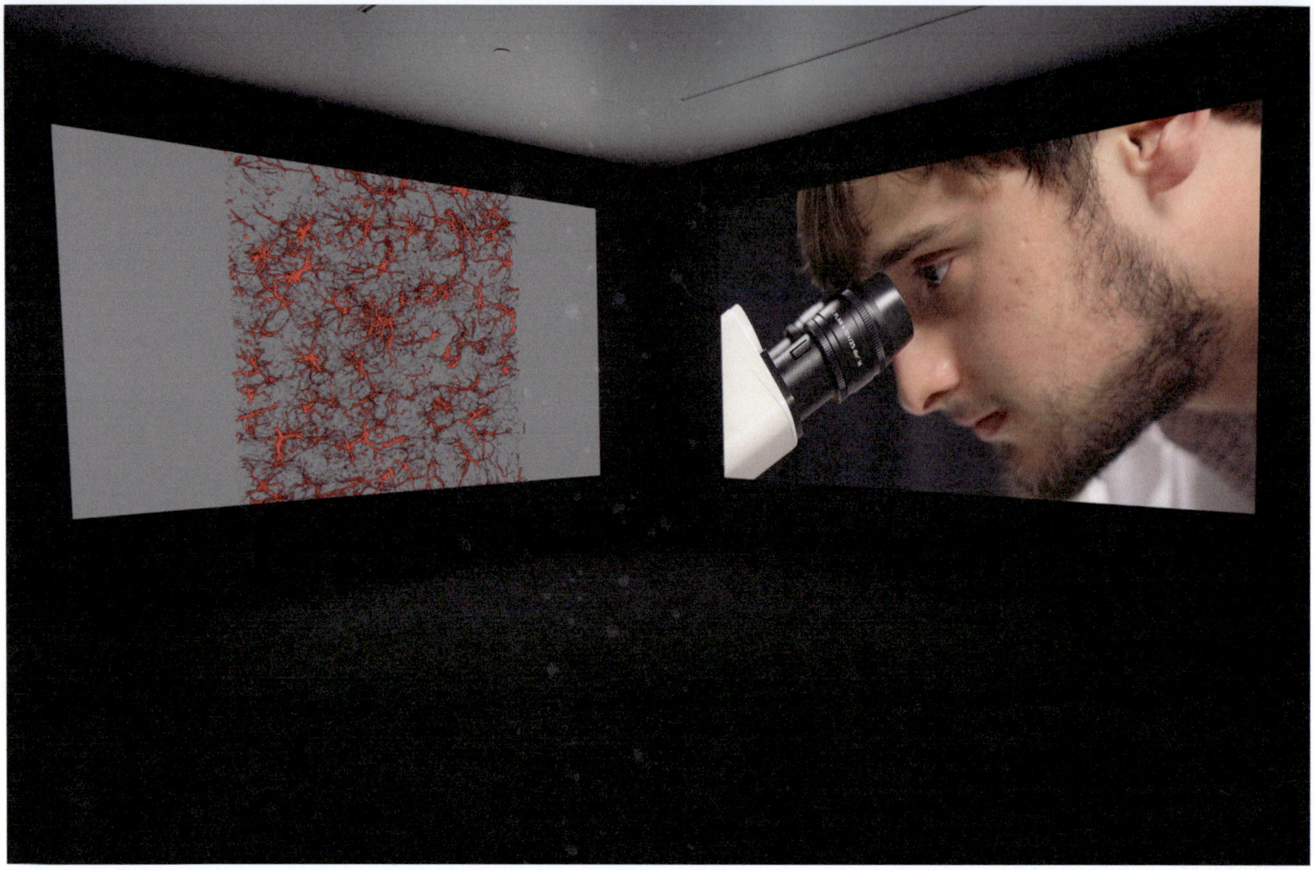

Fig. 3 The brain is impacted by many different pathological events, such as the development of damaged blood vessels (shown in red) in Alzheimer's disease. Janet Biggs, *The Persistence of Hope* (installation view, Blaffer Art Museum), 2015. Two-channel HD video installation with sound. Courtesy of the artist, Cristin Tierney Gallery (New York, NY), Analix Forever (Geneva, Switzerland) and CONNERSMITH (Washington, DC)

they tell do not completely recapitulate what we see in Alzheimer's, and what we prove in the lab may not, and often does not, translate to success in the clinic [4]. As Thomas Huxley wrote, "*the great tragedy of science—the slaying of a beautiful hypothesis by an ugly fact*" [5]. Although the many clinical failures have been challenging, the discovery of the limitations of these models has spurred research groups to formulate new visions that can be used to explore Alzheimer's disease.

Science continues to remain a collaborative effort around a shared vision. Artistic inspiration has been particularly useful in shaping science, as these imaginative constructions are a combination of intuition and craftsmanship that are not subject to the limitations of current models. Consequently, artistic representations have the potential to exhibit disproportionate impacts on the process of discovery, leading to new scientific breakthroughs.

Jokūbas Žiburkus—Just like crystals and minerals in the earth's deepest crevasses and highest peaks, the brain's connections are changing, evolving with age, and adapting to the environment and technology. The brain, comprised of billions of neurons and trillions of connections, is a collection of dynamic networks that act as weakly connected oscillators. These constant activity oscillations are called the brain rhythms or electrochemical waves and are representations of thoughts, emotions, and actions.

The brain rhythms come in a variety of frequencies, where each dominant frequency represents different perception and behavior. These numerous rhythms are created by the underlying architecture of the brain networks and synaptic communications between interconnected neurons. Architectural (anatomical) features of neuronal networks are plastic, adaptive, and entrained by the external environment. The cytoarchitecture of the networks supports formation of the brain rhythms and allows them to spread, creating multidimensional, dynamic maps of electrochemical activity. If the architecture is imbalanced or broken, the brain rhythms and the brain activity maps transform. The cytoarchitecture and proper rhythmicity of the brain can be permanently damaged or lost in neurological conditions, like Alzheimer's disease or epilepsy.

My deep interests of music and art, brain architecture and electrochemical brain rhythms, is what led me to seek to record and image electrochemical oscillations in the developing and aging brains. Undertaking the most significant challenges in science, art, and humanity and stepping outside of the accepted comfort zones and scientific or religious dogmas, is the beauty of both science and art. The boldest scientific discoveries often were not based on the calculations, but rather intuition, dreams, visions, and creativity. Exploring the uncharted territories takes psychological and physical determination.

To chart new territories, one has to understand and challenge the accepted paradigms, listen to their gut, and seek the unknown. I challenged myself to think how Alzheimer's plaques may interfere with electrochemical brain waves, how their proximity to individual neurons can impede with their electrochemical signaling. We set off to record and image oscillations in aged brains of transgenic mice that develop amyloid pathology, a key feature associated with Alzheimer's disease. Because of the old age of the tissue and neurons, months of our work went into getting skilled enough to do what would be a basic experiment in the young brain tissue. As a result, we spent months setting up experiments, spending 10–12 h/day in trying to move the needle forward often without any results. We finally obtained recordings from rare cells in the oldest neurons recorded to date.

Janet Biggs—My intent was to create an immersive experience for the museum visitor through video installations, objects, and sound work that address memory, it's creation and loss, from biological foundation through personal histories, all the while allowing for metaphoric meandering, journeys through the past, present, and into possible futures.

For a two-channel video installation, titled *The Persistence of Hope* (Fig. 4), I originally envisioned portraying one of my memories of my uncle, an avid bird watcher with a garden full of hummingbird feeders. After his funeral, the family gathered back at his house. My cousin, asking if I knew about my uncle's freezer, showed me to a large, freestanding freezer in their kitchen. Opening the lid, the frost dissipated, revealing dozens of tiny hummingbirds, lovingly wrapped and placed side by side. Hummingbirds have an extremely high heart rate and are prone to heart attacks. As my uncle felt his life slipping away, he began collecting and preserving the hummingbirds he found that died in his garden.

During my early research, I visited Dr. Moses V. Chao's lab at the Skirball Institute, NYU Langone. As we toured his lab, Dr. Chao opened one of the lab's freezers which

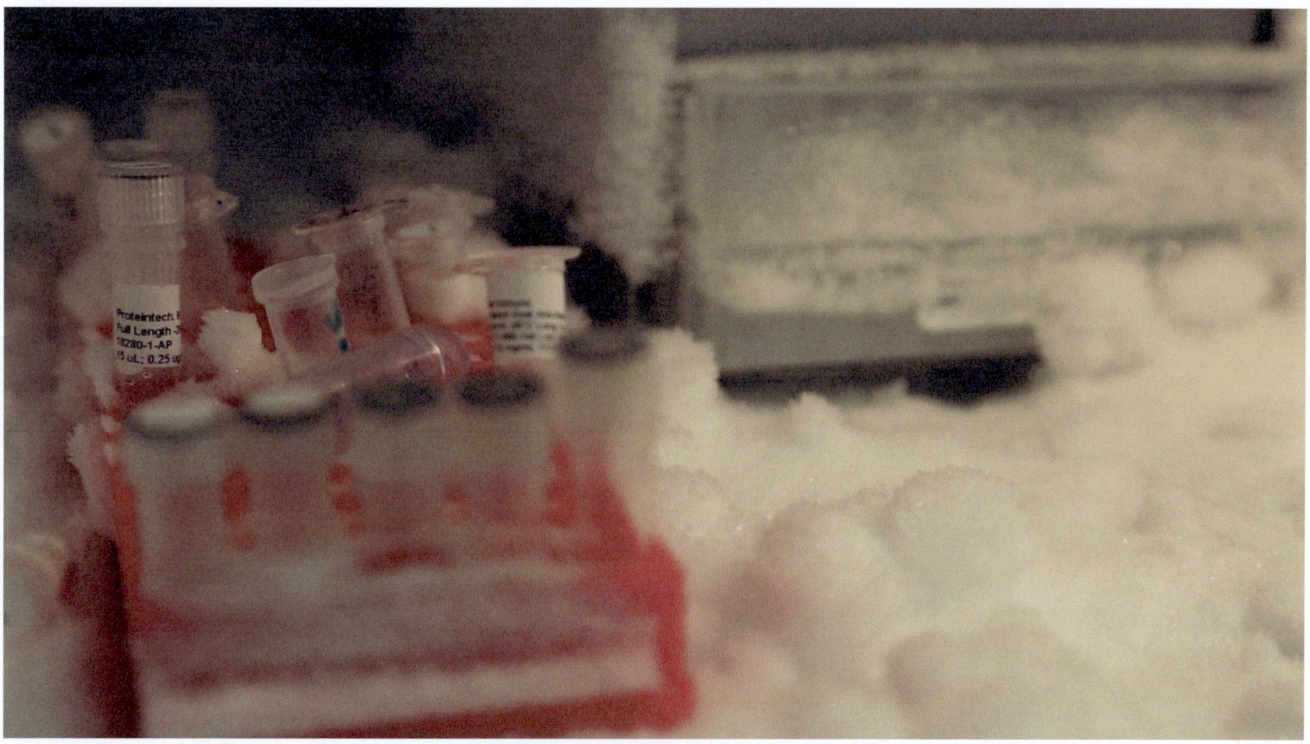

Fig. 4 Some of the frozen antibodies and other chemicals that are used to visualize changes in brain structure. Janet Biggs, *The Persistence of Hope* (detail), 2015. Two-channel HD video installation with sound. Courtesy of the artist, Cristin Tierney Gallery (New York, NY), Analix Forever (Geneva, Switzerland) and CONNERSMITH. (Washington, DC)

contained everything from single cell organisms to human brains. The visual parallel between Chao lab's freezer and my uncle's was striking. The desire to realize my vision in a video work led me to Dr. Eriksen at the University of Houston. Dr. Eriksen's research, insights, and generosity became instrumental to the physical and conceptual formation of this project, both broadening and augmenting my initial intent.

Our interaction opened doors for me in terms of specific images of the brain, expanding my perceptions and interpretations of the brain's biological functions, while still recognizing the inherent poetry of the brain and its many states, as so beautifully stated in the quote above by neuroscientist Charles S. Sherrington.

I am always conscious about authorship in art and science collaborations. In this case, I initiated the interaction with a specific goal of producing a body of work for the Blaffer exhibition. This set fairly tight parameters in terms of authorship and perhaps even on the exchange itself. We set clear ground rules from the start that you would review and approve all footage I took in your lab so I didn't unintentionally reveal any of your research prior to publishing.

I often start projects from a documentary point of view, recording a process, event, or location. This documentary approach gives me a clear path to follow as I gather information and imagery (Fig. 5), but at some point, I need to push myself off that path, to slide sideways. For me, this is where the art happens, where the project broadens allowing for unexpected juxtapositions or convergences. If I allow myself to follow distant threads, a kind of poetry can be created in the work, allowing the viewer to make their own interpretations. There are many different kinds of shared authorship that can happen throughout a project, from interactions in the creation of the work to the audience's interaction in its reception.

For a second installation in the project, I recalled another specific memory, the time of my grandfather's struggle with Alzheimer's. My grandfather was an amateur mineral collector. He could remember detailed information about samples in his collection, their geologic names, the places of extraction, long after he lost the ability to identify the names and faces of family members and friend. I wanted to both physically and metaphorically place myself inside one of his geodes, inside one of his moments of presence within the vast sea of loss that is Alzheimer's disease. My interaction with Dr. Žiburkus and his students helped shaped my four-channel video installation titled *Can't Find My Way Home*, both conceptually and visually, and defined the work in terms of sound.

Jason Eriksen—During the development of this project, I was particularly struck by Janet's use of footage of a deep mine filled with crystalline quartz of all different shapes and sizes. Janet wanders around the cavern, offering the viewer a

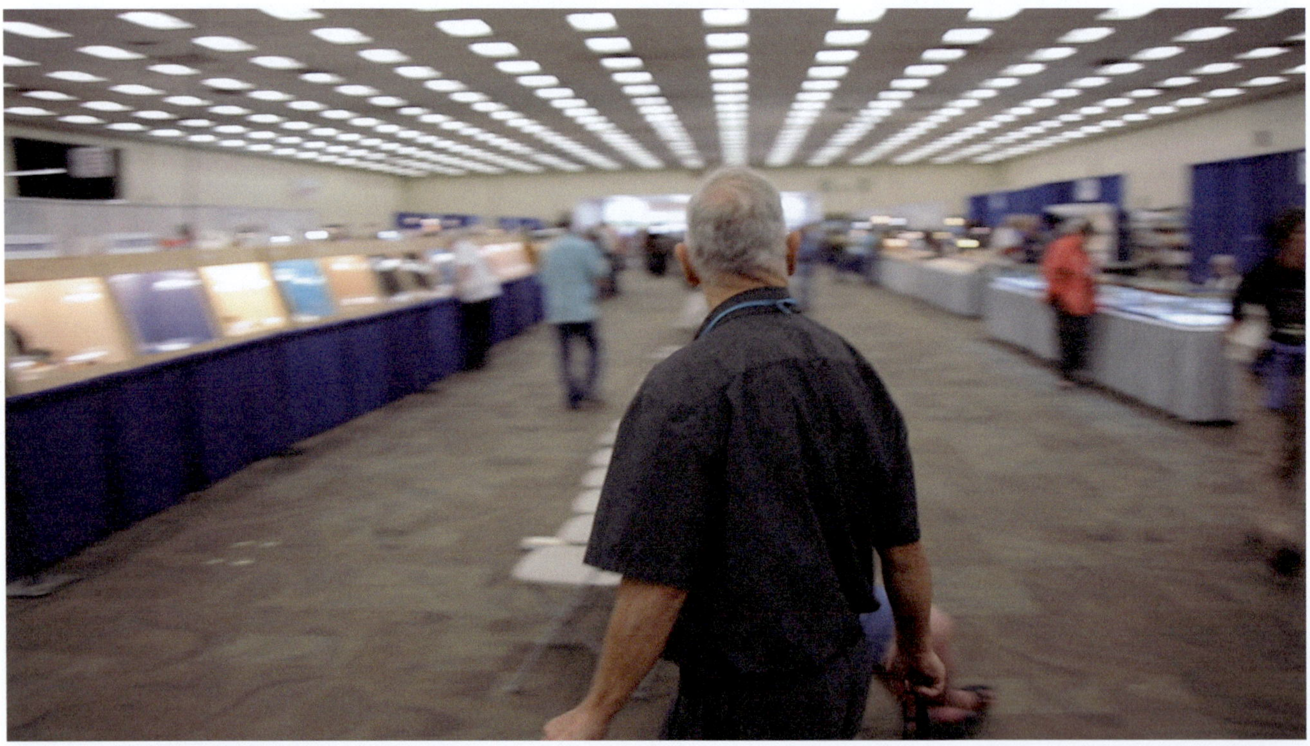

Fig. 5 Janet Biggs, *Breathing Without Air* (detail), 2015. Single-channel HD video with sound. Courtesy of the artist, Cristin Tierney Gallery (New York, NY), Analix Forever (Geneva, Switzerland) and CONNERSMITH (Washington, DC)

brief, illuminating, tantalizing, but partial glimpse of her surroundings. The dust mask and protective orange suit remind the viewer that the conditions of the mine—the heat, temperature, and pressure—are elements that are inhospitable to life. From my perspective as a scientist, this visual metaphor dovetails beautifully with our current understanding of Alzheimer's disease. It has been known for over a century that as the disease progresses, the brains of patients begin to fill with highly ordered, crystalline arrays of proteins known as amyloid (Fig. 6); a diverse array of amyloid beta plaques have been discovered. Due to their prevalence and their association with disease, the Alzheimer's disease community has strongly held the belief that these amyloid plaques are intrinsically hostile to the brain and drive the disease process. However, studies of human patients have increasingly suggested that plaques are byproduct of processes that accompany the development of disease but are not the primary drivers of the disease itself [6]. We now suspect, but have not entirely defined, the fundamental processes that drive the disease.

Janet Biggs—One thing I wasn't expecting was how my interaction with Dr. Jokūbas Žiburkus would influence and alter the sound of my installation (Fig. 7). Dr. Žiburkus' research involves recording the sound created by electrical activity of two cells "talking" to each other in the brain of someone suffering from Alzheimer's disease. Naively, I had expected the percussive sound to slow down in the brain of someone with Alzheimer's, but the opposite happens. The brain frantically tries to find connections that are disappearing. I used recorded sound of the electrical activity as part of the soundtrack, which altered my original trajectory of the work, dictating the mood and heightening the drama of the journey depicted.

Jason Eriksen—Far from the cessation of activity, the electrical chaos and repetitive seizures that develop in the brains of Alzheimer's patients is a highly counterintuitive observation, one that has come about over the last few years through careful scientific exploration [7]. These electrical storms interfere with the normal communication within the central nervous system, analogous to the impact of voltage

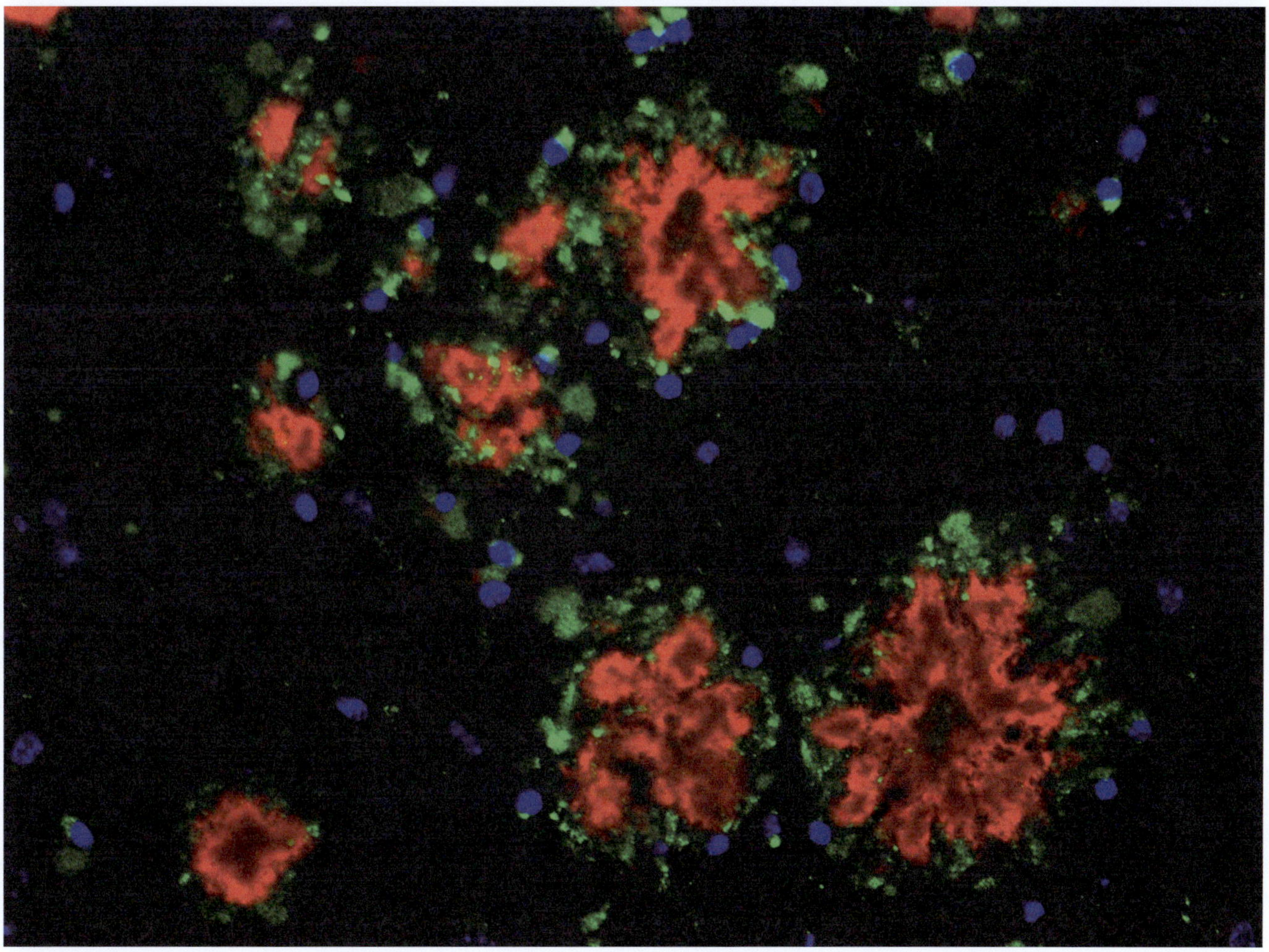

Fig. 6 A microscopic cross-section showing a diversity of amyloid plaques (red) in the brain of a genetically engineered mouse. Image provided by Jason Eriksen

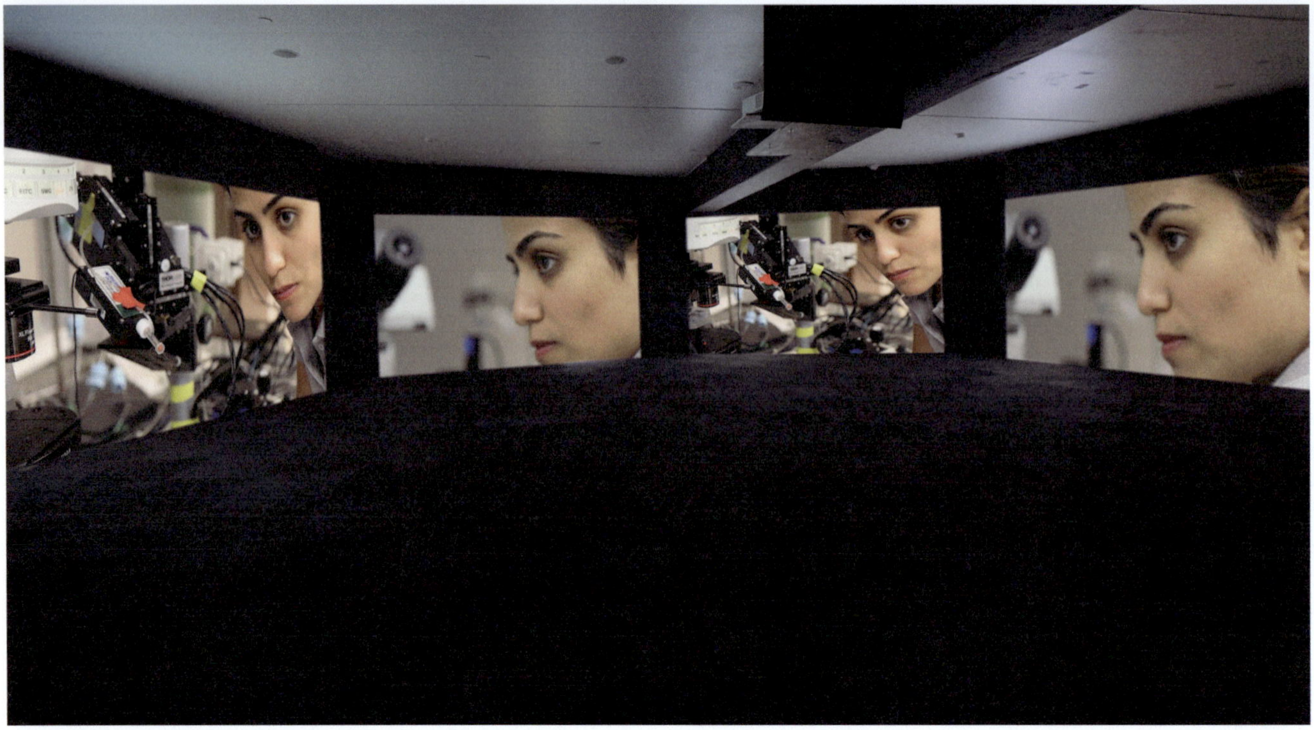

Fig. 7 Janet Biggs, *Can't Find My Way Home* (installation view, Blaffer Art Museum), 2015. Four-channel HD video installation with sound. Courtesy of the artist, Cristin Tierney Gallery (New York, NY), Analix Forever (Geneva, Switzerland) and CONNERSMITH (Washington, DC). Pictured is Dr. Mahshid Hosseini-Zare. In addition, not pictured are Drs. Anupam Hazra and Feng Gu, and Jokūbas Žiburkus who performed the unprecedented brain tissue recordings

spikes in the power line on household electronics. While not all portions of the brain are equally affected by discharges across the cortical network, these events can potentially have devastating impacts on the ability to recall and to develop new memories. These electrical waves may potentially lead to long-term changes in behavior by disrupting existing networks and creating new ones over time. Alzheimer's patients often experience periods of memory loss and periods of lucidity, abrupt changes in mood, periods of agitation followed by periods of calmness. Even in relatively early stages of disease, the increased electrical activity of the Alzheimer's brain can lead to progressive changes in memory, cognition, and behavior. This discovery raises many interesting questions of how these altered electrical networks ultimately impact the minds of those who are afflicted with the disease.

Jokūbas Žiburkus—When Janet walked into the electrophysiology recording room in my laboratory, she quickly adapted to the seriousness of our efforts. Searching for and then recording from these old neurons was a painstaking task. During the process, we remained completely focused on the sights and sounds of the brain tissue, while navigating through the complex architecture of the brain. The final approach into a neuron is crucial. The tension is so high that the air can almost be cut with a knife. Then we hear an unusual "pop"—electrochemical activity transferred into audio sound—indicating a successful contact with the neuron. There is a window of only ten minutes for the experimenter to capture as much information from the "talking" neuron, before the short, fleeting contact is lost. Entering the world of the dark room, microscope, traces on the computer screen, crackling sounds from the speakers connected to the amplifiers is challenging and fascinating. Most of us cherish the ability to exit out from these distinct worlds and dissociate distinct environments, events, and people the same day. That precious ability is lost in Alzheimer's disease, the maps are rigid, and the brain architecture is collapsing, often hopelessly beyond repair.

As humans, we are endowed with seamless abilities to interconnect art, music, math, and science. These are irreplaceable gifts, yet we often take these fleeting favors for granted or, even worse, place ourselves within boxes. Blurring the lines between statistical and metaphorical, between science and art, and thinking outside the box secures progress and reminds us of our dynamic and evolving human fate at the intersection of the spiritual and the physical.

3 Impact

Janet Biggs—*Echo of the Unknown* (the overarching title of my exhibition at the Blaffer Museum of Art) confirmed my belief that interactions and exchanges of information and methodologies between diverse communities create new vision and opportunities. The accumulated information used to produce this exhibition was gathered along an evidentiary path of new discovery, combining both the sensory and the empirical (Fig. 8).

The impact of this project has been broad, both in terms of identifying challenges and opportunities, in lessons learned and future ground to be explored. The concrete has been easy to chart. In conjunction with *Echo of the Unknown*, the Blaffer collaborated with more than a dozen University of Houston colleges and Houston institutions on programming through its *Innovation Series*, an ambitious slate of lectures, gallery talks, and panel discussions, enhancing the exhibition's role as a catalyst for cross-disciplinary learning. The exhibition and *Innovation Series* programming became one of the inspirations for the Brain on Art Conference(s). My exhibition has since traveled to museums and institutions in Europe and Canada and received numerous articles and reviews in publications ranging from the *Houston Chronicle* to the *New Yorker*, including a major article in *Art in America*.

An unexpected occurrence from the project was its use in a green card application by a Ph.D. candidate. The candidate's research was highlighted in one of my video installations, generating extensive conversations throughout the exhibition and educational programming at the Blaffer and at subsequent exhibitions. The candidate was able to demonstrate the impact and breadth of her research through documentation and press about the exhibitions.

The success of this project has been further acknowledged and supported by outside institutions including the John Simon Guggenheim Memorial Foundation through a 2018 fellowship that supports production of my new project that will continue exploring art and science perspectives and interactions.

Collaborations between different disciplines, between art and science, increases fluidity and the potential for new discovery. Taking the hand of another is a responsibility, but within each grasp there is hope. By overcoming our

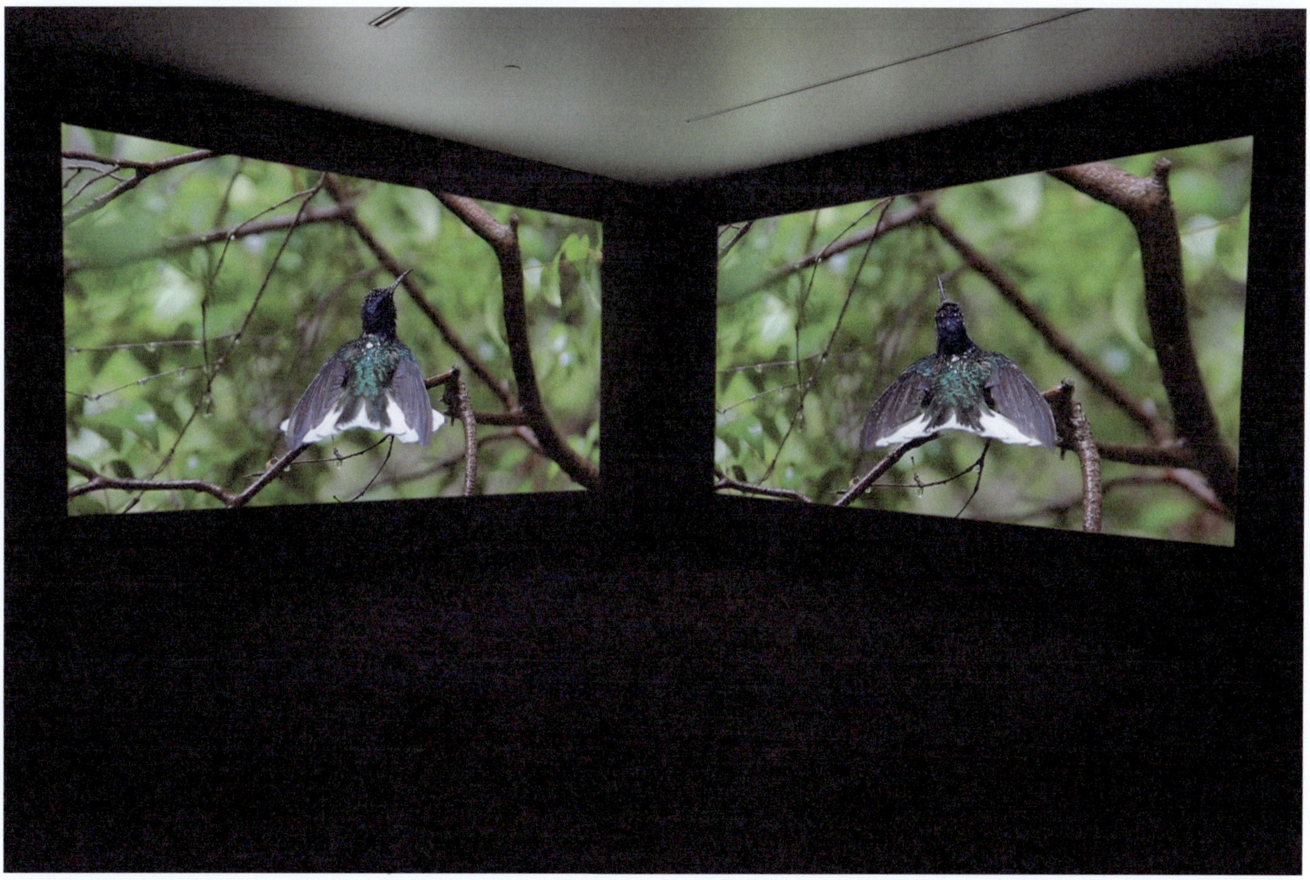

Fig. 8 Janet Biggs, *The Persistence of Hope* (installation view, Blaffer Art Museum), 2015. Two-channel HD video installation with sound. Courtesy of the artist, Cristin Tierney Gallery (New York, NY), Analix Forever (Geneva, Switzerland) and CONNERSMITH (Washington, DC)

vulnerabilities, we learn to ask new questions, to embrace possibilities. Interactions with scientists like Jason and Jokūbas shook many of my preconceived ideas and approaches to their core. These seismic shifts have replaced the static bedrock of working within comfortable confines and were indispensable to this project, which will alter my process in the future. For me, it was the more ephemeral aspects of this project's impact that promise to reverberate the most. The resonance and revelatory nature of the project has encouraged renewed attentiveness to the voice and perspective of others. The confluence of art and science demands increased attentiveness so that we might all better understand and be more acutely attuned to perceptions around us, both given and received.

References

1. Rosenberg, H.: The De-Definition of Art [1972], p. 10. Colliers, New York (1972)
2. Sherrington, C.S.: Man on His Nature, p. 142. Cambridge University Press (1942)
3. McKhann, G., Drachman, D., Folstein, M., Katzman, R., Price, D., Stadlan, E.M.: Clinical diagnosis of Alzheimer's disease: report of the NINCDS-ADRDA Work Group under the auspices of Department of Health and Human Services Task Force on Alzheimer's Disease. Neurology **34**(7), 939–944 (1984). PMID: 6610841
4. Cummings, J.L., Morstorf, T., Zhong, K.: Alzheimer's disease drug-development pipeline: few candidates, frequent failures. Alzheimers Res. Ther. **6**(4), 37 (2014). https://doi.org/10.1186/alzrt269. eCollection 2014. PMID: 2502475
5. Huxley, T.H.: Collected Essays: Volume 8, Discourses: Biological and Geological, p. 229. Cambridge University Press (2012)
6. Nelson, P.T., Alafuzoff, I., Bigio, E.H., Bouras, C., Braak, H., Cairns, N.J., Castellani, R.J., Crain, B.J., Davies, P., Del Tredici, K., Duyckaerts, C., Frosch, M.P., Haroutunian, V., Hof, P.R., Hulette, C.M., Hyman, B.T., Iwatsubo, T., Jellinger, K.A., Jicha, G.A., Kövari, E., Kukull, W.A., Leverenz, J.B., Love, S., Mackenzie, I.R., Mann, D.M., Masliah, E., McKee, A.C., Montine, T.J., Morris, J.C., Schneider, J.A., Sonnen, J.A., Thal, D.R., Trojanowski, J.Q., Troncoso, J.C., Wisniewski, T., Woltjer, R.L., Beach, T.G.: Correlation of Alzheimer disease neuropathologic changes with cognitive status: a review of the literature. J. Neuropathol. Exp. Neurol. **71**(5), 362–381 (2012). https://doi.org/10.1097/nen.0b013e31825018f7. Review. PMID: 22487856
7. Amatniek, J.C., Hauser, W.A., DelCastillo-Castaneda, C., Jacobs, D.M., Marder, K., Bell, K., Albert, M., Brandt, J., Stern, Y.: Incidence and predictors of seizures in patients with Alzheimer's disease. Epilepsia **47**(5), 867–872 (2006). PMID: 16686651

Janet Biggs is a 2018 Guggenheim Fellow, known primarily for her work in video. Solo exhibitions include the Neuberger Museum of Art, SCAD Museum of Art, Blaffer Art Museum, and d'art contemporain de Montréal. Reviews include the New York Times, the New Yorker, ARTNews, Art in America, and many others.

Jokūbas Žiburkus Ph.D. is an Associate Professor at the University of Houston's Department of Biology and Biochemistry researching epilepsy and dementias. Along with academic achievements and peer-reviewed publications, he is an entrepreneur, innovator, and public speaker with a widely watched TEDx talk on epilepsy and cannabis.

Jason L. Eriksen is an Associate Professor of Pharmacology at the University of Houston. He is a specialist in Alzheimer's disease and other neurodegenerative disorders. He received his Ph.D. in Neuroscience from Loyola University and served as a postdoctoral fellow at the Mayo Clinic in Jacksonville, Florida.

PARIESA: Practice and Research in Enactive Sonic Art

Lauren Hayes

1 Introduction

> The nature from which man has selected his musical styles is not only external to him; it includes his own nature—his psychophysical capacities and the ways in which these have been structured by his experiences of interaction with people and things, which are part of the adaptive process of maturation in culture.
>
> [1, p. 25]

In his book, *How Musical Is Man?*, John Blacking gives some suggestions about the biological and social origins of music. Blacking's experiences with the Venda people of South Africa led him to reject his former beliefs and strongly held assumptions about the function and nature of music, up until that point formed through a narrow, predominantly Western classical musical perspective. From this, he imbued musicological research with methodologies from social anthropology, rather than being constrained by theories of music based only around notions of pitch, rhythm, harmony, melody, and so on. In a similar vein, Christopher Small has offered a useful definition of music-as-human-action, or rather, as a verb: *musicking* [29]. Small emphasises the relational elements between sounds, people, and spaces, and the behavioural rituals that are bound up within the various practices of listening and music making, such as the tacit etiquette that accompanies attending a concert. Small labours the point that in any musical activity, the contributions of many more people than just the 'key players'—composers, performers, and audiences—are involved in bringing about a musical event.

These themes align with the emerging paradigm of enactive music cognition, a field borne out of radical and non-traditional cognitive science research and philosophy of mind, which offers a compelling framework for musical research. This is based on the idea that individuals are autonomous agents for whom cognition is formed in the embodied process of living within their socio-cultural and physical environments. This developmental process of identity and sense making depends on the repeated and ongoing sensorimotor and affective coupling between individuals and their world [30]. Recent developments in the field have highlighted that many activities—musicking could be included here—involve the conservation of group dynamics within the maintenance of such identities [21]. This field of research has emerged as a challenge to the traditional cognitivist and even certain embodied approaches to music cognition which rely on the role of mental representations in understanding musical creativity and experience (see [22]). It suggests that music cannot simply be abstracted into notated forms, or frameworks such as harmonic structures, but is wholly bound up within our embodied living. Even the practice of listening becomes an active, bodily endeavour.

The enactive approach to cognition has been suggested as a possible candidate for contextualising recent mirror neuron research within a broader, phenomenologically-compatible understanding of how humans interact with each other in the world [19]. The human mirror neuron system has been the focus of several recent neuroimaging studies examining musical experiences, concerning factors such as 'pitch memory, beat detection, [and] emotional response' [24, p. 489]. It is specifically the coupling of action and perception that is central to this phenomenological view of how

Electronic Supplementary Material
The online version of this chapter (https://doi.org/10.1007/978-3-030-24326-5_7) contains supplementary material, which is available to authorized users.

L. Hayes (✉)
Arts, Media and Engineering, Arizona State University, Tempe, AZ, USA
e-mail: Lauren.S.Hayes@asu.edu
URL: http://pariesa.com

J. L. Contreras-Vidal et al. (eds.), *Mobile Brain–Body Imaging and the Neuroscience of Art, Innovation and Creativity*, Springer Series on Bio- and Neurosystems 10, https://doi.org/10.1007/978-3-030-24326-5_7

we form musical meaning out of sequences of sounds produced by others over time: 'action understanding through the mirror mechanism is a *direct* activation of motor representation. It does not require a cognitive simulation of others' behaviour as suggested by simulation theory of action understanding' [7, p1]. While advances in neuroimaging techniques have led to a rapid increase in the number of studies related to music and cognitive processes, in the majority of cases, functional magnetic resonance imaging (fMRI), for example, tends to be used to explore computational models of the mind [8]. Yet, there exist several studies at the edges of neuroscientific research which challenge these traditional views and point towards the enactivist emphasis on embodiment, suggesting a holistic view in which 'the brain participates in a system, along with eyes and face and hands and voice, and so on, that enactively anticipates and responds to its environment' [8, p. 421].

My own work within the fields of music and sonic art has explored these threads from multiple perspectives, often through the use of electronic and digital technologies. Through the extended applications of enactive and embodied music cognition, performance, technology, and design, and the connections between these topics, I have developed a practice that has fed into a range of interdisciplinary collaborations from pedagogy to health and well-being, in addition to performance and improvisation. By viewing musical engagement—both perceiving as well as creating—as a relational, evolving, and embodied process, it can be demonstrated that the relevance of technological developments in the field of live electronic and digital musical practice lies not necessarily within the material aspects per se. But rather, an important consequence is the potential for individualised practices to emerge, where each musical agent (perhaps a performer) enacts a unique musical world in coordination with their physiological, socio-cultural, and musical histories. By viewing creative practices in this way, we are afforded the opportunity to view musical activity in general as—what socialist Tia DeNora has proposed—a 'medium of social relation', in various contexts [4, p. 14].

In what follows, I provide a selection of brief accounts of how these research themes have been addressed through the development of creative technologies used within my music and sonic art practice. In this work, I explore how the relationships that Small points to are forged over time through the lens of practice-based and ethnographic methodologies, but also within my more recent interdisciplinary collaborations, which integrate scientific and quantitive research. This activity is curated and documented as a collection of collaborative research projects, presented as Practice and Research in Enactive Sonic Art (PARIESA) [11]. In addition to the textual output, this research can be listened to, viewed on the stage, and at times it is even palpable. As such, links to representative media have been included where relevant. PARIESA involves working with performers, sonic artists, sound designers, composers, improvisers, as well as researchers and artists from other academic disciplines who have a shared interest in sound and music-driven creative practice research. Owing both to its similarities and differences to language, and its pervasiveness across cultures, music has been identified as a unique field for the investigation of human cognitive processes. PARIESA provides a rich milieux from which to explore the question: 'what do brains do in the complex and dynamic mix of interactions that involve moving, gesturing, expressive bodies, with eyes and faces and hands and voices; bodies that are gendered and raced, and dressed to attract, or to work or play; bodies that incorporate artifacts, tools, and technologies, that are situated in various physical environments, and defined by diverse social roles and institutional practices?' [8, p. 421].

2 Background

When I started performing live electronic music in the mid-to-late 2000s, I found that micro-gestures could often lead to macro-sonic events: a small finger movement could produce eruptions in the sound; a single key press could trigger a complex sequence of tones, or a pre-recorded audio sample. In spite of the ease of movement, these performance gestures contained none of the effort or struggle that I was accustomed to making use of as a classically trained pianist. Furthermore, I found that although I could hear the result of digital signal processing (DSP) through loudspeakers around a stage, I felt physically disconnected from the sounds that I was producing. This led me to explore more deeply the links between action and perception, specifically for the performer.

An analysis of the software involved in my performance system—which has been iterated through numerous performances and diverse collaborative situations—would certainly evidence different types of mapping strategies, which link the parameters of physical controllers to DSP. However, this has been far less important in the instrument's evolution than how it feels to perform with it. Sile O'Modhrain's research has focused extensively on the feel of digital musical instruments (DMIs) (see, for example [25]). Similarly, Kristina Andersen, along with collaborators, has been working for several years with 'non-functional physical mock-ups and prototypes' [10, p. 2347] to allow instrument designers to imagine affordances and potential functionalities of their instruments before becoming overly concerned with technical issues. Over time, I have developed metaphors and imagined agencies [6] within my system that have undoubtedly shaped both my experience of performing with it, and the actions and gestures I use when doing so.

Importantly, these are key aspects of my musicking which cannot be understood by quantitative analysis alone.

From the phenomenological perspective, I have worked extensively with the perception of sound as vibration, using customised vibrotactile devices [12, 13]. While digital technologies offer a host of new sonic possibilities—and the means to explore ideas that could only be laboriously achieved in the analogue or acoustic domains—we are no longer dealing with the physical vibrations of strings, tubes, and solid bodies as the sound source. Rather, our material is the impalpable numerical streams of digital signal processing and control data. As a result, when we perform with DMIs, we can no longer make use of vibrational feedback provided through the body of the instrument itself [18]. These haptic devices allow me to feel aspects of the electronic sound I am making, even when it has no real world resonating physical source, such as the wood and strings of the piano. It is perhaps not surprising that my training as a pianist has led to an exploration of musical human-computer interaction (HCI) that is largely focused around the expressive capacities of the hands and fingers.

Hungarian psychologist Géza Révész first introduced the word haptic, from the Greek *haptesta* (to touch), in 1931 [3]. It was used to describe the process of actively exploring a shape, or spatial dimension, with the hands, discussed in the context of his research into blindness and its effects on the other sense modalities. He contrasts this process with the sensation of indirectly perceiving something via the skin [3], such as experiencing differences in temperature, or feeling a feather brush against one's arm. However, when discussed in terms of HCI, 'haptic sensation' is often used as an umbrella term, encompassing both the active information gathering that Révész describes, as well as the passive tactile sensations that he classes separately. It is also sometimes used to refer to kinaesthetic sensory information, which deals with the relationship between the body and limbs, and their position and movement in space. The term haptic is often used in relation to both the somatosensory system—dealing with the perception of sensations on the surface of, or under the skin—as well as in relation to the proprioceptive system—involving intentional or active touching, and the actions and movements of the hands and body.

3 Bodily Hearing

The link between sound and touch is inherent: hearing is, essentially, 'a specialized form of touch' [9]. Profoundly deaf musician, Evelyn Glennie asserts a view that is the grounding premise for much of my work. Her statement could be explained in physical terms by the fact that sound is the rapid vibration of molecules in the air, or oscillations of pressure, which excite the membranes, hair and fluid inside of our ears, allowing us to hear. Moreover, our perception of sound goes beyond just the penetration of the auditory canal, and in fact is felt by our whole body, through vibrations within the organs and the bones. This engagement with the somatosensory system is something that Glennie affirms, claiming that she can sense, repetition of perceive/perceptable as vibrations, even those higher frequency sounds, which we may not have considered to be tangibly perceivable [9].

Of course, it is well known that below around twenty hertz, sound passes out of audible range, into palpable sensation. While many of us are familiar with the physical thumping of a bass line in a nightclub, we are not necessarily aware of our body's ability to haptically perceive higher pitched sounds. Glennie claims that is this simply because the auditory modality is more efficient in these lower ranges, and so becomes more prominent:

> If you are standing by the road and a large truck goes by, do you hear or feel the vibration? The answer is both. With very low-frequency vibration the ear starts becoming inefficient and the rest of the body's sense of touch starts to take over. For some reason we tend to make a distinction between hearing a sound and feeling a vibration, in reality they are the same thing. It is interesting to note that in the Italian language this distinction does not exist. The verb 'sentire' means to hear and the same verb in the reflexive form 'sentirsi' means to feel. Deafness does not mean that you can't hear, only that there is something wrong with the ears. Even someone who is totally deaf can still hear/feel sounds [9].

Glennie's account is a subjective and phenomenological one, as described from the unique perspective of a profoundly deaf virtuosic musician. We should pay serious attention to her comments, given her lifetime of experience and heightened awareness of a finely tuned sensory system. The idea of hearing through physical sensation is a sentiment that is echoed by other deaf musicians [28], as well as elsewhere in anthropological literature (see [20] for further discussion). We might start to wonder to what extent the body makes use of this sensory information in building up our impression of the world around us without, necessarily, our active awareness. Indeed, my most recent interdisciplinary art-science research—an ongoing project in collaboration with speech and hearing scientist Xin Luo—explores how haptic technology can improve music perception and enjoyment for people with cochlear implants.

This question of extending the sonic through touch has impacted two research areas within my work. First, in the creation of audio-haptic experiences in which audiences can both hear and feel musical material. For example, the *Skin Music* series [15] explores techniques for developing combined audible and tangible musical experiences. In the works that make up the series, participants experience a piece of music audibly, but also palpably through an arrangement of physical actuators embedded into furniture (see Fig. 1). A fixed media composition plays through loudspeakers,

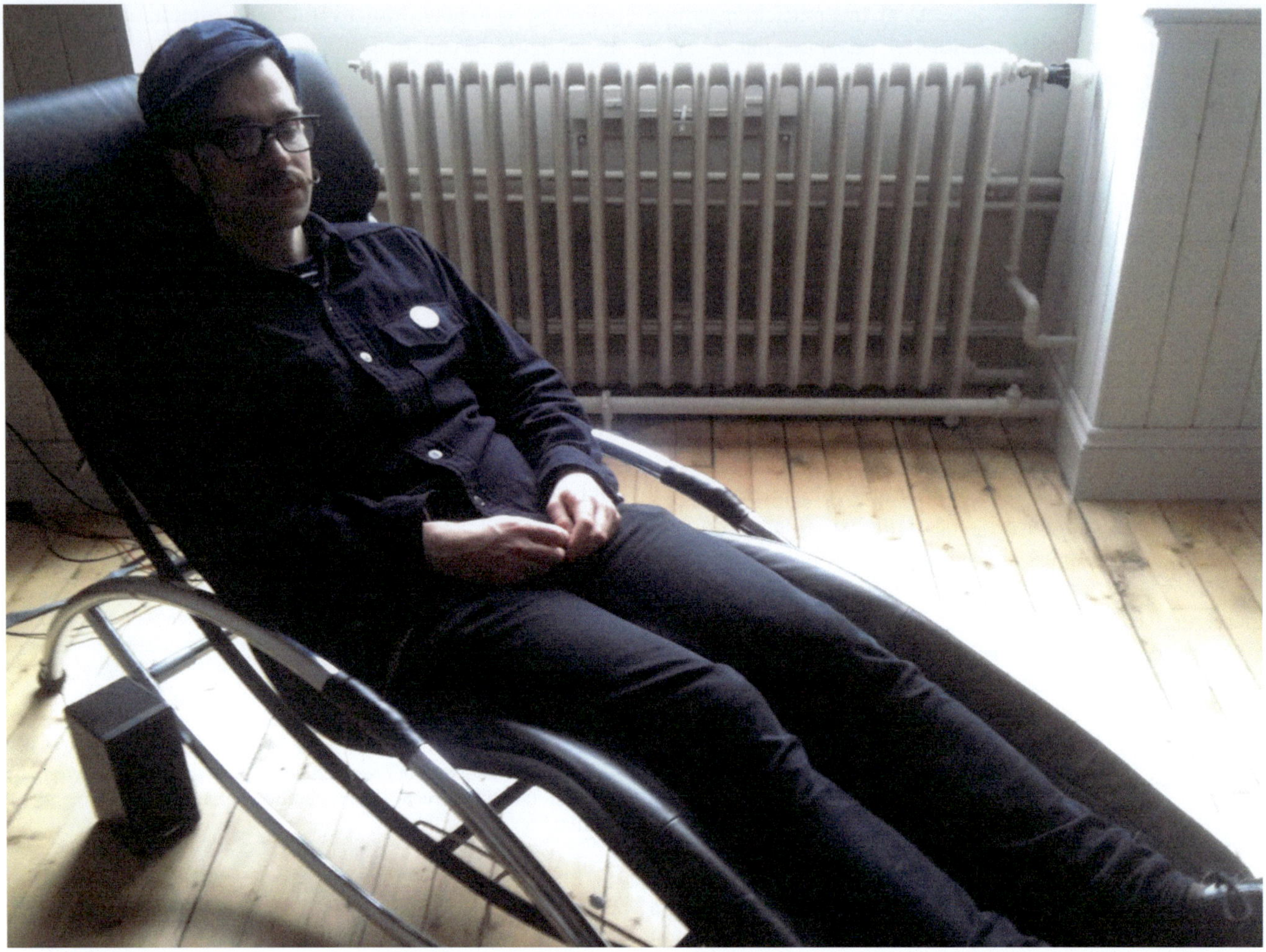

Fig. 1 A participant experiencing music through ears and body in *Skin Music, 2012*. Vibration motors are embedded into a chaise longue and tactile transducers are fixed onto under side of chair, proximal to the lower spine and feet

while different automated haptic patterns can be felt through vibration motors positioned within the structure of the chair, along with tactile transducer loudspeakers which are located proximally to, for example, the spine and feet, offering a range of different sensations (see Media Example 1). In a more recent performance work, in collaboration with somatic practitioner and interdisciplinary artist, Jessica Rajko, I performed the haptics in real-time, extending my existing live electronic musical performance environment. By attaching tactile transducers to the seating area of the performance space, I was able to shape different types of sensations being felt by audience members through their chairs [17].

Touching or feeling can bring about a sense of realism or truth, this sentiment being expressed in the idiom, 'seeing is believing, but feeling's the truth' (Thomas Fuller, quoted in [26, p. 73]). However, it is interesting to note that only the first part of this dictum is commonly used. Mark Paterson seems to highlight such phenomena as evidence to support his claim that popular media has an 'infatuation with visuality', and that we live in an 'academic climate that celebrates visual cultures' [26, p. 1]. Indeed, it is only within the last hundred years, or so, that theories of sound studies and audio culture have started to emerge. Research into haptic aesthetics is even more in its infancy. Yet anthropologist Tim Ingold provides a rejection of 'the thesis that attributes the dominance of objective thinking in the West to an obsession with the eye' [20, p. 245]. Ingold suggests that the problem with such criticism lies fundamentally in the 'reduction of vision… to its construal as a sensory modality specialised in the appropriation and manipulation of an objectified world' [20, p. 287]. We might rather understand the active processes of looking, hearing, and so on, as inseparable aspects of perception.

4 Digital Musical Instrument Design

This detached, or observational perspective is often exemplified within the more traditional aspects of HCI, where interaction is based around graphical user interfaces and onscreen icons: the visual representation of the metaphorical desktop. Yet touch brings us into direct contact with the objects that are within our visual field. Touching can also renew our relationship with a person or object, and if we have not picked up or practiced our instrument for a significant length of time, we may say that we are 'out of touch'. But the role of touch within musical practice goes much deeper than this.

The second avenue within my research that is concerned with how to extend the sonic through touch involves the design of new DMIs. Many of my first works using technology involved digital augmentation of acoustic pianos (see Fig. 2). While, as a 4-year-old child, I may have originally been drawn to the piano simply due to its ubiquity as a traditional Western instrument, through repeated engagement with the instrument from this young age, by way of lessons, exercises, and the sort of experimentation that I much later learned was called improvisation, I enacted my musical environment based around a very specific type of tactile engagement. I learned to make use of both the vibrational feedback of the resonating body of the piano, as well as the particular resistances that it offered me as a physical instrument.

Over the last decade, I have undertaken an approach to DMI design that focuses specifically on these relationships between the sonic and the tactile. This explores the double aspects of Maurice Merleau-Ponty's notion of embodiment—of the body as at once biological and phenomenological [23]. I have explored ways of enfolding physical resistances into my instruments. This has involved force-feedback haptic technology, often repurposing low-cost games controllers, and using, for example, physical models within haptic design to offer different types of palpable feedback to play against. While the technical aspects of my DMI design philosophy have been described elsewhere (see, for example, [14]), this process was heavily influenced by Claude Cadoz's notion of instrumental gesture, where energy is transduced from the physical world into the digital domain [2]. This is fundamentally distinct from the commonly accepted

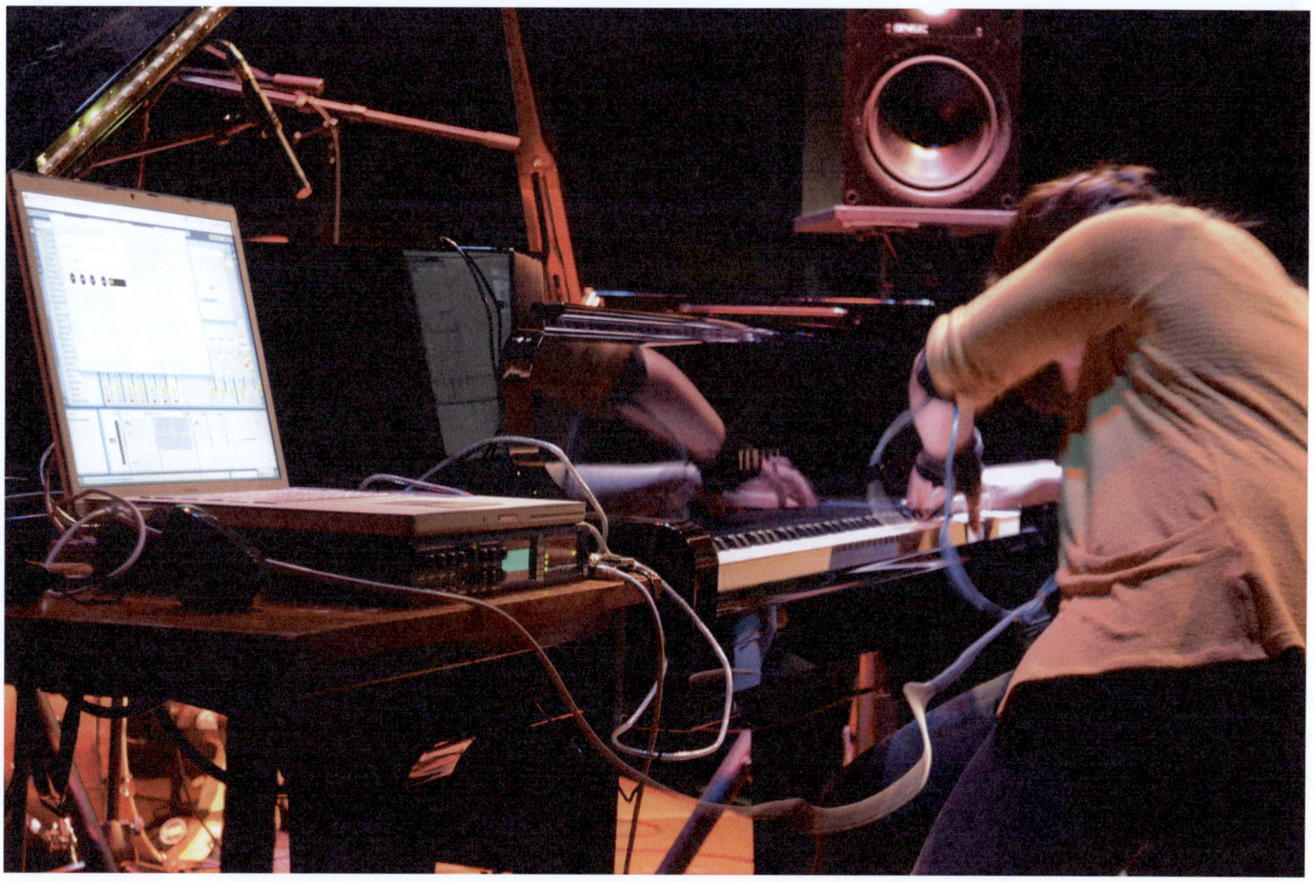

Fig. 2 Performing on a digitally-augmented hybrid piano, using an early version of the vibrotactile-feedback glove. The acoustic sound of the piano is amplified and processed using custom software. The digital audio signals are analysed and converted into haptic information, which is sent to the hand of the performer

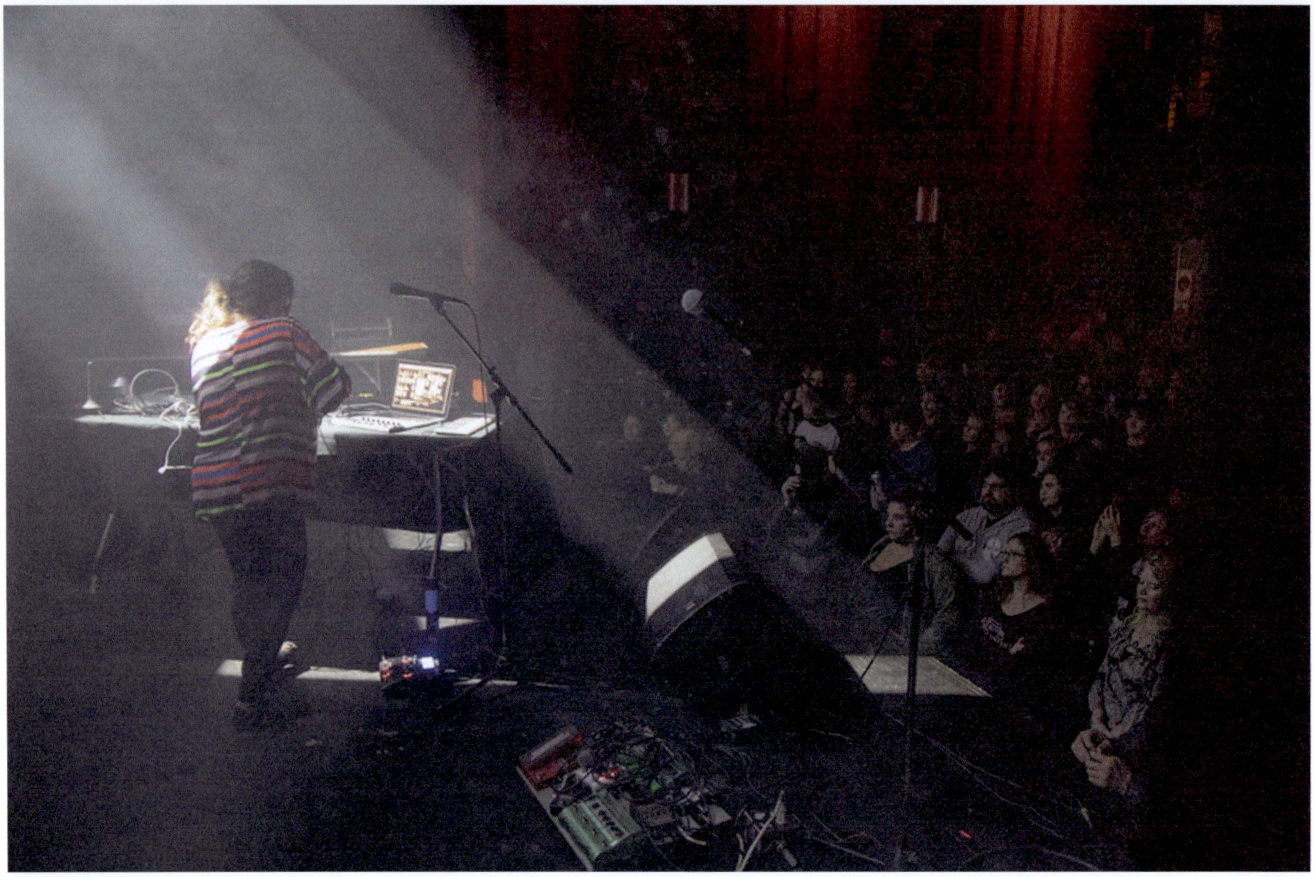

Fig. 3 Improvising with the hybrid analogue-digital performance system comprising commercial hardware and bespoke software. Gestural energy is transduced into the digital domain and used to affect DSP. *Photo credit* Jason Thrasher

paradigm of mapping where the onus is on the engineer or instrument designer to successfully build in functional relationships between input gesture and sonic result. As Cadoz demonstrated, working with virtual-physical models, which are excited using haptic interfaces, is one possible alternative approach.

Accounting for the unique physiology of the performer is crucial. I have often aligned boundaries within the physical world, such as, for example, where my hand falls upon a piano keyboard, or where my furthest reach inside the piano on the soundboard might be, with areas of instability within the digital world—such as the point at which a short looped sound speeds up until it becomes a sustained tone (see Media Example 2). Many DMIs are derived from interfaces designed for effortlessly smooth human-computer interaction. But as Pedro Rebelo and Richard Coyne note: 'there [is] no impetus to develop a violin that blends ergonomically with the player' [27, p. 2]. Whether I am building a performance system for myself, or for someone with very specific physiological requirements—perhaps due to sensory impairment or a learning difficulty (see [16])—I seek to find meaningful points of resistance or friction within these systems, which tend to be the places where the most potential for expressive musical engagement lies.

The instrument I play presently is an evolving hybrid assemblage which comprises commercial hardware such as analogue synthesizers, voice processors, and drum machines; bespoke software which I continue to develop incrementally over time; and repurposed games controllers (see Fig. 3). In my most recent hybrid analogue/digital performance system an excessive number of components mutually affect each other through an ecological network of sound analysis and DSP. Engaging with different parts of the instrument through tangible and haptic controllers, I bring a sense of immediacy into my hands: the slightest movement may trigger a mechanical relay bank, which in turn may active digital processes. The resistances in my performance environments lie within the extreme potential for activity through interconnections within the audio signal path. Yet, a joystick-centred controller is so easy to move—a movement of even one millimetre can drastically alter the sound—that musicality and expression come from resisting this.

The idea of sound sculpting [5] suggests an active process of deliberately shaping sonic material through tangible interactions. As a performer, not only do I want to be able to manipulate the material that I create, but I want to be able to feel this sense of the malleability of sound through my audio-tactile interactions, and to be able to sense that I am approaching the thresholds of my electronic processes both with my hands, as well as my ears (see Media Example 3). Approaching this as an improviser, I navigate my participation as a human in the ongoing and dynamic interactions between my body, hardware, software, the loudspeakers, the space in which we are situated, and the audience. This performance work explores how musical experiences which are emergent, unpredictable, and non-linear can be created, and how new instruments which are individually engaging, yet allow collaborative creativity, and are challenging to play can be designed.

5 Conclusion

Being a time-based media, musical performance occurs in the present moment of the historical unfolding of material, socio-cultural, and sensorimotor interactions that have led to a particular aesthetic experience. An enactive understanding of music cognition acknowledges the importance of the repeated and ongoing sensorimotor action in the world where musical activity arises out of perceptually guided and situated action. It is important to consider the role of these processes in shaping musical activity in order to develop new creative practices beyond the paradigm of human-computer interaction, as well as suggesting how we might conduct neuroscience research that reaches beyond computational models of mind.

While the benefits of using haptic and enactive technology for improving certain aspects of instrumental skill acquisition are well documented, research in this area tends to be focused around technical development alone. My own research has attempted to provide an in-depth, practice-based perspective in this field. Much of this work is situated within the realm of performance practice, yet working as an improviser and technologist has allowed me to navigate interdisciplinary collaborations including those between artists working in various disciplinary fields, hearing scientists working in areas such as music perception, and with many publics including children, adults who have not been trained as musicians, and people with profound and complex learning difficulties. By understanding brain activity in 'nonrepresentational, integrative and dynamical terms' [8, p. 421] there lies significant potential to develop creative, therapeutic, and rehabilitative technologies which acknowledge the importance of sensuous feedback, while simultaneously incorporating it in their design.

Acknowledgements Images courtesy of Craig Jackson and Jason Thrasher.

References

1. Blacking, J.: How Musical Is Man?. University of Washington Press, Seattle (1974)
2. Cadoz, C.: Instrumental gesture and musical composition. In: Proceedings of the 1988 International Computer Music Conference, San Francisco (1988)
3. Davidson, P.W.: Haptic perception. In: S. of Pediatric Psychology (ed.) J. Pediatr. Psychol. **1**(3), 21–25. Oxford University Press, Oxford (1976)
4. DeNora, T.: Music in Everyday Life. Cambridge University Press, Cambridge (2003)
5. Emmerson, S.: Music imagination technology. In: Proceedings of the 2011 International Computer Music Conference, Huddersfield (2011)
6. Ferguson, J.R.: Imagined agency: technology, unpredictability, and ambiguity. Contemp. Music Rev. **32**(2–3), 135–149. Taylor & Francis, London (2013)
7. Ferrari, P.F., Rizzolatti, G.: Mirror neuron research: the past and the future. Philos. Trans. R. Soc. Lond. B Biol. Sci. **369**(1644), 20130169 (2014). https://doi.org/10.1098/rstb.2013.0169
8. Gallagher, S., Hutto, D.D., Slaby, J., Cole, J.: The brain as part of an enactive system. Behav. Brain Sci. **36**(4), 421–422 (2013)
9. Glennie, E.: Hearing essay. https://www.evelyn.co.uk/hearing-essay/ (1993). Accessed 23 Nov 2018
10. Grote, F., Andersen, K., Knees, P.: Collaborating with intelligent machines: interfaces for creative sound. In: Proceedings of the 33rd Annual ACM Conference on Human Factors in Computing Systems. Seoul, 18–23 April 2015 (2015)
11. Hayes, L.: Practice and research in enactive sonic art. http://www.pariesa.com (n.d.). Accessed 23 Dec 2018
12. Hayes, L.: Vibrotactile feedback-assisted performance. In: Proceedings of the International Conference on New Interfaces for Musical Expression. Oslo (2011)
13. Hayes, L., Michalakos, C.: Imposing a networked vibrotactile communication system for improvisational suggestion. Organ. Sound **17**(1), 36–44. Cambridge University Press, Cambridge (2012)
14. Hayes, L.: Haptic augmentation of the hybrid piano. Contemp. Music Rev. **32**(5), 499–509. Taylor & Francis, London (2013)
15. Hayes, L.: Skin Music (2012): an Audio-haptic composition for ears and body. In: Proceedings of the ACM SIGCHI Conference on Creativity and Cognition. Glasgow, 2012 (2015)
16. Hayes, L.: Enacting musical worlds: common approaches to using NIMEs within both performance and person-centred arts practices. In: Proceedings of the International Conference on New Interfaces for Musical Expression. Baton Rouge (2015)
17. Hayes, L., Rajko, J.: Towards an aesthetics of touch. In: Proceedings of the 4th International Conference on Movement Computing. ACM, London (2017)
18. Hayes, L.: Live electronic music performance: embodied and enactive approaches. In: Proceedings of the 5th International Conference on Movement Computing. ACM, Genoa (2018). https://doi.org/10.1145/3212721.3212891
19. Hayman, G.: Mirror neurons, Husserl, and enactivism: an analysis of phenomenological compatibility. Perspectives **6**(1), 13–23 (2016)

20. Ingold, T.: The Perception of the Environment: Essays on Livelihood, Dwelling and Skill. Routledge, London (2000)
21. Loaiza, J.M.: From enactive concern to care in social life: towards an enactive anthropology of caring. Adapt. Behav. (2018). https://doi.org/10.1177/1059712318800673
22. Matyja, J.R., Schiavio, A.: Enactive music cognition: background and research themes. Constr. Found. **8**(3), 351–357 (2013)
23. Merleau-Ponty, M.: Phénoménologie de la perception (1945). English edition: Merleau-Ponty, M.: Phenomenology of Perception (trans: Smith, C.). Routledge and Kegan Paul, London (1962)
24. Molnar-Szakacs, I., Overy, K.: Music and mirror neurons: from motion to'e'motion. Soc. Cogn. Affect. Neurosci. **1**(3), 235–241 (2006)
25. O'Modhrain, S.: Playing by feel: incorporating haptic feedback into computer-based musical instruments. Ph.D. Dissertation. https://ccrma.stanford.edu/~sile/thesis.html (2001). Accessed 23 Dec 2018
26. Paterson, M.: The Senses of Touch: Haptics, Affects and Technologies. Berg, Oxford (2007)
27. Rebelo, P., Coyne, R.: Resisting the smooth: time-based interactive media in the production of distressed space. In: Digital Design: 21st eCAADe Conference (2003)
28. Roebuck, J.: I am a deaf opera singer. http://www.theguardian.com/theguardian//sep/29/weekend7.weekend2 (2007). Accessed 1 Dec 2018
29. Small, C.: Musicking: The Meanings of Performing and Listening. Wesleyan University Press, Middletown (2011)
30. Varela, F.J., Thompson, E., Rosch, E.: The Embodied Mind: Cognitive Science and Human Experience. MIT Press, Cambridge (1991)

Lauren Sarah Hayes is a Scottish musician and sound artist who creates physical live electronic music. She is a regular improviser on both augmented piano and analogue/digital hybrid electronics. Her research examines the links between sound and touch, and she has developed and performed with bespoke haptic and vibrotactile devices over a prolonged period of creative practice. Her work has been presented globally at festivals and within publications such as Organised Sound and the Computer Music Journal. She is currently Assistant Professor of Sound Studies at Arizona State University and an associate of the New BBC Radiophonic Workshop. In 2018, she became Director at Large of the International Computer Music Association.

Into the Mind of an Artist: Convergent Research at the Nexus of Art, Science, and Technology

Jesus G. Cruz-Garza, Anastasiya E. Kopteva, Jo Ann Fleischhauer, and Jose L. Contreras-Vidal

We can try to use machines just as machines or as an extension of the body. It's a question of attitude.
—Pipilotti Rist

1 Introduction

Understanding the creative brain in action and in context remains one of the grand challenges in human neuroscience. However, neuroscience studies of the human creative artistic process have typically been constrained to laboratory settings or brain imaging scanner suites that are far removed from the artist's usual work environment and tools. As discussed extensively in Part I of this book, the experimental paradigms that have been used to study creativity have failed to explain creativity (see Chap. 4), in part because studies of creativity have used highly parameterized tasks (e.g., counting the number of different uses for a predetermined object as an index of creativity). Moreover, from these parametrized tasks, researchers have attempted to quantify brain activity associated with a proposed creativity metric often within a single experimental session. These methods constrain both the artist and the measurements from investigating important components of the creative process, which is a highly dynamic, evolving process that can take days, weeks, months or even years to be completed.

The creative process therefore cannot be considered a sequence of isolated moments of inspiration, but the incremental progress of successes and failures that lead to an envisioned result. This demands a period of time where ideas mold [24], which often surpasses the duration allotted to most experiments, or even precedes the identified initial conceptualization. The creative components typically include researching and conceiving the problem or goal of the artwork or art commission; researching and selecting tools, materials and suitable locations for conducting the work; planning, exploration, and ideation leading to prototypes or precursors (studies) of the ultimate artwork; body movement; deliberation and feedback; spontaneity, and revision of the work, to name a few. Indeed, the process to create an artistic installation is often nonlinear: the projected vision may change over time, the artist adapts the materials and presentation to the exhibition space requirements, or new knowledge and interests emerge. Da Vinci, Dali, and Picasso did not experience their Aha! moment *sans* effort; instead their work was fraught with repetitive successes and failures over an extended period of time. Furthermore, it is very likely that the situational context and the individual experiences, intentions, rivalries, and goals of the artists contributed to their creative output. These factors have not been considered in prior studies of the creative process.

Studying the creative process of the artist "in action and in context" has been additionally constrained by mobility limitations of neuroimaging technology. Protocols had to be constrained to laboratory settings [1] and other artificial environments where subjects were connected to bulky and restrictive technology such as magnetic resonance imaging (MRI), wired scalp electroencephalography (EEG), positron

Electronic Supplementary Material
The online version of this chapter (https://doi.org/10.1007/978-3-030-24326-5_8) contains supplementary material, which is available to authorized users.

J. G. Cruz-Garza (✉) · A. E. Kopteva · J. L. Contreras-Vidal
IUCRC BRAIN, University of Houston, Houston, TX, USA
e-mail: jgcruz@central.uh.edu

J. L. Contreras-Vidal
e-mail: pepeum@gmail.com

A. E. Kopteva
Department of Theater and Dance, University of Houston, Houston, TX, USA

J. A. Fleischhauer
Cullen College of Engineering, University of Houston, Houston, TX, USA

J. L. Contreras-Vidal et al. (eds.), *Mobile Brain–Body Imaging and the Neuroscience of Art, Innovation and Creativity*, Springer Series on Bio- and Neurosystems 10, https://doi.org/10.1007/978-3-030-24326-5_8

emission tomography (PET), or magnetoencephalography (MEG). The advent of context-aware mobile brain–body imaging (MoBI) technology [14] addresses the mobility limitations in neuroimaging studies; further allowing for potential development of new tools to study and potentially support the creative process as in the case of artistic brain–computer interfaces (BCIs) where artists can use brain reading technology to control and augment their surroundings based on the state of their brain activity (e.g., thoughts or feelings) [15, 21, 29].

Thus, context-aware MoBI technology provides the opportunity to study longitudinally the creative process of artists at their own studio and other complex natural settings without movement, spatial or time constraints [8, 9, 20]. Synchronized video cameras and motion sensors, as well as manual annotations from the artist and the researchers, enable the labeling of actions of freely behaving artists as they work on an art piece from inception to production. Further, the fast setup time, ease of use, and increasing affordability of MoBI technology (e.g., dry electrode EEG headsets integrated with accelerometers) may allow for wide adoption in the future. Such a dataset would contain highly valuable, context-dependent, and timely information about the human creative process as it happens naturally in real complex settings. It would capture brain activity related to project ideation, planning, and execution, including changes in brain activity patterns due to shifts in internal states, elusive "Aha!" moments (provided these can be identified and annotated effectively by the artist, or inferred from the artist's data), all while documenting the dynamic creation of an artistic installation as a whole; including life situations artists encounter over extended periods of time such as discovery, failures, successes, revisions, external feedback, life events, etc. The richness of such a real-world MoBI dataset, complemented with environmental, location, and demographic information, poses its own inherent challenges. Emotional considerations, medication status, neurological and mental conditions, family events, executive decisions, changes in the project, deadlines, prosperous and arduous periods of production, and the very nature of the specific tasks undertaken in an artistic installation (materials, development of skills, etc.), are all variables of potential interest to track in the data.

In this chapter, we discuss the inception of the first context-aware MoBI long-term study of the creative process of an artist working on her own creative practice at her home and studio, and elsewhere. The MoBI data collection was planned for a period of one year of work, eventually extended to 18 months, starting on January 18, 2016. We present early behavioral findings from the first 9 months of data with a focus on the artist's perspective on transforming her artwork into an experiment. The artist, also a research member of the team, envisioned an installation where the spectators' multisensory perception played a central component in their experience of the space. Her work explores the interaction of the human body and cognition (knowledge, memory, evaluation, reasoning) to interpret sensory input and build experiences. Refer to Online Resource 10[1] for a video interview about the longitudinal study with remarks by the authors.

One expected outcome of the convergent art-science collaboration was that MoBI recordings would provide insights into the artist's creative process, while also making visible work-related habits, and unknown correlations with environmental variables. Moreover, the data would allow the researchers to assay the evolution of internal states across days, weeks, and months, while cataloging the patterns of brain activity, and characterizing their individuality and variance associated with various stages of her work. Here, we focus our discussion on the identified challenges, opportunities, and initial findings from this case study, the artist's experience, and the construction of a labeled MoBI dataset.

The processing of MoBI, and in particular scalp EEG data, is on-going given the massive data acquired and the multiple steps required to process the data, such as labeling, denoising, detecting neural patterns using machine learning techniques, visualization, statistical analyses, and interpretation of findings [10, 20, 30]. The discussion aims to assess the usability of context-aware MoBI technology during the evolving process of artistic creation. We also suggest an approach to annotate behavioral data in this context, and automatically extract useful data from the MoBI data collected.

1.1 Authenticity and Privacy in Science-Art-Engineering Collaborations

To ensure authenticity from the point of view of both the artist and the experimenter(s), it is highly desirable to include the artist in the conception and implementation of experimental protocols to minimize disruption of the artist's creative process and to facilitate integration of the MoBI technology into the artist's everyday work activities. Engaging the artist early into the planning of the study provides an equal consideration in the experimental design and evaluation process to best assess the creative process in a minimally intrusive way [6]. In this regard, the artist becomes an integral member of the research team and provides valuable top-down information regarding the artist's thought process, while annotating the data with important and timely events for the analysis phase. This collaboration on experimental protocol articulation allows the artist to create authentically and on her own schedule, while respecting her privacy. Indeed, in our case study and after a

[1] Video credit: Carlos Landa, University of Houston Cullen College of Engineering.

period of training with the MoBI technology, the equipment set-up, and data acquisition was carried out and controlled by the artist herself at times and places of her own choosing (e.g., home, study, studio, gym, etc.). Moreover, it allowed her brain and body to assimilate the instrumentation as an extension, or even an ornament or tool, of her body, just as when one wears a new pair of amplifying glasses, rides a new bicycle, or wears a new hat.

2 The Artist

Fleischhauer [11], a Houston-based sculptor and installation artist and Artist-in-Residence in Neuroaesthetics at the Cullen College of Engineering's Industry-University Cooperative Research Center for Building Reliable Advances and Innovations in Neurotechnologies (IUCRC BRAIN), at the University of Houston, volunteered to participate in the study while she worked on her next major art installation project. The longitudinal study included periods of conception, research, planning, prototyping, and production of the installation's components, as well as novel skill development, and time periods where the artist was relaxing or thinking. The artist was technically trained by laboratory members in donning the MoBI headset independently, identifying potential recording errors, ensuring the care and maintenance of the device, and in performing an electrode impedance check prior to start of a recording session. The artist incorporated the technology into her daily life and work and was an integral member of the research team with constant communication about the project's progress and direction, as well as in monitoring the state of the recording equipment and suggesting best practices for its portability and usability.

The artist began data collection early in the conception of her project after receiving approval from the Institutional Review Board (IRB) at the University of Houston. The start date for MoBI data collection was January 18, 2016. For the first four months, she primarily spent her time conducting research on topics of interest for her new installation, mostly within her home and studio. As a sculptor and installation artist, her work is site responsive, and often site-specific. Her practice is research-oriented; she spends much of her project time reading and researching, experiencing the site, and manipulating and investigating materials.

3 Assaying the Artist's Creative Mind

Context-aware MoBI technology was deployed to track Jo Ann Fleischhauer's brain activity (Fig. 1) as she worked through ideas and possibilities—walking on a treadmill, researching the botanical aspects of pollen, and using a kit of essential oils and other scents to spark ideas about incorporating the sense of smell into the multidimensional project. The MoBI headset was a wireless, mobile, 20-channel dry EEG headset (Cognionics Inc., San Diego, CA) that provided continuous EEG recording of her brain activity synchronized with a motion sensor that monitored her head movements at a 500 Hz sampling frequency. A smartphone app for journal annotations was used to link the brain activity with physical location, weather information, and other variables. The artist took notes to track her thoughts and feelings, medications taken, and caffeine consumption; the weather information was logged automatically. She donned the headset at home, where a network of video cameras provided contextual cues as she pondered various aspects of the project across locations such as at her studio just east of downtown Houston, and at home. The headset was also used on the treadmill at the gym and while walking dogs, as she would actively think of her project's direction during these activities. The MoBI data was transferred from the headset to a Microsoft Surface Pro 3 tablet, held on a lightweight backpack, via Bluetooth wireless transmission. The total weight of the MoBI (EEG and accelerometer) headset was 340 g.

In her experience, going out in public wearing the 3D printed headset seemed awkward at first. According to the artist, "*in the beginning, it was almost like getting used to wearing a prosthetic limb.*" Over time, it grew more comfortable, both physically and mentally. Most people did not inquire about it when she was out in public. Those that did were intrigued about the Art-Science collaboration and the research being conducted. Wearing the headset changed how she thought and worked, to an extent. The artist remarked that "*it is making me much more conscious of what I'm thinking. I am much more analytical, more conscious about remembering the threads*" [16, p. 63].

The artist noted that ideas often began to flow after about 30 min into running on the treadmill, so she started to use a stopwatch, clicking it when an idea arrived and, once off the treadmill, rushing to match and notate those time points to the specific idea. For Fleischhauer, inspiration is not one discrete moment in time. "*I'm finding my [Ah-ha!] moments; they're accumulations of lots of different things,*" Fleischhauer said. Anything—from detailed research or a trip to the museum to watching television and listening to music—can trigger an idea. And when it happens, the headset records her brain activity, in action and in context, something that cannot be done in a laboratory setting.

For privacy and comfort reasons, while Fleischhauer provided the data for the study, she also controlled what she provided, deciding when, where, and for how long to wear the headset. The length of this study provides the data for researchers to learn about the dynamics of the evolving creative process as the artist created an installation project -

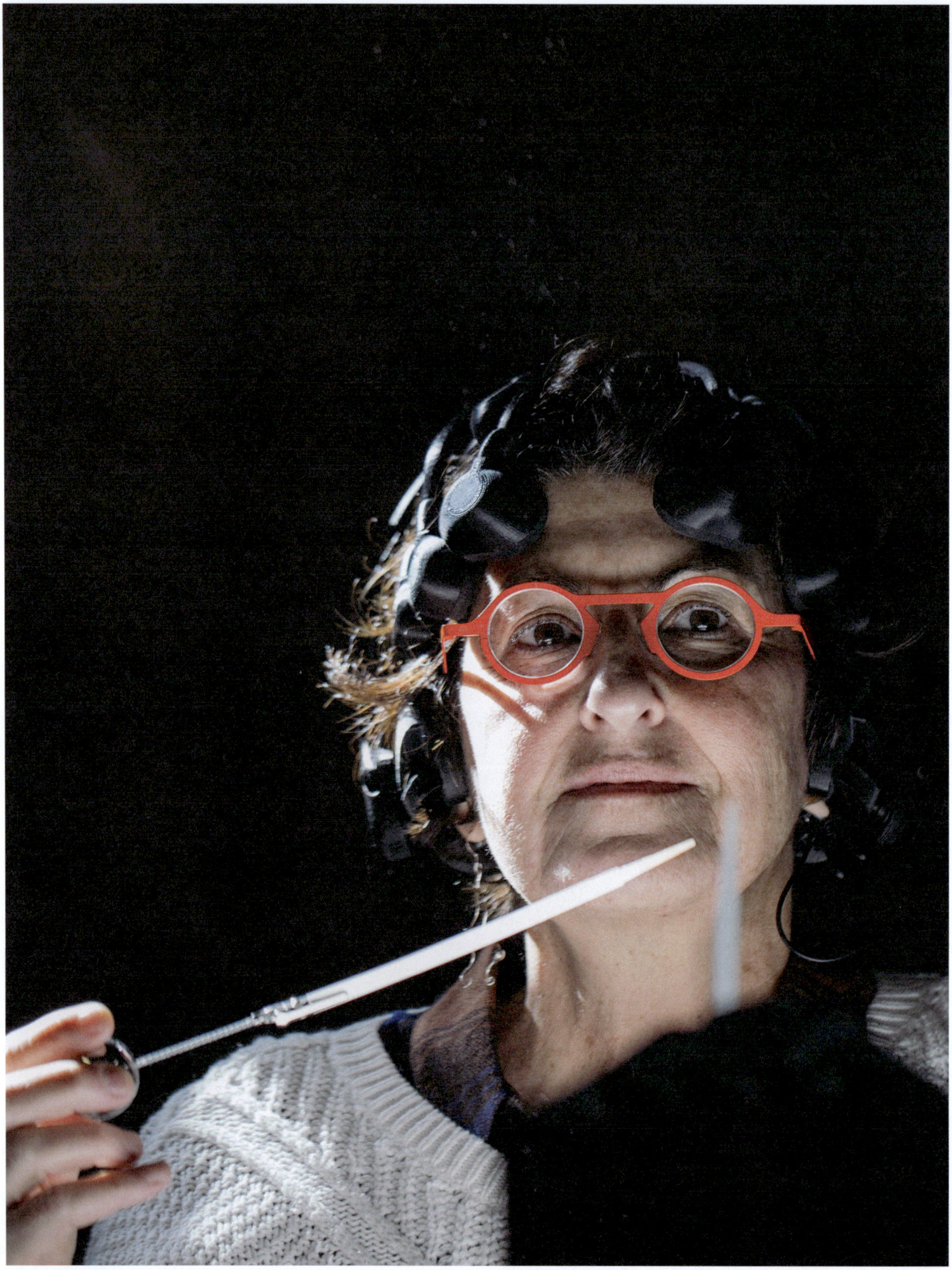

Fig. 1 The artist is shown wearing the MoBI headset while training to identify and differentiate scents. *Photo credit* Carlos Landa, Cullen College of Engineering at University of Houston

from conception to research, exploring new ideas, developing new skills (olfactory training, working with new material), daily life events, experimenting with materials, and prototyping in preparation of a final installation. These tasks are described in the next sections.

3.1 Pollen Preparation

Pollen was collected from two sources: a large grocery store and a small florist shop in Houston, Texas. The florists collected the immature anthers from Oriental Lilies and put them in small plastic containers. The artist gathered the collected anthers once or twice a week.

The anthers were spread out on a black piece of paper left to dry and open at the artist's home. This process exposed the pollen (Fig. 2). After several days, the artist gathered up the pollen and sifted it to separate the dried anther from the pollen. The artist then put the sifted pollen in small plastic bags: labeling the day that it was collected from the store and the day that it was sifted. The pollen taken from the two sources were always kept separate. The plastic bags were wrapped in tin foil to block out any light, taped closed, and put in the freezer until required, as the pollen is very susceptible to light and heat. When ready, the artist used the pollen as pigment to infuse the Japanese paper, experimented with it (Fig. 3), and created drawings. After completing each drawing, the artwork was wrapped in a foil-lined wrapping and placed in cold storage. The artist wore the MoBI headset while preparing the pollen in her home and studio: laying out the collected pollen, sifting, labeling, and freezing. The headset was not worn when collecting the pollen from the florists.

3.2 Olfactory Training

Another aspect of the art installation was to introduce scent as a component to the project. The artist researched the topic and received training on scent identification skills so that she could collaborate with a professional perfumer to develop and construct a scent, which would be a central part of the installation. The following materials were used to train her scent identification skills:

- 48 small vials of natural essential oils as well as synthetic molecules, which are used in the perfume industry
- Blotting paper strips—thin strips of absorbent blotting paper that is used to dip into the vials of scent and then lightly smelled
- Holders to hold the strips of paper after dipping into the vials
- Notebook to write impressions
- Reference books
  - One pamphlet- source to identify the smells
  - In-depth reference book [2] where the artist read about the corresponding plants, manufacturing processes and uses of the scent that had been identified.

There was a small table where all of the above olfactory-training components were laid out (Fig. 4). The artist practiced smelling 5–10 different scents during each session, while wearing the MoBI headset (Fig. 5). The vials are "blind", meaning that the artist did not know a priori what scents she was testing. Each smelling trial consisted of taking the scent vial, dipping a smelling paper strip into the liquid, closing up the vial, and taking short whiffs of the paper strip. The artist would write down her impressions of the scent: her recollections, what the smell reminded her of, and any impression that she had at the time of smelling. These notes would help her remember the scent the next time she tried it. At the end, after smelling all of the vials, she would go back and smell each one again. Her perception of the smell of the vials changed over time: they would get more diffused, softer, and sometimes they would have different characteristics that were not noticed initially. The artist was continuously annotating her responses. Sometimes she could identify the smell immediately and it could be generally associated with something familiar, i.e., a holiday, an experience, sometimes the association was not clear, and sometimes the smell triggered a very specific memory for her. The artist wore the MoBI headset during all of her scent and olfactory training sessions.

3.3 Book and Internet Research

The artist wore the headset when she was reading. She read via a variety of mediums- either sitting at the computer, holding a book, or a photocopied article. She used a yellow highlighter and took notes in several notebooks.

3.4 Treadmill Workout

The artist reported that exercising on the treadmill was a prosperous introspective time in which she would think about her project and conceive her creative ideas. She wore the MoBI headset during these times. The artist typically walked on the treadmill for 1 h and at the maximum incline of 15° at 4.8 miles per hour. She put the headset on prior to getting on the treadmill. She used a stopwatch that was initialized when the workout session on the treadmill started. She actively thought about her project and documented

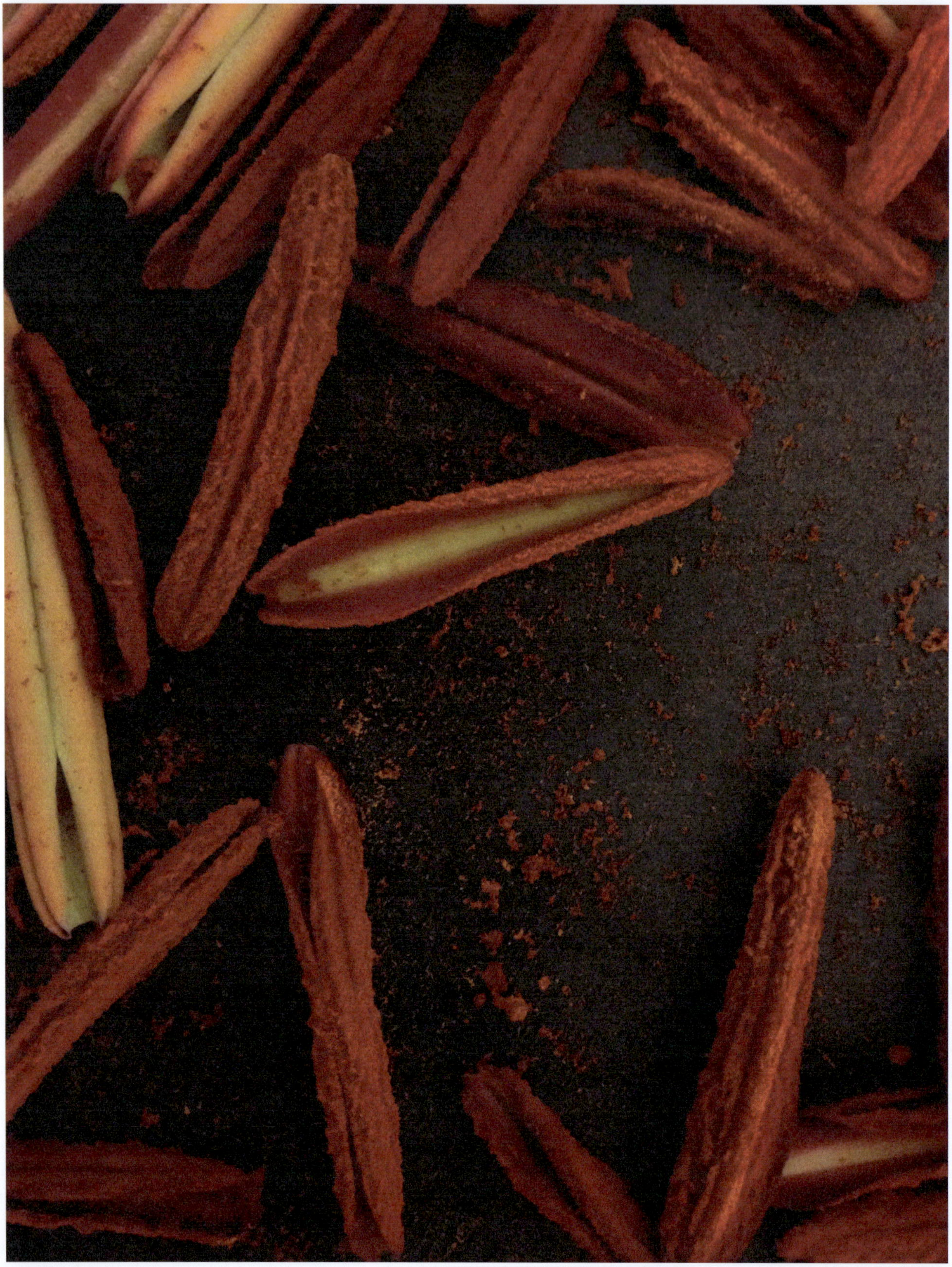

Fig. 2 Lily anthers opening and exposing pollen. Pollen was a central material of the installation project. *Photo credit* Jo Ann Fleischhauer

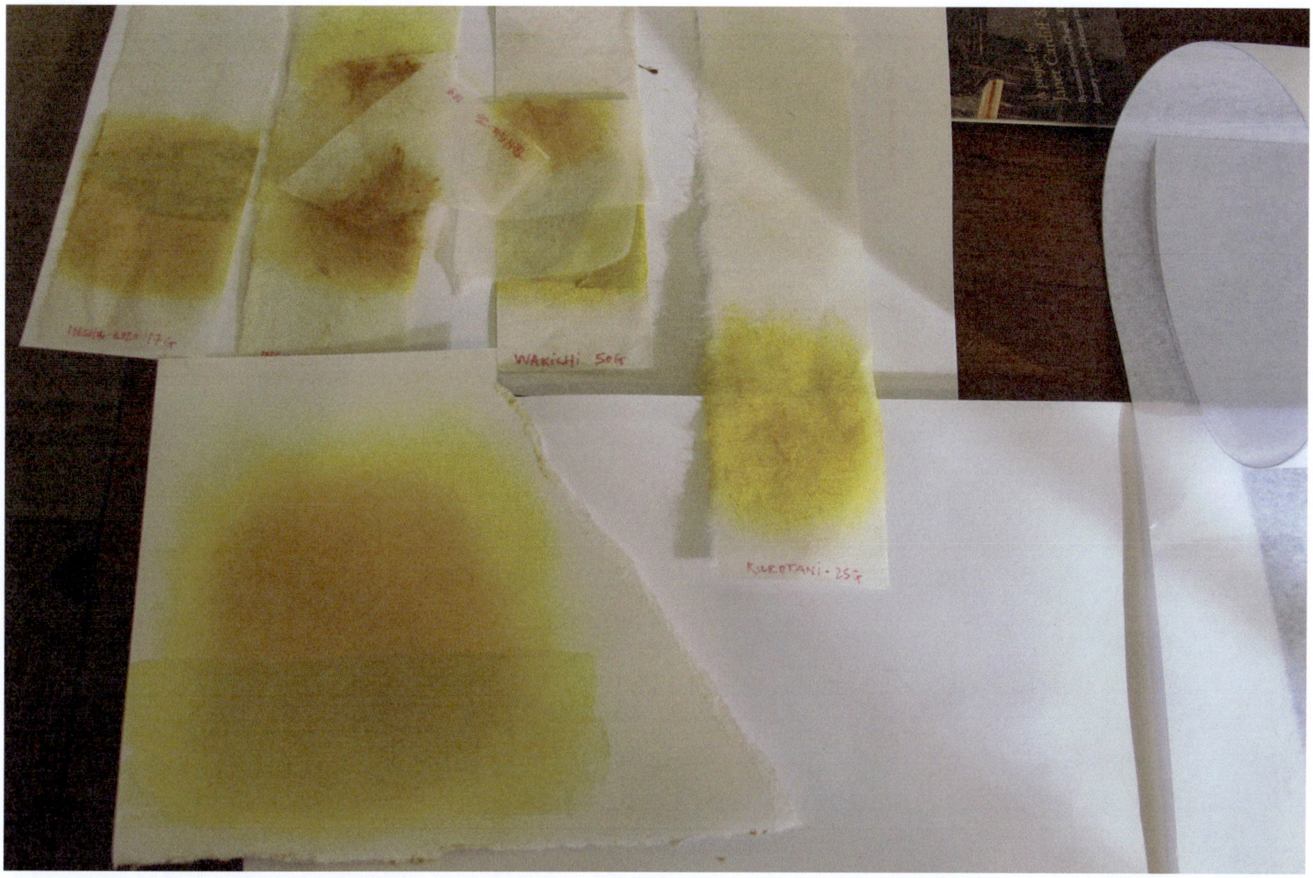

Fig. 3 Handmade Japanese paper impregnated with lily pollen. Experiment looking at oxidation and degradation of yellow pollen pigment on different types of handmade Japanese paper. *Photo credit* Jo Ann Fleischhauer

creative ideas that she had during the workout sessions, even if they were not used later on. When she thought of an idea, she would click the stopwatch once. After the workout session, she would retrieve the times where she had creative ideas, annotate those times, and write down as much as she could remember about such ideas.

4 Headset Usability Metric: Hours of Use Per Day

The artist set up the system herself independently, using a mirror to align and adjust the headset's electrodes correctly. The setup was complete once all electrodes were in place and showed acceptable levels of impedance ($<15\ \mathrm{k}\Omega$) in the company-provided software on the tablet. Video cameras were installed in the artist's workplaces (house, studio, workshop), sampling at 30 fps and 1080 ppi to enable the MoBI recordings to be context-aware. This allowed for video review and tagging of the EEG and motion sensor activity, and together with denoising algorithms [4, 18], assisted in the identification of artifacts in the MoBI recordings [8]. The artist took notes that included: task times, headset usability and comfort, prescriptions taken, coffee and alcohol intake, specifics on the tasks, and creative ideation.

Factors such as the hours of use per day and week give us insight on the level of acceptance of the user, comfort levels and ease of adoption of the technology. Maintaining data integrity and data security for privacy reasons are also necessary to better facilitate research and future BCI applications. Additionally, the large amount of data collected allows for examination of how factors such as location, length of use, time of day, temperature, and weather conditions impact signal quality and usability daily and seasonally.

5 Results

In the following sections, we present early findings from this longitudinal study.

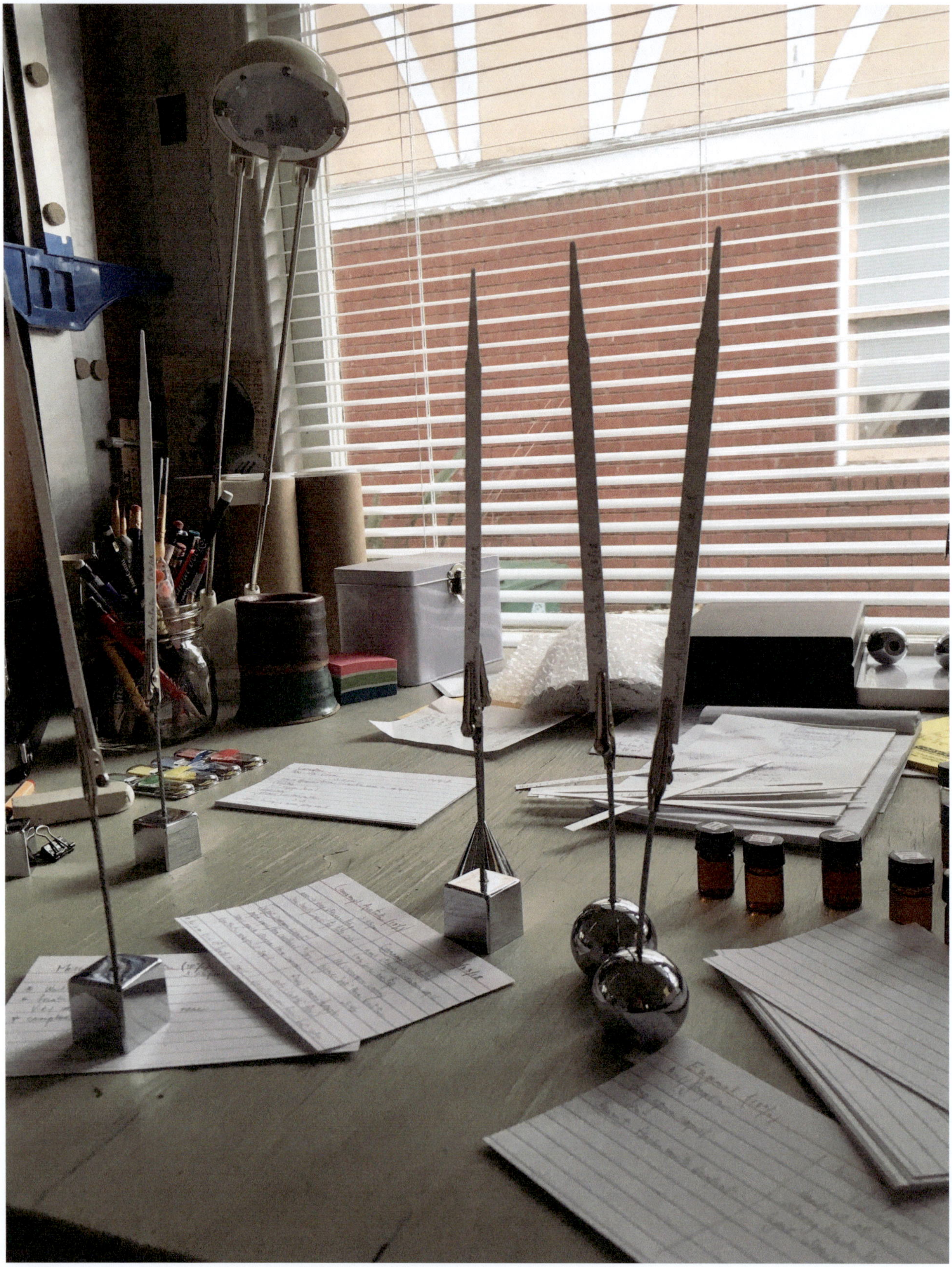

Fig. 4 Olfactory work table. Vials of different essential oils, paper smelling strips dipped in different essential oils to identify, note cards used to jot down impressions of different scents. *Photo credit* Jo Ann Fleischhauer

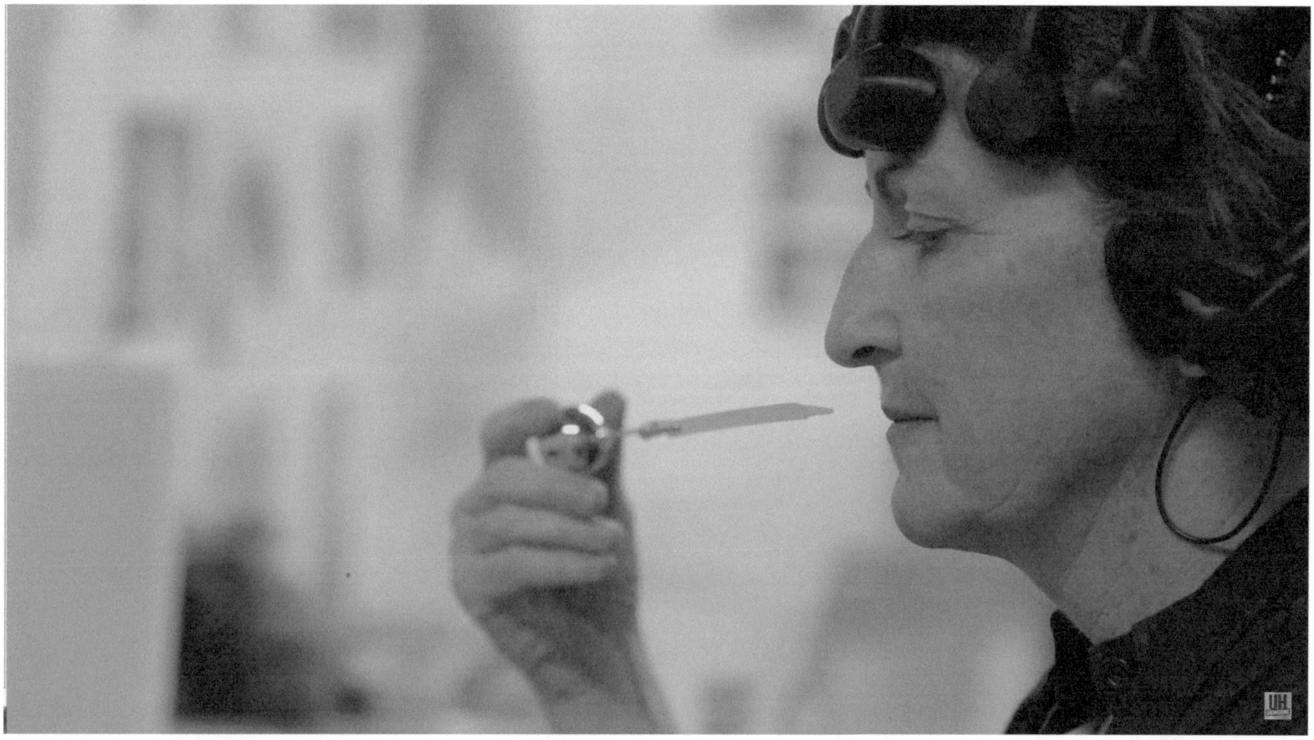

Fig. 5 The artist trained her sense of smell to identify and describe natural and synthetic molecules. The multisensory installation included an olfactory component. *Photo credit* Carlos Landa, Cullen College of Engineering at University of Houston

5.1 Long-Term Usability of MoBI Device in Artwork Practice

To the best of our knowledge, this is the first long-term use of a MoBI headset to study creativity in a real-world setting. As such, questions about the long-term usability and signal quality of MoBI technology, the level of subject's familiarization and assimilation of the technology and the possibility of using the subject-specific data collected for automatic pattern identification in real-world settings has not been explored.

The artist used the headset as she worked on her installation, from conception to planning to implementation. The artist wore the headset at her discretion, aiming to provide the most data collection without compromising her comfort. The headset was also not used when its' integrity could be compromised, such as in situations where the artist would require using aerosols or work in extremely confined spaces. The artist reported spending an estimated half of her time working on project-related tasks while wearing the headset. In the first 34 weeks, she wore the headset for a total of 323 hours at an average of 1.36 hours per day. There was a clear preference for Sundays, Tuesdays, and Thursdays for data recording over the 34 consecutive weeks, representing the artist's work schedule (Fig. 6). The device usability results, in terms of hours of use per day, are shown in Fig. 6.

When the headset caused any discomfort, the artist would remove it. The artist reported mild discomfort after more than 3 h of continuous use of the headset, primarily due to extended pressure from the temporal electrodes. The artist did not wear the headset for several extended periods of time for a variety of reasons including maintenance (week 2), sickness (week 7), and traveling (week 24). After longer periods of time without use, the artist reported discomfort earlier into the recording session when she resumed data recording. Overall, the headset proved durable and reliable, with consistent-quality data provided.

5.2 Annotating MoBI Datasets

The artist's actions were labeled by visual inspection of the video recordings: The artist started with research on the computer during the conception of her project, transitioned gradually to research in books and printed articles, prototyping in the third month of data collection, performing olfactory training, and spending additional time prototyping at later stages.

From the video cameras placed at her studio and home, we identified visually the following tasks performed by the artist: (a) conducting research on her computer, "Researching in Computer", (b) reading printed books and articles,

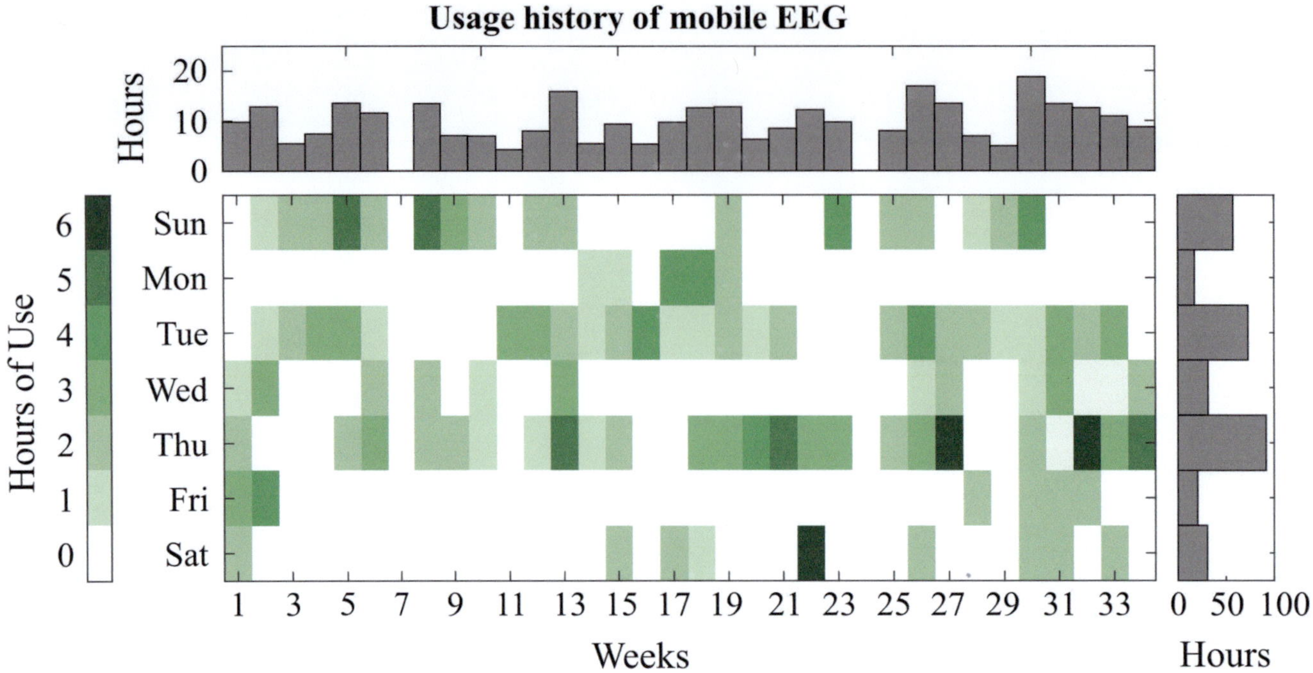

Fig. 6 Usage history of EEG cap during the first 34 weeks of the artist's work. The total recording time collected for the 36 weeks is 322.64 h of use representing 130 GBs of MoBI data: 105 Gb of EEG and acceleration data, 25 Gb of video. A. The gray histograms indicate the cumulative number of hours of use across week (vertical) for EEG data, and per day (horizontal). The green gradient shows the number of hours the artist wore the headset each day (average of 1.36 h/day) with video data

"Researching in books", (c) developing scent discrimination, "Olfactory Training", (d) manually creating components of the artwork, "Prototyping", (e) walking during creative ideation around week 22 after olfactory training and prototyping, "Walking", and (f) times when the artist was taking a break and g) doing noncreative tasks such as cleaning. A "Prototyping" task is shown in Fig. 7. We are currently deploying computer vision, image pattern recognition and deep networks to mine and learn the artist's various types of behavioral actions from the videorecording, which should accelerate the discovery process.

Figure 8 shows a histogram of the first four classes described here, for the 34 weeks of data collection, where each "sample" is a 4 s window of EEG data, with 25% overlap. The histogram shows an imbalance of classes, with "Research in Computer" as the most performed task with the mobile EEG headset, followed by "Olfactory training", "Research in Books", and finally "Prototyping". In the final stages of the installation project, it is expected to acquire more EEG data from the "Prototyping" stage.

6 Discussion

This study was designed to investigate the behavioral and neural correlates of a real world creative process of a skilled artist over several months, from conception to implementation. This work is considered a ground-breaking feasibility study. This chapter addressed issues of usability of the MoBI technology, experience of the artist, and context-awareness of the MoBI data for event tagging and annotations. Early findings in regard to usability were presented which suggest this type of study is feasible. However, there are some potential pitfalls, which are discussed next.

7 Labeling of the Artist's Actions

The artist's creative actions were labeled by manual scanning of the video recordings and labeling overall actions that could be discerned from the video recordings. This process is tedious and time-consuming. Automated computer vision techniques could be deployed to speed up the annotation of the video to link to the remaining MoBI data as discussed below. The artist started with research on the computer during the conception of her project, transitioned gradually to research in books and printed articles, prototyping in the third month of data collection, performed olfactory training, and then spent time prototyping at later stages. The progression of tasks was a real-world example of an artist's creative process, with heavy research-oriented practices at the start of the project, to skill development, and finally installation prototypes. This process was highly nonlinear,

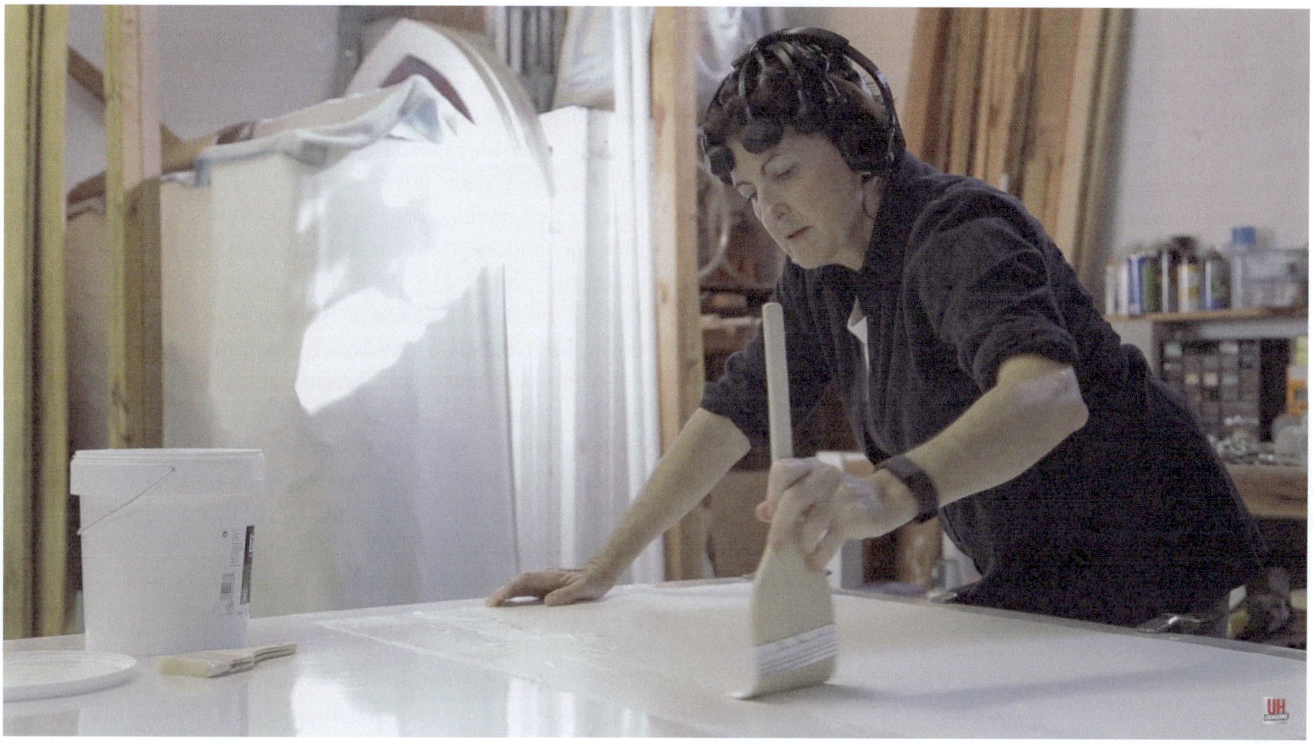

Fig. 7 The artist is shown wearing the MoBI headset while preparing the Japanese paper for later work. *Photo credit* Carlos Landa, Cullen College of Engineering at University of Houston

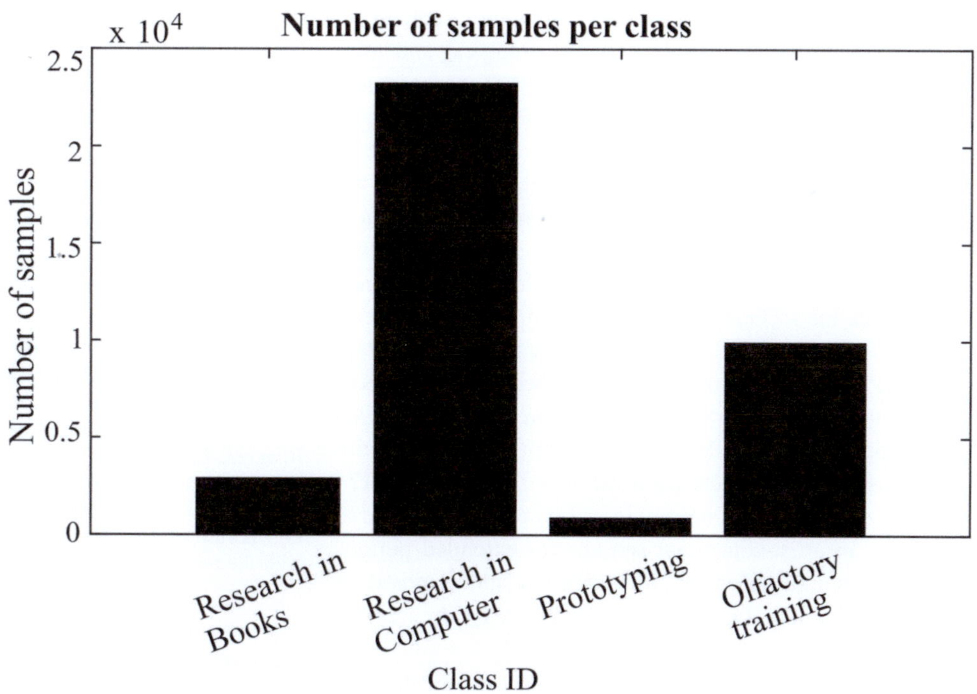

Fig. 8 Number of samples per type of activity (Class ID) during the first 34 weeks of data collection, illustrating four classes. Each "sample" is a 4 s window of EEG data with 25% overlap

with many iterations and revisions on research questions for the creative output, prototype experiments, and reframing and revision being integral parts of the process. For example, for the first 9 months, a significant class imbalance from labeled actions was obtained (Fig. 8), as expected in an unconstrained real-world study of the human creative process. The class imbalance poses additional challenges in the functional and statistical analysis of the data.

The behavioral annotations were performed by human visual inspection, based on clear actions discernible from the video recordings. The context-awareness component of MoBI in this study was performed manually. If context-aware MoBI technology is to be used in future unconstrained experimental settings (e.g., tracking biological data from humans during daily living), the labeling of the data through manual inspection will certainly be a limiting bottleneck. Advances in machine vision are expected to help annotate human behavior from one or more simultaneous subjects, and coupled with MoBI technology, large neural datasets from populations are certainly a possibility. Real-world data in complex environments will need to involve the collection and analysis of a large number of subjects, environments, synchronized measuring devices, massive-EEG preprocessing tools [5], and automatized annotation tools for effective implementation.

7.1 Automatic Artifact Identification and Denoising of EEG Signals

Dry-electrode, context-aware mobile EEG headsets allow for neural data collection in real-world settings [20, 27, 26]; opening the door for the study of natural cognition, "in action and in context" [6, 7]. However, the acquisition of EEG brain waves in unconstrained situations such as in MoBI studies requires handling of physiological and non-physiological artifacts associated with the measurement modalities. Scalp EEG measurements are often prone to excessive motion artifacts and other types of artifacts such as eye blinks, eye movements, electromyographic (EMG) activity from scalp and neck muscles, amplifier's voltage shifts, changes in electrode impedance due to humidity, temperature and movement, and other artifacts that may contaminate the EEG recordings. Although the magnitude of such artifacts heavily depends on the task and the setup, complete minimization or isolation of such artifacts is generally not possible. Fortunately, there are powerful signal processing and machine learning algorithms that can identify and remove such artifacts from the raw EEG signals [17–19]. Utilizing MoBI technology in real-world settings, and in particular, in artist-contextual settings for extended periods of time, requires automatic identification and removal of noise from the data. In a long-term MoBI study, the number of recordings is expected to be large, therefore supplying a large quantity of examples to identify motion, EMG, and ocular-related artifacts. This work is computational intensive and it is currently ongoing in this study.

7.2 Challenges of the Study

Given the nature of this study, which was conducted in natural complex settings over a period of several months, outside variables need to be taken into account for data analysis and data interpretation. For example, it is well known that some medications and other substances such as nicotine and alcohol can affect brain activity over hours, days and weeks [12, 25]. Thus, it is expected that these factors will have some impact on several features throughout the day. Moreover, additional factors that may affect internal states and brain activity include the amount of time spent awake, circadian rhythms, and changes in daily routine. Other factors such as temperature and humidity, and their effect on longitudinal mobile EEG data collection are yet to be analyzed using clustering and regression techniques.

Ongoing analyses of this dataset, to be reported in future publications, investigate the neural stages of the creative process. Alongside the artist, we aim to integrate data-driven (bottom-up) and high-level model-based (top-down) methods to identify the neural networks and stages of the artist's creative process as she worked on the installation. For example, spectral estimation methods are used to identify the frequency, prevalence and distribution of brain rhythms across the data set [22]. Functional connectivity analyses are used to quantify the level and direction of cortico-cortical communication across brain areas related to aesthetic tasks [3, 20, 28]. Source analyses methods can also be deployed to understand the cortical locus of the uncovered signals about intentionality [23], emotion [13], and preference during art creation, uncovered by machine learning algorithms. These methods can also be used to track changes in the internal states of the artist over time to understand the evolving dynamic nature of art making.

Acknowledgements The authors thank the support of this research by the Cynthia Woods Mitchell Center for the Arts at the University of Houston for providing the artist with a research grant to conduct olfactory research and consult with orchid bee researcher, Dr. Santiago Ramirez, The Ramirez Lab, University of California Davis, Davis, CA; and the Institute of Art and Olfaction in Los Angeles, CA. Research funds for this project were provided by the National Science Foundation Award #BCS 1533691. Jeannie Kever conducted the interview with the artist, transcribed and adapted with permission.

References

1. Abraham, A.: The Neuroscience of Creativity. Cambridge University Press, Cambridge (2018)

2. Arctander, S.: Perfume and Flavor Materials of Natural Origin. CreateSpace Independent Publishing Platform, Elizabeth, NJ (1960)
3. Beaty, R.E., Benedek, M., Wilkins, R.W., et al.: Creativity and the default network: a functional connectivity analysis of the creative brain at rest. Neuropsychologia (2014). https://doi.org/10.1016/j.neuropsychologia.2014.09.019
4. Bigdely-Shamlo, N., Mullen, T., Kothe, C., et al.: The PREP pipeline: standardized preprocessing for large-scale EEG analysis. Front. Neuroinform. **9**, 16 (2015). https://doi.org/10.3389/fninf.2015.00016
5. Bigdely-Shamlo, N., Touryan, J., Ojeda, A., et al.: Automated EEG mega-analysis I: spectral and amplitude characteristics across studies. BioRxiv 409631 (2018)
6. Contreras-Vidal, J.L., Cruz-Garza, J., Kopteva, A.: Towards a whole body brain-machine interface system for decoding expressive movement intent challenges and opportunities. In: 2017 5th International Winter Conference on Brain-Computer Interface (BCI), IEEE, pp. 1–4 (2017)
7. Cruz-Garza, J.G., et al.: Your brain on art: a new paradigm to study artistic creativity based on the 'exquisite corpse' using mobile brain-body imaging. In Brain Art, pp. 283–308. Springer, Cham (2019). https://doi.org/10.1007/978-3-030-14323-7_10
8. Cruz-Garza, J.G., Brantley, J.A., Nakagome, S., et al.: Deployment of mobile EEG technology in an art museum setting: evaluation of signal quality and usability. Front. Hum. Neurosci. **11** (2017a). https://doi.org/10.3389/fnhum.2017.00527
9. Cruz-Garza, J.G., Brantley, J.A., Nakagome, S., et al.: Mobile EEG Recordings in an Art Museum Setting. IEEE Dataport (2017b). https://doi.org/10.21227/h2tm00
10. Cruz-Garza, J.G., Hernandez, Z.R., Nepaul, S., et al.: Neural decoding of expressive human movement from scalp electroencephalography (EEG). Front. Hum. Neurosci. **8**, 188 (2014). https://doi.org/10.3389/fnhum.2014.00188
11. Fleischhauer, J.A.: Media Sculpture & Installation Artist (2019). http://jofleischhauer.com/
12. Gilbert, D., Dibb, W.D., Plath, L.C., Hiyane, S.G.: Effects of nicotine and caffeine, separately and in combination, on EEG topography, mood, heart rate, cortisol, and vigilance. Psychophysiology **37**, 583–595 (2000)
13. Goshvarpour, A., Abbasi, A., Goshvarpour, A.: Combination of sLORETA and nonlinear coupling for emotional EEG source localization. Nonlinear Dyn. Psychol. Life Sci. **20**(3), 353–368 (2016)
14. Gramann, K., Ferris, D.P., Gwin, J., Makeig, S.: Imaging natural cognition in action. Int. J. Psychophysiol. **91**, 22–29 (2014)
15. Kever, J.: Art and Science Come Together on the Dance Floor. TMC News (2015)
16. Kever, J.: Contemplating creativity. Univ. Houst. Mag. **62**. Online PDF: https://issuu.com/uhmagazine/docs/uhmag_s2017_final.compressed?e=3877706/49165470. Online resource: https://www.uh.edu/magazine/2017-spring/contemplating-creativity/ (2017)
17. Kilicarslan, A., Contreras-Vidal, J.L.: Full characterization and removal of motion artifacts from scalp EEG recordings. In: 2017 International Symposium on Wearable Robotics and Rehabilitation (WeRob). IEEE, p. 1 (2017)
18. Kilicarslan, A., Grossman, R.G., Contreras-Vidal, J.L.: A robust adaptive denoising framework for real-time artifact removal in scalp EEG measurements. J. Neural Eng. **13**, 026013 (2016). https://doi.org/10.1088/1741-2560/13/2/026013
19. Kleifges, K., Bigdely-Shamlo, N., Kerick, S.E., Robbins, K.A.: BLINKER: automated extraction of ocular indices from EEG enabling large-scale analysis. Front. Neurosci. **11**, 12 (2017)
20. Kontson, K.L.K.L., Megjhani, M., Brantley, J.A.J.A., et al.: Your brain on art: emergent cortical dynamics during aesthetic experiences. Front. Hum. Neurosci. **9**, 626 (2015). https://doi.org/10.3389/fnhum.2015.00626
21. Kovacevic, N., Ritter, P., Tays, W., et al.: 'My Virtual Dream': collective neurofeedback in an immersive art environment. PLoS ONE **10**, e0130129 (2015)
22. Lotte, F., Bougrain, L., Cichocki, A., et al.: A review of classification algorithms for EEG-based brain-computer interfaces: a 10 year update. J. Neural Eng. **15**(3), 031005 (2018)
23. Luu, T.P., Brantley, J.A., Nakagome, S., et al.: Electrocortical correlates of human level-ground, slope, and stair walking. PLoS One (2017). https://doi.org/10.1371/journal.pone.0188500
24. Mace, M.A., Ward, T.: Modeling the creative process: a grounded theory analysis of creativity in the domain of art making. Creat. Res. J. (2002). https://doi.org/10.1207/s15326934crj1402_5
25. Pickworth, W., Herning, R., Henningfield, J.: Spontaneous EEG changes during tobacco abstinence and nicotine substitution in human volunteers. J. Pharmacol. Exp. Ther. (1989). https://doi.org/10.1111/gbb.12420
26. Sujatha Ravindran, A., et al.: Assaying neural activity of children during video game play in public spaces: a deep learning approach. J. Neural Eng. **16**(3), 036028 (2019). https://doi.org/10.1088/1741-2552/ab1876
27. Sujatha Ravindran, A., Cruz-Garza, J.G., Kopteva, A., et al.: Multi-modal mobile brain-body imaging (MoBI) dataset for assaying neural and head movement responses associated with creative video game play in children. IEEE Dataport (2017). https://doi.org/10.21227/H23W88
28. Takeuchi, H., Taki, Y., Hashizume, H., et al.: The association between resting functional connectivity and creativity. Cereb. Cortex (2012). https://doi.org/10.1093/cercor/bhr371
29. Todd, E., et al.: Self-conscience/physical memory: an immersive, kinetic art installation driven by real-time and archival EEG signals. In Brain Art, pp. 309–323. Springer, Cham (2019). https://doi.org/10.1007/978-3-030-14323-7_11
30. Zhang, Y., Prasad, S., Kilicarslan, A., Contreras-Vidal, J.L.: Multiple kernel based region importance learning for neural classification of gait states from EEG signals. Front. Neurosci. (2017). https://doi.org/10.3389/fnins.2017.00170

Jesus G. Cruz-Garza is a Ph.D. candidate in Electrical Engineering at the University of Houston and Graduate Fellow on Translational Research with Houston Methodist Research Institute and University of Houston. He graduated from BS Engineering Physics at Tecnologico de Monterrey, Mexico.

Anastasiya E. Kopteva is a Los Angeles-based innovator, dancer, and researcher. She investigated how physical orientation, artistic creation, and perception are manifested in the brain in her roles at the Non-Invasive Brain–Machine Interface Systems Laboratory, the Houston Methodist Research Institute, and her artist in residency with ResidenceSEA in Heraklion, Greece.

Jo Ann Fleischhauer was born in Brooklyn and raised in Buffalo, New York. She received her BA in Art from Bates College and BFA from San Francisco Art Institute. Based in Houston, she has completed multiple large-scale public art installations, both nationally and internationally, that interface art, science, architecture, and history.

Jose Luis Contreras-Vidal Ph.D., IEEE Fellow, is Cullen Distinguished Professor of Electrical and Computer Engineering, and Director of the NSF Industry-University Collaborative Research Center on Building Reliable Advances and Innovations in Neurotechnologies (IUCRC BRAIN) at the

University of Houston. He has pioneered the development of noninvasive brain–machine interfaces to control robots, exoskeletons, virtual avatars, and prosthetic limbs to restore motor function in individuals with spinal cord injury, stroke, and limb amputation. He has also pioneered the concept of "The Museum and the Classroom as Laboratories" to understand the brain in action and in context of free-behaving individuals, with applications to medical devices, art therapy, aesthetics, neuroscience, and regulatory science. His work has been supported by the NIH, NSF, VA, DARPA, ONR, Foundations, donors, and industry. Dr. Contreras-Vidal's research has been highlighted by The Economist, Nature, Science, Science News, Der Spiegel, NSF, Wall Street Journal, SFN, O&P, Scientific American, NPR's Science Friday, and Neurology Today among others. His career development in biomedical engineering has been highlighted in the magazine Science.

Webpage: http://www.ece.uh.edu/faculty/contreras-vidal.

Webpage: https://www.houstonmethodist.org/faculty/joseluis-contreras-vidal/.

IUCRC BRAIN: http://brain.egr.uh.edu/.

Special Feature: Interdisciplinary Mobile Brain–Body Imaging Art-Science Demonstration

María José Delgadillo

Interdisciplinary work, at its core, aspires to understand the relationships and ties between what different approaches to ideas may offer, and how these relationships can further develop research and advancement of any given topic under diverse perspectives. In this performance specifically, the interdisciplinary work helps us see in real time the importance of creating links between creativity and research on neurosciences. With the participation of Rebecca B. Valls, a dancer, Dario Robleto, a conceptual visual artist, Woodrow Witt, a jazz musician, and Jose L. Contreras-Vidal and his team from the University of Houston's IUCRC BRAIN Center; this experiment is centered on the experience of performance and improvisation. Through data visualization, gathered by the use of EEG technology and inertia measurement units, we become witnesses to the relationships formed instantly by the performers who are, in real time, creating a unique type of collaboration. Each one of the artists is creating, in their own area of expertise with attention to each other. In this experiment, movement is what brings the experience together: it is through experiencing the moving of Vall's body that the strokes and choices of Robleto become intertwined with this specific scenario, it is also through her body that music performed by Witt moves and, more importantly, the quality of the sound is moved and shifted through the connection with both the body of Witt and the strokes from Robleto. For the duration of the experiment, the audience also becomes witness to the specific movement that occurs inside the brain activity of the performers. In this sense, the audience can experience both the process in which the intent to move is visible through the changes in the EEG, and the tangible result of this thought process in the experiences of music, visual art creation, and dance responding to each other simultaneously. The relevance of this experiment and its interdisciplinary nature lies not only in the wonder of creation when artists come together to generate an experience in collaboration; but it also offers a window into the possibilities of what both art and neuroscience can achieve when their processes of research intertwine and, moreover, can expand the scope of our own understanding of movement, creativity, intent, and ultimately, human curiosity.

Electronic Supplementary Material
The online version of this chapter (https://doi.org/10.1007/978-3-030-24326-5_9) contains supplementary material, which is available to authorized users.

M. J. Delgadillo (✉)
Department of Hispanic Studies, University of Houston, Houston, TX, USA

J. L. Contreras-Vidal et al. (eds.), *Mobile Brain–Body Imaging and the Neuroscience of Art, Innovation and Creativity*, Springer Series on Bio- and Neurosystems 10, https://doi.org/10.1007/978-3-030-24326-5_9

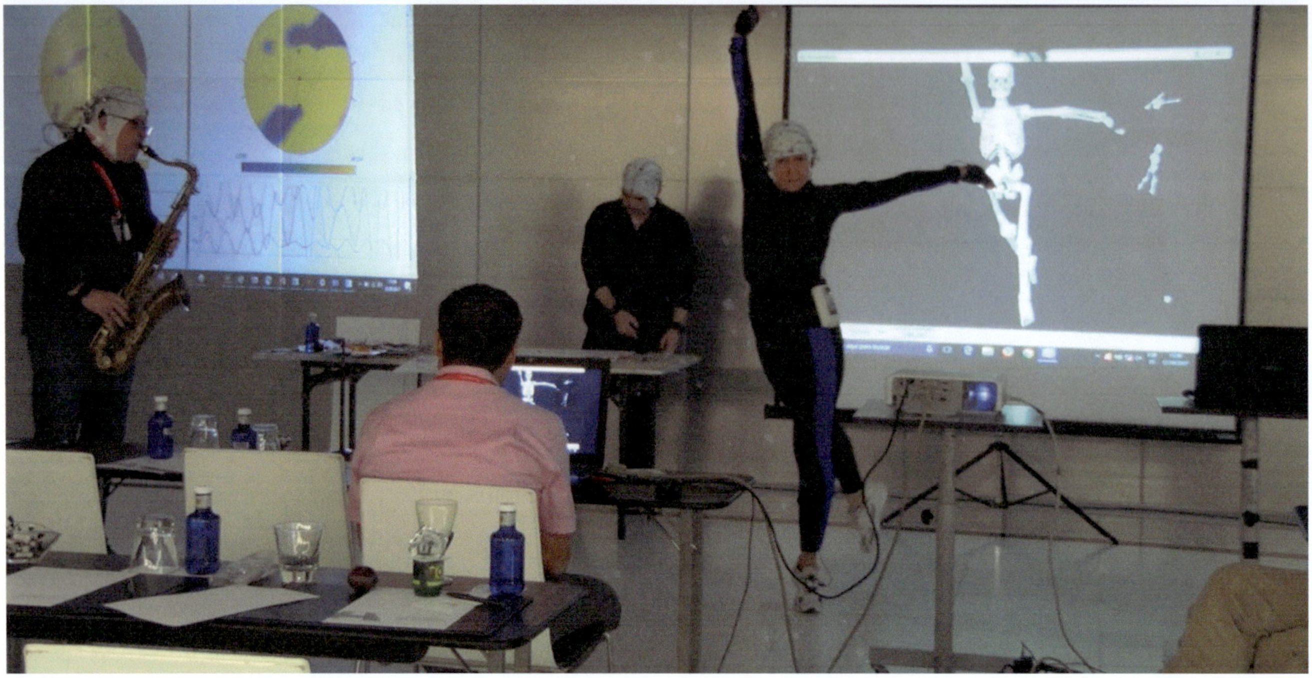

Interdisciplinary performance featuring Woodrow Witt (left), Dario Robleto (center), and Rebecca Valls (right). EEG caps measured brain activity and inertial measurement units tracked motion data of the dancer. Refer to Online Resource 11 for a video of the performance.

María José Delgadillo holds an MFA in Creative Writing from the University of California San Diego and is currently a Ph. D. student in Hispanic Studies with concentration in Creative Writing in Spanish at University of Houston. She is currently the Student Artist-in-Residence in Neuroaesthetics at the IUCRC BRAIN Center at University of Houston.

Part III
Brain Mechanisms of Aesthetic Perception

Introduction

Jose M. Azorin and Chang S. Nam
Universidad Miguel Hernández de Elche, Elche, Alicante, Spain
Department of Industrial and Systems Engineering, North Carolina State University, Raleigh NC, USA

One of the main goals of the International Conferences on mobile brain–body imaging (MoBI) and the Neuroscience of Art, Innovation, and Creativity, the so-called Brain on Art Conferences, held in Cancun, Mexico (2016) and Valencia, Spain (2017), was to improve the understanding of the brain mechanisms implied in the aesthetic perception, since aesthetic reflection has been traditionally a profound mystery of the human brain. This part of the book addresses this challenge, showing two different scientific approaches that provide a better understanding of the brain mechanisms of aesthetic perception.

In a comprehensive chapter "How we see art and how artists make it," the author (Stephen Grossberg) presents an overview of how the paintings of visual artists activate multiple brain processes that contribute to the conscious perception of their paintings, based upon his recent publication in Art and Perception. Supplemented by a set of PowerPoint presentation slides, his chapter also illustrates that different artists and artistic movements may focus on different combinations of brain processes to achieve their aesthetic goals. Finally, Dr. Grossberg highlights two contemporary research issues through two types of paintings. First, how neural models of how advanced brains consciously see have characterized these processes was explained by paintings of a group of artists (e.g., Jo Baer, Ross Bleckner, Gene Davis, etc.). In addition, paintings by Rembrandt, Graham Rust, and Sean Williams are explained to illustrate brain processes that range from discounting the illuminant and lightness anchoring, to boundary and texture grouping and classification, through filling-in of surface brightness and color, to spatial attention, conscious seeing, and eye movement control.

In chapter "Is Beauty in the Eye of the Beholder or an Objective Truth? A Neuroscientific Answer", Hassan Aleem and colleagues provide a neuroscientific approach to answer a question that has captured scientists' and philosophers' attention since antiquity: "Is beauty subjective or objective?". Authors provide support for both point of views. On the one hand, authors describe a cognitive psychology theory that proves the existence of objective aspects of beauty. On the other, they show how subjectivity arises from the networks in the brain responsible for learning and motivation.

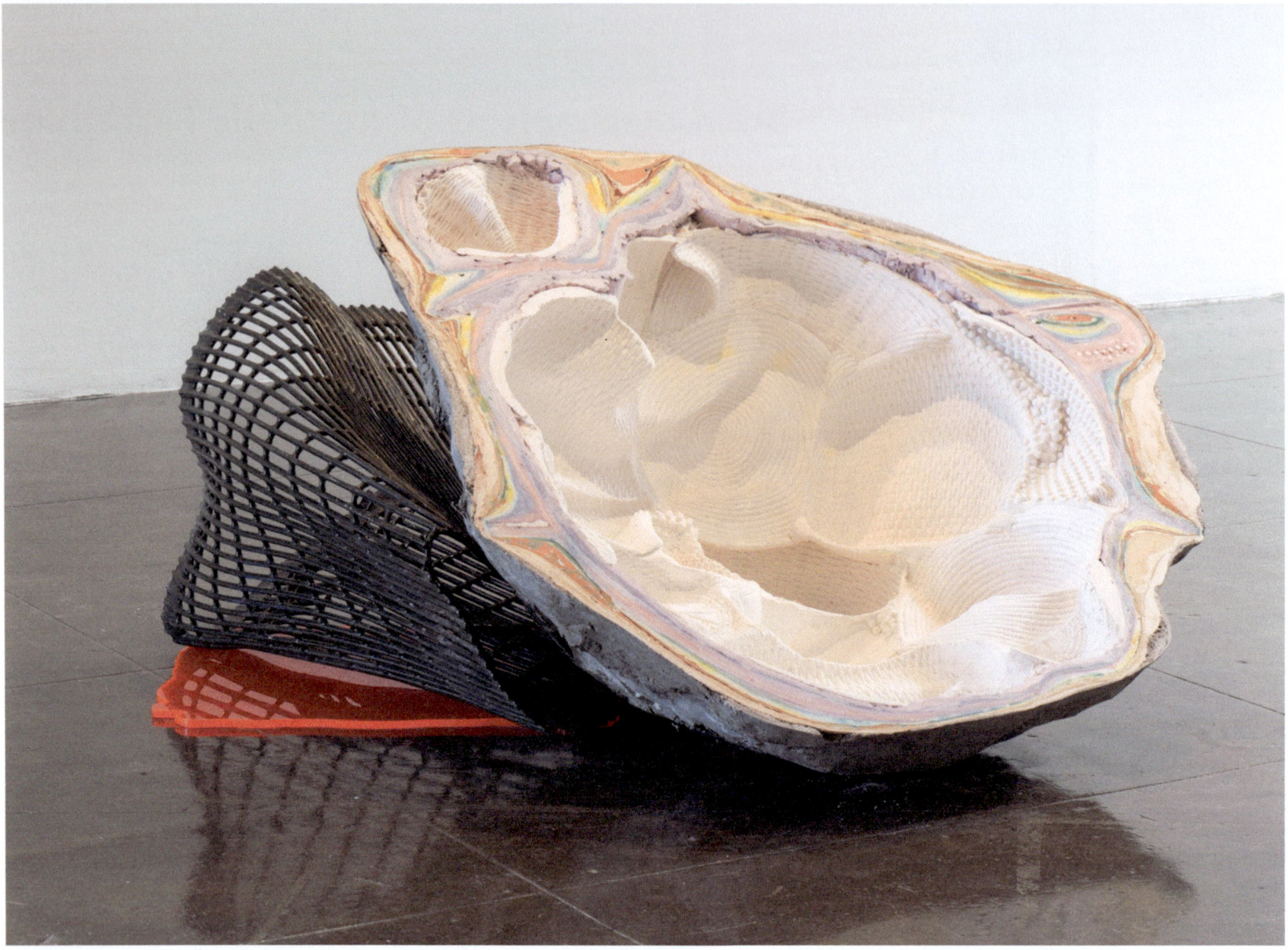

Lily Cox-Richard
Thunder Egg
Gypsum cement, concrete, pigment, trashcan, acrylic
30 × 60 × 56 inches
2016

Photo credit: Adam Schreiber
In the sculpture Thunder Egg, an aggregate of woven baskets is positioned as a geological formation
Originally commissioned and produced by Artpace San Antonio

How We See Art and How Artists Make It

Stephen Grossberg

1 Introduction: From Strokes to Conscious Percepts and Back

Whenever an artist manipulates a canvas, say by applying a dab of color to a canvas, he or she immediately experiences a conscious percept of the result. This percept emerges from all the brain machinery whereby we consciously see and know about our visual world. Artists typically have no explicit knowledge about the brain processes that mediate between painterly manipulations and the resulting conscious percepts. Yet despite this intellectual chasm between action and percept, the particular interests and aesthetic sensibilities of different artists have led each of them to emphasize different combinations of these brain processes, and to thereby create their own artistic style. In the hands of a master, the results can be both astonishing and transcendently beautiful.

The corpus of works of art on two-dimensional surfaces, across time and culture, provide an incredible richness of issues that paintings elicit, both scientific and aesthetic. This chapter reviews several of these issues through a discussion of specific paintings by well-known artists that have been chosen to illustrate how different combinations of brain processes were used to achieve their aesthetic goals. Illustrative paintings or painterly theories by nine artists were given a unified analysis in Grossberg and Zajac [44] using neural design principles and mechanisms that have been articulated and computationally characterized by the most advanced neural models of how advanced brains consciously see. This article also summarized, where possible, descriptions of an artist's stated goals, or summarized reviews of the artist's work written by art historians, curators, or critics.

The current chapter does not attempt to scientifically explain why a painting looks beautiful, or how it may arouse strong emotions. Such an analysis would require the study of how perceptual, cognitive, and *emotional* processes interact. Some promising approaches have been described whereby to understand aesthetic emotions by using mathematical models of the mind (e.g., Perlovsky [54]). The current goal is to first try to better understand the brain mechanisms of perception and cognition whereby humans consciously see paintings, and whereby painters have achieved their aesthetic goals. Further studies of beauty and of aesthetic emotions may benefit from the considerable neural modeling literature about the brain processes that create coordinated conscious experiences of seeing, knowing, and *feeling* (e.g., Grossberg [32, 34]). These more comprehensive theoretical insights would, in any case, need to build upon insights such as those described herein.

In addition, Grossberg [34] summarizes some (but not all!) of the basic brain processes that are needed to understand how we perceive and recognize music.

The current summary will provide comments about the numbered powerpoint slides in the lecture with the same title as the current article that can be found at Online Resource 12 and Online Resource 13.

2 A Step-by-Step Theory of How We See Art and How Artists Make It

Let's begin by raising the basic question of how various painters struggled to intuitively understand how they see in order to generate desired aesthetic effects in their paintings (Slides 1–3). Answering this question is made possible due to neural modeling work that clarifies what goes on in each brain as it consciously sees, hears, feels, or knows something. In Grossberg [34], I provide a self-contained, non-technical

Electronic Supplementary Material
The online version of this chapter (https://doi.org/10.1007/978-3-030-24326-5_10) contains supplementary material, which is available to authorized users.

S. Grossberg (✉)
Departments of Mathematics & Statistics, Psychological & Brain Sciences, and Biomedical Engineering, Center for Adaptive Systems, Boston University, Boston, MA 02215, USA
e-mail: steve@bu.edu

J. L. Contreras-Vidal et al. (eds.), *Mobile Brain–Body Imaging and the Neuroscience of Art, Innovation and Creativity*, Springer Series on Bio- and Neurosystems 10, https://doi.org/10.1007/978-3-030-24326-5_10

summary of current modeling knowledge about how this happens. The current article focuses only on one aspect of how we consciously see. It also summarizes a claim concerning *why* evolution was driven to discover conscious states in the first place. This analysis begins with Slide 136 in the Supplementary Materials to this book. It proposes how conscious perception is used to close the loop between perception and action, in this case between manipulating a painting, seeing it, and then manipulating it again.

In brief, the chapter and its Supplementary Materials will explain how multiple processing stages overcome the incompleteness and ambiguities of the raw sensory data that reaches our brains. These sensory data are hopelessly inadequate for triggering effective actions that can enable us to survive in a changing world that is filled with potentially life-threatening challenges. After these processing stages do their work, the result is sufficiently complete, context-sensitive, and stable perceptual representations upon which to base effective actions. In civilized societies, these actions include the strokes that create a painting. The article hereby proposes that evolution discovered conscious states in order to mark, or "light up", the sufficiently complete, context-sensitive, and stable perceptual representations that can support effective actions, notably feature-category resonances for consciously knowing about objects, and surface-shroud resonances for consciously seeing them and triggering actions based upon them. These resonances will be defined and discussed below.

Slide 5 summarizes some of the painters whose work will be discussed. They include Jo Baer, Banksy, Ross Bleckner, Gene Davis, Charles Hawthorne, Henry Hensche, Henri Matisse, Claude Monet, Jules Olitski, and Frank Stella. These painters were chosen to demonstrate how the paintings of different artists, and even of different artistic movements, can often be easily recognized due to their emphasis on different combinations of brain processes. Works of several other artists, such as Rembrandt, Graham Rust, Georges Seurat, and Sean Williams, will also be briefly mentioned to make specific points.

A reader can rightly ask: How can this kind of insight about paintings be discovered in the first place. In order to understand this, one needs to appreciate how scientists have been discovering and developing brain models of psychological processes, including artistic processes like painting. Slides 6–9 emphasize that, since "*brain* evolution needs to achieve *behavioral* success," neural models that hope to link brain to mind need to discover and model the level of brain processing that governs behavioral success. A half-century of modeling has consistently shown that these are network and system levels, which is why we study neural networks.

In order to complete such a model, individual neurons must be designed and connected in networks whose emergent, or interactive, properties give rise to successful behaviors. Keeping all these levels in mind at once—behavior, network, neuron—requires an appropriate modeling language whereby to link them. Such a mathematical model makes it much simpler to understand how brains give rise to minds, not only by articulating appropriate brain design principles and mechanisms, but also by explaining the emergent properties that they generate when they interact together in response to a rapidly changing world. Unaided intuition cannot, by itself, understand these emergent properties.

Although rigorous mathematical modeling and computational analyses are needed to understand how brains give rise to minds in a way that feels inevitable, it is nonetheless possible to explain the ideas upon which these models are based using simple, self-contained, and intuitively understandable stories. That is what these articles try to illustrate. In so doing, they clarify that perhaps the hardest obstacle to understanding mind and brain is to know how to think about each problem. Once one is on the right path, the technical details can then often follow in a natural way. Finding such paths requires guidance from lots of data.

This perspective argues that, as illustrated in Slides 10 and 11, to deeply understand how brains work, you need to understand how evolution selects brain designs based on their behavioral success. That is why the modeling method and cycle that I have developed with many colleagues over the past 50 years always starts with behavioral data, often scores or even hundreds of experiments in a given area of psychology. Having lots of data to guide one's thinking helps to rule out incorrect, but initially appealing, ideas.

The Art of Modeling consists in large part of figuring out how to understand behavioral data, which one receives as static curves that plot one variable against another, as interactive, or emergent, properties of individual behaviors as they adapt autonomously in real time to a changing world. For example, one might be trying to understand why the curve that summarizes the number of correct responses at each position in a list after a fixed number of learning trials has the shape that it does, with more correct responses at the beginning and the end of the list than in its middle. This kind of *bowing* effect occurs during essentially every experience we have when we are trying to remember sequences of events that we have experienced. If you look at these data in the right way, you can see that they embody lots of exciting philosophical paradoxes.

The results of such top-down analyses from behavioral data have always been the discovery of brain design principles that are translated into the simplest possible mathematical models (Slide 11). Then mathematical and computational analyses of these models are used to generate emergent behavioral properties that explain much more behavioral data than went into the hypotheses from which the model was derived. In this way, the modeling loop between behavior-to-design-to-model-to-behavior is closed.

In addition, and of critical importance, is the fact that the mathematical models always look like part of a brain. As a result, despite using no facts about the brain to derive these models, they explain a body of known brain data, as well as predict as yet unreported new brain data. Because this derivation proceeds from behavior-to-design-to-model-to-brain, it often proposes novel functional explanations of both known and unknown brain data.

Once the connection is made between behavior and brain, one can explain and predict lots of behavioral and brain data using the currently derived model. After the explanatory and predictive range of the model in its current form is understood, one can press both top-down from behavioral data, and bottom-up from brain data, to identify an additional design principle that the model does not currently embody. Then this new design principle is consistently added, "embedded", of "unlumped" into an expanded model, and the cycle begins again, leading to a broader range of interdisciplinary data that can be explained and predicted.

This cycle has been repeated many times during the past 50 years. As a result, we now have models that can *individually* explain and predict psychological, neuroanatomical, neurophysiological, biophysical, and even biochemical data. In this sense, the classical mind/body problem is incrementally being solved.

After going through this modeling cycle, what is the result? Is the brain just a "bag of tricks" as even famous neuroscientists like my colleague V. S. Ramachandran have claimed in the past (Slide 12)? If that were the case, true theories would be impossible.

Instead, as illustrated in Slide 13, a small number of fundamental equations have sufficed to explain thousands of interdisciplinary experiments, just as in physics. A somewhat larger number of modules, or microcircuits, that are defined using these fundamental equations, are used in specialized forms to compute useful, but not universal, combinations of properties. These modules, in turn, are assembled into modal architectures for carrying out different kinds of biological intelligence. The word "modal" stands for different modalities of intelligence, such as vision, audition, cognition, emotion, and action. None of them computes all possible computable functions in the manner of a modern von Neumann computer. However, each of them is general-purpose within its own modality of intelligence, can respond adaptively to wide range of environmental challenges, and can seamlessly interact with other modal architectures to generate autonomous adaptive intelligence as we know it.

What principles determine how modal architectures are designed (Slide 14)? It is here that the novel computational paradigms, and corresponding design principles that underlie brain computing play a critical role in ensuring that we can autonomously adapt to rapidly changing environments that are filled with unexpected events. Two of these paradigms are called Complementary Computing and Laminar Computing (Slide 15). Together they also imply a third fundamental brain design that I call the Hierarchical Resolution of Uncertainty. It is this latter design that requires multiple processing stages before our brains can compute perceptual representations that are complete, context-sensitive, and stable enough to be used to generate effective actions. It is because only such complete representations can be selectively used to generate effective actions that conscious states "light them up" to use them, and not earlier representations, for this purpose. These are the processing stages that enable a painter to apply paint to a canvas and consciously see and appreciate his or her handiwork.

Complementary Computing asks what is the nature of brain specialization (Slide 18). It provides an alterative to the earlier idea that brains compute using independent modules (Slide 17). There are lots of specialized brain regions in the visual cortex, and at least three parallel cortical processing streams with which to activate them. However, independent modules should compute each property—such as luminance, motion, binocular disparity, color, and texture—independently of the others. In reality, huge perceptual and psychophysical databases show that there are strong interactions between these various perceptual qualities.

Complementary Computing explains how such specialization coexists with, and indeed requires, these interactions by providing a very different answer to the question: What is the nature of brain specialization? Complementary Computing identifies new principles of uncertainty and complementarity that clarify why multiple parallel processing streams exist in the brain, each with multiple processing stages to realize a hierarchical resolution of uncertainty (Slide 19).

There are analogies to computationally complementary properties, such as a key fitting into a lock, and puzzle pieces fitting together (Slide 20), but these analogies do not explain the dynamism that is required to carry out Complementary Computing. In particular, computing one set of properties at a processing stage prevents that stage from computing a complementary set of properties. These complementary parallel processing streams are balanced against one another. This kind of balance is reminiscent of classical ideas about Yin and Yang, but again not explained by them. Instead, prescribed interactions between these streams, at multiple processing levels, overcome their complementary weaknesses and support intelligent and creative behaviors. They do so, in particular, by creating conscious visual states that can be used to guide looking and reaching behaviors, including those used to create and see paintings.

Each row in Slide 21 summarizes a pair of computationally complementary processes and the cortical streams in which they are proposed to occur. This list is not, however, exhaustive of all the complementary processes in our brains (Figs. 1 and 2).

When one puts together the first four of them (Slide 22), one is led to an emerging unified theory of visual intelligence, starting at our photosensitive retinas and ending at the prefrontal cortex, or PFC (Slide 23). Each box in the slide functionally describes a basic process that occurs in the corresponding part of the brain, and both the What and Where cortical streams are included. The What, or ventral, cortical stream carries out processes of perception and recognition, whereas the Where, or dorsal, cortical stream carries out processes of spatial representation and action. The modeling work that I and my colleagues have carried out over the years to explain hundreds of interdisciplinary experiments support my hypothesis that the bottom-up, horizontal, and top-down interactions between these various processes help to overcome complementary processing deficiencies that each process would experience if it had to act alone.

Slides 24–26 begin to show what it means for visual boundaries and surfaces to be complementary. Much psychophysical evidence has supported my prediction that 3D boundaries and surfaces are the basic functional units in natural vision. This prediction was first made in Grossberg [25] and was supported by computer simulations of perceptual and psychophysical data in Grossberg and Mingolla [39, 40] and Grossberg and Todorovic [43]. I began to extend it in Grossberg [26, 27] to explanations and simulations of data about 3D vision and figure-ground perception using the Form-And-Color-And-DEpth (FACADE) model of 3D vision and figure-ground separation, and its 3D LAMINART model extension to simulate identified cell types within the laminar circuits of visual cortex. This major research program was carried out with multiple Ph.D. students and postdoctoral fellows, including Rushi Bhatt, Yongqiang Cao, Nicolas Foley, Gregory Francis, Alan Gove, Simon Hong, Piers Howe, Seungwoo Hwang, Frank Kelly, Levin Kuhlmann, Jasmin Leveille, John Marshall, Niall McLoughlin, Steven Olson, Luiz Pessoa, Rajeev Raizada, William Ross, Aaron Seitz, David Somers, Karthik Srinivasan, Guru Swaminathan, Massimiliano Versace, James Williamson, Lonce Wyse, and Arash Yazdanbakhsh. The vision models were complemented by the SACCART, SAC-SPEM, TELOS, and lisTELOS models of the saccadic and smooth pursuit eye

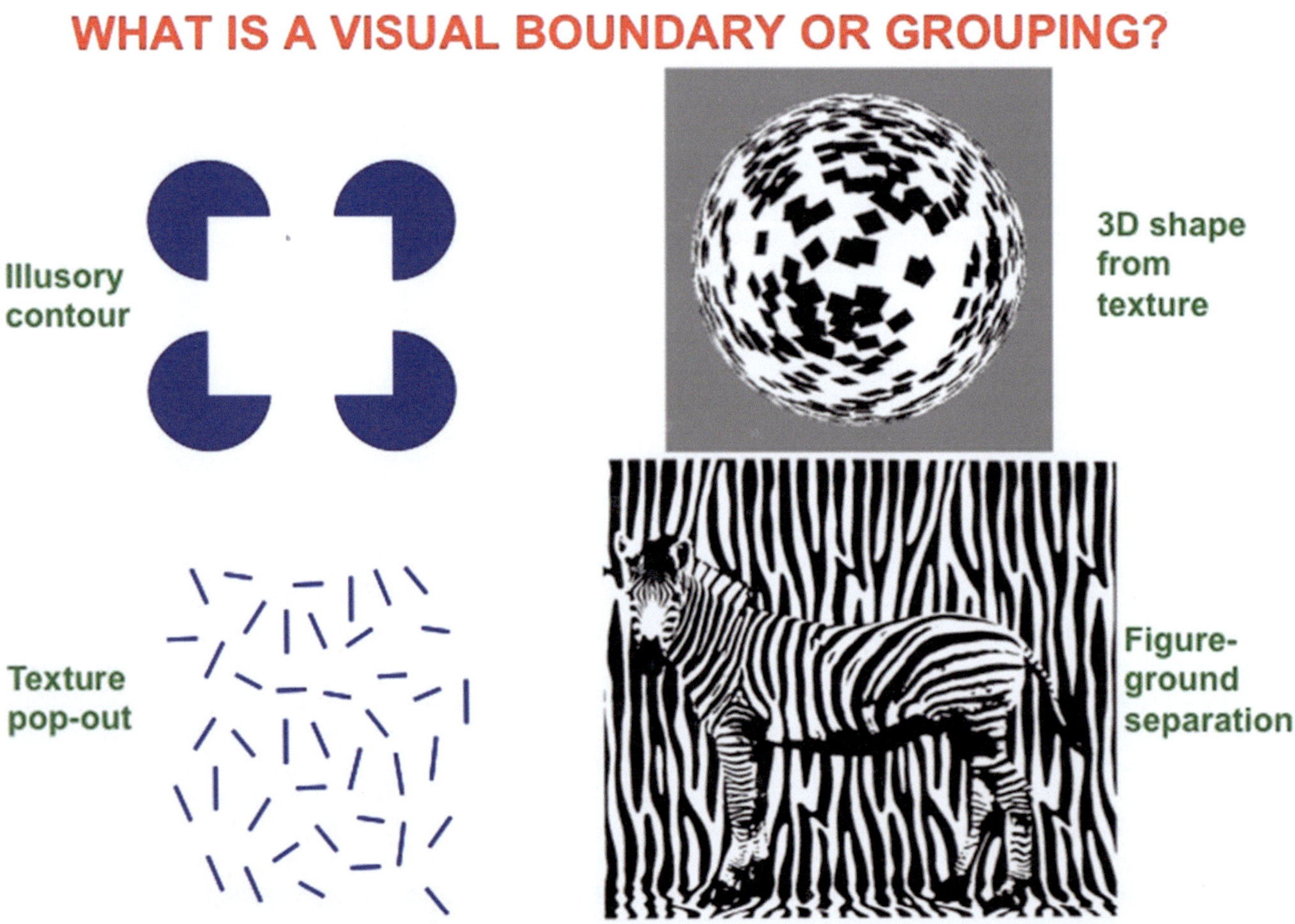

Fig. 1 What is a visual boundary or grouping? (Slide 25)

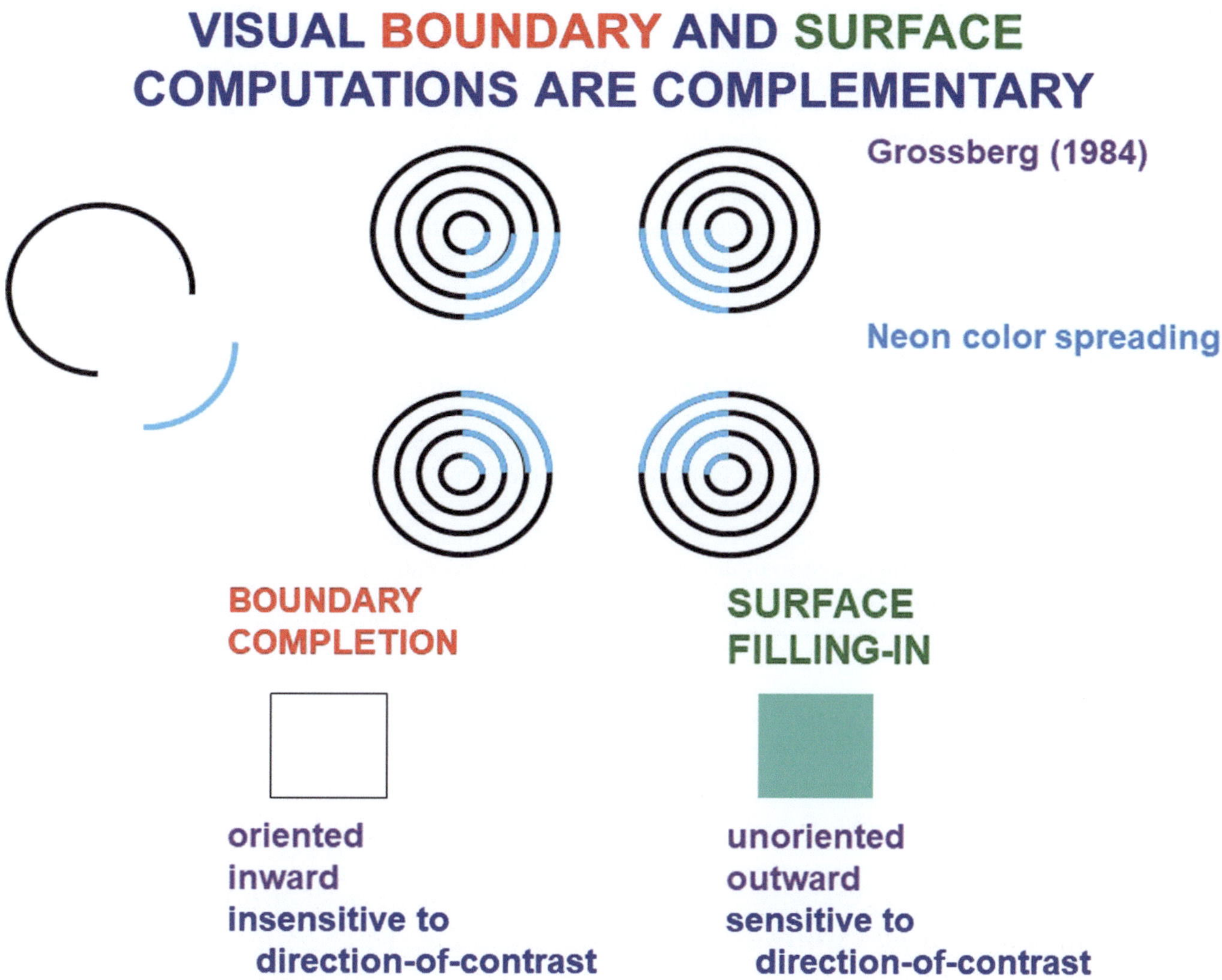

Fig. 2 Visual boundary and surface computations are complementary (Slide 26)

movements that occur during visual perception and planning, and invariant object category learning. A parallel but distinct line of work also developed the 3D FORMOTION model of visual motion perception, with its extensions to visually-based navigation and target tracking. See my personal web page sites.bu.edu/steveg for many such archival articles and https://en.wikipedia.org/wiki/Stephen_Grossberg for a list of the names of available models and the areas of biological intelligence to which they contribute.

Visual boundaries are emphatically not just edge detectors. Rather, boundaries can form in response to many different kinds of images and scenes. Boundaries hereby give rise to properties of texture pop-out, 3D shape from texture, figure-ground separation, and visual illusions, among others (Slide 25). This versatility spares our brains from having to use specialized detectors for each of these types of stimuli, only to have to figure out at a later processing stage how to put all the information together. Such specialization cannot, in any case, work in response to natural scenes if only because edges, shading, texture, and figure-ground properties are often overlaid at the same perceptual positions in a scene.

Neon color spreading is one of the visual illusions that provides lots of useful information about the complementary properties of visual boundaries and surfaces (Slide 26). A typical neon-inducing image is constructed of black and blue arcs, where the blue contrast relative to its white background is smaller than that of the black contrast. When these arcs are properly arranged, both boundary completion and surface filling-in of a neon color spreading illusion are caused. The boundary completion generates the illusory square that passes through the positions where the blue and black arcs touch. The surface filling-in causes the square to be filled with a bluish hue.

Three properties of boundary completion and surface filling-in are illustrated by neon color spreading (see the bottom of Slide 26). The first two boundary properties are that boundaries are completed between pairs of inducers in an *oriented* and *inward* fashion. If outward completion were possible, then a single dot in an image could cause a radial proliferation of boundaries that could seriously obstruct vision. By comparison, the spread of the blue color through the square is generated by small breaks in the blue boundaries where they touch the more contrastive black boundaries. The blue color can then spread in an *unoriented* manner *outward* in all directions until it hits the square illusory boundaries. These boundary and surface properties are manifestly complementary: oriented versus unoriented; inward versus outward.

Where do these boundaries and surfaces form? Slide 27 shows that boundaries are completed within several processing stages of the interblob cortical stream from the lateral geniculate nucleus, or LGN, through V1 interblobs, V2 interstripes, and V4. The surfaces are completed in the parallel blob cortical stream processing stages of the V1 blobs, V2 thin stripes, and V4. These are two of the brain's computationally complementary processing streams (Fig. 2).

What does the third boundary completion property of "insensitive to direction-of-contrast" mean in Fig. 2 (Slide 28)? This has to do with the classical distinction between seeing versus knowing, or seeing versus recognition. For example, in Fig. 3 (Slide 29), the lower left image shows an Ehrenstein Figure that is generated by blue lines pointing toward the center of an imagined disk. One can both see and recognize this disk because its interior is brighter than its background. This brightness difference is a visual illusion that is due to filling in of "brightness buttons" that are generated just beyond each of the line ends, whence this brightness spreads within the illusory circle that is also generated through the line ends.

In contrast, in response to the Offset Grating to the right of the Ehrenstein Figure, a vertical boundary is generated that passes through the line ends of the horizontal blue lines. We can *recognize* this vertical boundary, but we cannot *see* it: It is not brighter or darker, or nearer or further, from the rest of the background. This percept shows that one can consciously recognize objects that one cannot see. There are hundreds of such amodal percepts.

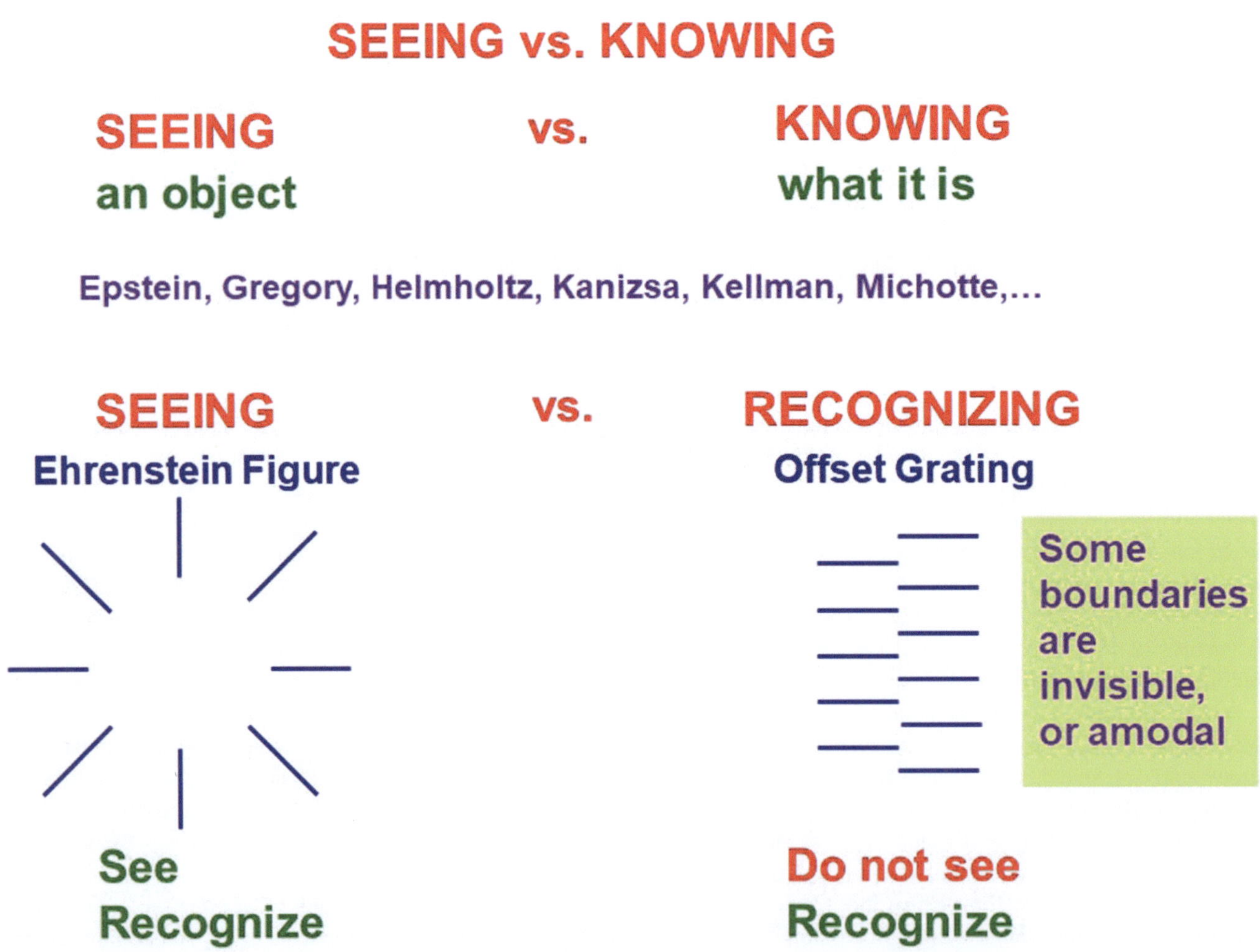

Fig. 3 Seeing versus knowing (Slide 29)

One plausible answer to the question "Why do we see?" is that "We see things to recognize them". However, we can recognize the vertical boundary that is generated by the Offset Grating without seeing it. This is thus a counterexample to the hypothesis that we see things in order to recognize them, because we can recognize this vertical boundary without seeing it. This conclusion does not deny that seeing objects does often help to recognize them, but it shows that there must be a different answer to the question "Why do we see?"

I earlier noted that, due to hierarchical resolution of uncertainty, our brains seem to have created conscious states of seeing so that we can selectively use those perceptual representations upon which to base actions like looking and reaching.

Slide 29 shows that some boundaries are invisible. Slide 30 provides one of several reasons why *all* boundaries are invisible, at least within the interblob cortical stream that generates boundaries. In particular, consider what happens if you move along the circumference of the gray disk in the right figure of this slide. One passes from gray-to-white, then gray-to-black, then grey-to-white, etc. contrasts all along the circumference. These reversals of relative contrast are often found when an object is seen in front of a textured background.

If our brains only had separate boundaries that compute dark-to-light contrasts (e.g., gray-to-white) or light-to-dark contrasts (e.g., gray-to-black), then each type of boundary would have big holes in it. Brightness and color could spread through these holes during the filling-in process and thereby seriously degrade vision.

Slide 31 shows that boundary computation does begin with oriented local contrast detectors, called simple cells, that individually can respond to either a dark-to-light oriented contrast, or a light-to-dark oriented contrast, but not to both. If boundary processing ended here, then there would be big holes in the resulting boundaries.

Instead, at each position, pairs of like-oriented simple cells that are sensitive to opposite contrast polarities input to cells at the next processing stage that are called complex cells. Each complex cell can respond to both dark-to-light and light-to-dark contrasts at, and close to, its preferred position and orientation. Thus, by the time complex cells respond at the circumference of the gray disk image in Slide 30, they would build a boundary at every position around its circumference.

It is precisely because they pool signals from both polarities—that is, are insensitive to direction-of-contrast—as noted in Slide 32, the complex cells cannot represent visual qualia like differences in relative luminance or color. Said in another way: *All boundaries are invisible!* We can experience how *salient* boundaries may be, but strong boundary salience does not imply a visible difference of qualia.

Despite being invisible, boundaries are extremely useful in helping us to recognize objects, especially objects that are partially occluded in a three-dimensional scene, as in Slide 33. The dashed red lines in Slide 34 illustrate where amodal boundaries of partially occluded objects may be created in order to help to recognize these objects. The abutting three rectangles in the right image of Slide 35 gives rise to a compelling 3D percept of a vertical rectangle that is partially occluding, and in front of, a horizontal rectangle. Even though we "know" that the horizontal rectangle is "behind" the vertical rectangle, we do not see it.

This property of figure-ground separation is exploited in all pictorial art, movies, and TV that use a 2D image to generate representations of 3D objects. For example, the face in the famous Mona Lisa painting of Leonardo da Vinci in Slide 35 partially occludes the background of the scene. The occluded collinear background boundaries can nonetheless be amodally completed behind her, at least in the upper part of the painting.

There are several basic reasons why boundary completion and surface filling-in occur. One of these reasons is clarified by inspecting Slide 36, which shows a side view of the interior of an eye. After light passes through the lens of the eye and the retinal fluid that helps to maintain the eye's shape, it needs to go past the nourishing retinal veins and all the other cell layers in the retina before it hits the photoreceptors. The photoreceptors that are activated by the light then send signals along axons via the optic nerve to the brain.

Slide 37 shows a top-down view of the retina. It includes the fovea, which is the part of the retina that is capable of high acuity vision. Our eye movements focus the fovea upon objects of interest several times each second. There is also a blind spot that is as big as the fovea. Here is where the axons from the photoreceptors are bundled together to form the optic nerve. No light is registered on the blind spot.

Even the simplest objects may be occluded by retinal veins and the blind spot at multiple positions before they can activate the retina. Slide 38 shows how this can happen to even a simple image like a blue line. This state of affairs raises several questions. For one, why do we not see retinal veins and the blind spot? This is true because our eyes rapidly jiggle in their orbits, even when we think that they are not moving. This jiggle generates transient visual signals from objects in the world. These transients refresh the neural responses to these objects. The veins and blind spot do not, however, generate such transients because they move with the eye. They are thus stabilized images. Hence, they fade. You may have noticed in an opthalmologist's or optometrist's office your own retinal veins or blind spot when he or she moves a small light alongside your eye in order to examine it. That motion can create transients with respect to the borders of the veins and blind spot and makes them momentarily visible.

Another important question is this: How do we see even images like a line if they can be occluded in multiple positions? Slide 39 shows that boundary completion completes boundaries within occluded regions and surface filling-in spreads colors and brightnesses from surrounding regions to complete the surface percepts of the occluded regions within these boundaries.

The percepts that are generated across the occluded regions are constructed at higher brain regions. Because they are not provided directly by visual inputs to the retinas, they are, mechanistically speaking, visual illusions. On the other hand, we often cannot tell the difference between the regions on the line that receive their signals directly from the retina, and those that have completed boundaries and filled-in colors and brightnesses. Both kinds of regions look equally "real". This raises the question in Slide 40: What do we call a visual illusion? I believe that we tend to call illusions those combinations of boundary and surface properties that look unfamiliar or unexpected, as in the case of the invisible vertical boundary that is generated by the Offset Grating in Slide 29.

If boundaries are invisible, then how do we consciously see? Slide 41 suggests that we see the results of surface filling-in after boundaries define the compartments within which lightness and color spread. Slide 42 summarizes the fact that the stimulus that generates the percept called the Craik-O'Brien–Cornsweet Effect has the same background luminance, but a less luminous cusp abutting a more luminous cusp in the middle of the image (see the red line labeled stimulus). These two regions are surrounded by a rectangular black frame. The percept is, however, one of two uniform gray regions (see the blue line labeled percept). This percept may be explained by the fact that the boundaries which surround the gray regions restrict filling-into each of them. Then filling-in of the less luminous cusp in the left region leads to the percept of a uniformly darker gray region than does the filling-in of the more luminous cusp in the right region. A more complete explanation, and simulations, of this percept is given in Grossberg and Todorovic [43], as well as of the very different percept that is seen when the black region is replaced by a gray region that matches the gray of the stimulus background. Many other brightness percepts are also explained and simulated within that article.

We can now understand the last computationally complementary property of boundary completion and surface filling-in that is shown at the bottom of Slide 43. As I earlier noted, "insensitive to direct-of-contrast" can also be summarized by the statement that "all boundaries are invisible". "Sensitive to direction-of-contrast" can be recast as "filling-in of visible color and lightness" since filled-in surfaces are what we can consciously see. Slide 44 can now summarize my prediction from 1984 that all boundaries are invisible in the interblob cortical stream, whereas all visible qualia are surface percepts in the blob cortical stream. I know many confirmatory experiments, but no contradictory ones, to the present time.

3 Toward a Mechanistic Understanding of the Aesthetic Struggles of Various Painters

We can now begin to apply these ideas to provide a better mechanistic understanding of the aesthetic struggles of various painters. Let us start with Henri Matisse. Slide 46 raises the provocative question: Did artists like Matisse know that all boundaries are invisible? Consider his painting, The Roofs of Collioure, from 1905 to understand a sense in which the answer to this question is Yes. Note that Matisse constructed much of this painting using patches of color to suggest surfaces. Slide 47 provides some quotations from Matisse about his life-long struggle to understand "the eternal conflict between drawing and color". He wrote that "Instead of drawing an outline and filling in the color…I am drawing directly in color".

The bottom image in this slide illustrates what this means. The color patches in this painting trigger the formation of amodal boundary webs in the cortical boundary stream. These boundary webs are then projected to the cortical surface stream where they organize the painting's color patches in surfaces. These surface colors are what we see in the painting. By not "drawing an outline" to define these surfaces, Matisse ensured that he did not darken these colors. Generating vivid colors in their paintings was one of the goals of the Fauve artistic movement to which some of Matisse's paintings contributed (Figs. 4 and 5).

Thus, as Slide 48 notes, when discussing The Roofs of Collioure with your friends, you can impress them by saying that this painting illustrates Complementary Computing in art because it generates so many invisible boundary representations to define its colorful surfaces.

Another Matisse painting from 1905, the Open Window, Colloure, is illustrated in Slide 49. This painting brilliantly combines surfaces that are created with sparse surface color patches, as well as surfaces that are rendered with continuously applied paint. Both types of surfaces blend together into a single harmonious scene.

Many artists have experienced Matisse's struggle to be "drawing directly in color", as noted in Slide 50. Slides 51 and 52 include quotes that summarize the approach to painting by two famous plein air painters who belonged to the Cape Cod school of art, including its founder, Charles Hawthorne, and his most famous student, Henry Hensche. Hawthorne wrote, in part, "Let color make form—do not

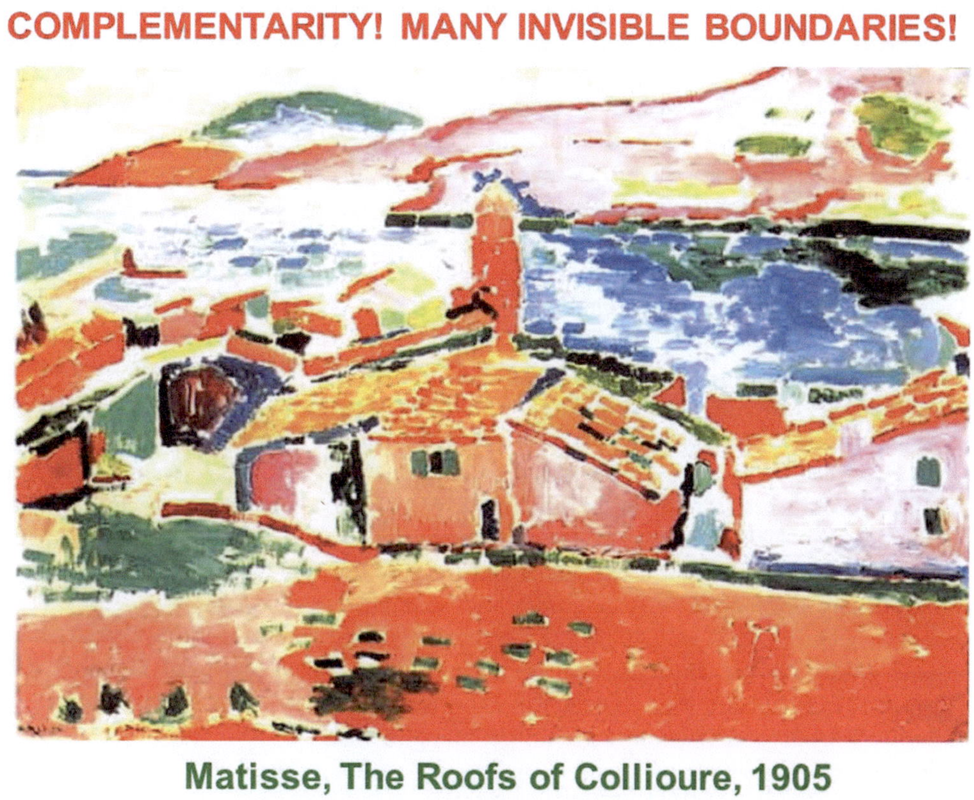

Fig. 4 Complimentarity! Many invisible boundaries! (Slide 48)

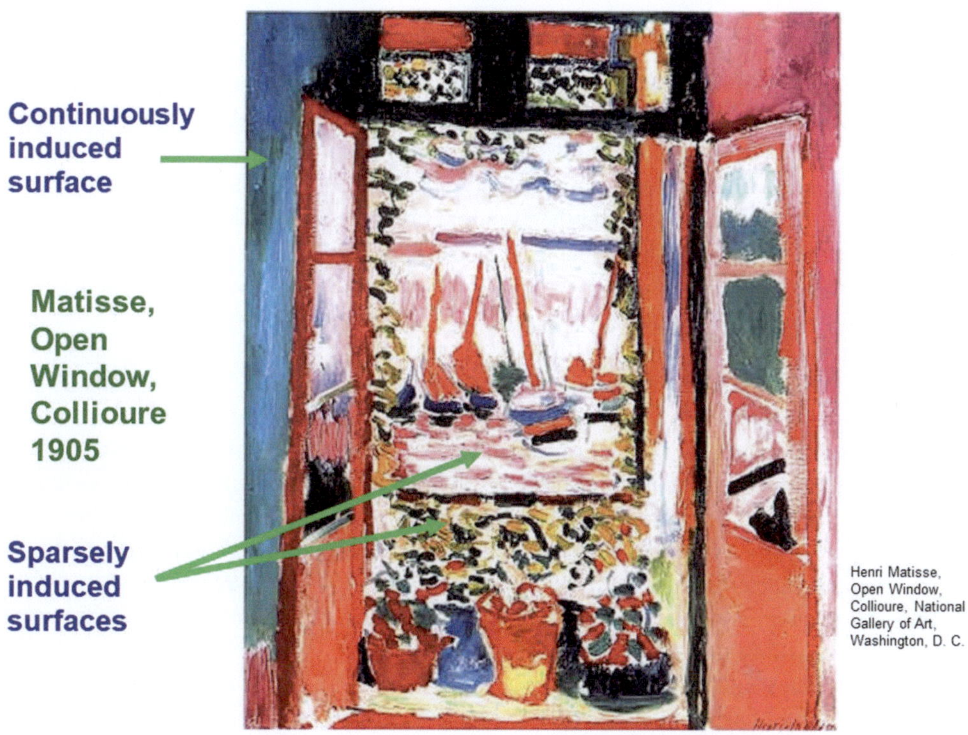

Fig. 5 Continously induced and sparsely induced surfaces (Slide 49)

make form and color it. Forget about drawing…" Hensche expressed his own approach by summarizing the view of the great Impressionist painter, Claude Monet, that "color expressing the light key was the first ingredient in a painting, not drawing…Every form change must be a color change…" Monet himself reduced this perspective to its essentials by writing, as summarized more fully in Slide 53, that "here is a little square of blue, here an oblong of pink…paint it just as it looks to you,…"

Slide 54 further illustrates this perspective using the famous painting *Femmes au bord de l'eau* of the French pointillist painter, Georges Seurat. Despite the fact that this painting is constructed from little spots, or "points", of color, it is consciously perceived due to the way in which boundaries complete between regions where feature contrasts change, and colors fill-in within these boundaries to form visible surface percepts. Slides 55 and 56 point out (in blue) that there are both large-scale boundaries that group regions of this painted scene, and small-scale boundaries that surround the individual color patches with which the painting was created. We can see both scales as our attention focuses upon different aspects of the painting.

It is all very well and good to discuss boundary completion and surface filling-in using words and images. But can we really understand these processes well enough to develop rigorous neural models that can process complex scenes? Slides 57–59 illustrate that the answer to this question is emphatically Yes. Indeed, the same brain processes of boundary completion and surface filling-in that enable use to appreciate Impressionist paintings also enable us to process natural images and images that are derived from artificial sensors.

Slides 57 and 58 illustrates this by showing how a Synthetic Aperture Radar, or SAR, image can be transformed by such a neural model into an image that can be easily interpreted by human observers. SAR is the kind of radar that can see through the weather, and is thus very useful in remote sensing and international treaty verification applications where SAR sensors in satellites and other airborne observers can observe activities on the ground even during bad weather conditions. The Input image in the upper left corner of Slide 57 contains five orders of magnitude in the radar return. This huge dynamical range is hard to represent on a powerpoint slide, and much of the image is darkened relative to the sparse, but very high intensity, pixels in it. The Feature image in the upper right corner of Slide 58 results from a process of "discounting the illuminant", or compensating for variable intensities or gradients of illumination that could otherwise prevent the extraction of information about object form. This process normalizes the Input image without distorting its relative intensities. Despite this normalization process, the resulting images still exhibit its individual pixels, just as in the painting by Seurat.

The Boundary image in the lower left corner of Slide 58 shows the completed boundaries around and between sets of pixels with similar contrasts. Finally, the Feature image fills-in within the Boundary image. The result is the Surface Filling-In image in the lower right corner of Slide 58. One can here see a road that runs diagonally downward from the middle of the top of the image toward its lower right. One can also see individual posts along this road, the highway that runs beneath it, and the trees and shadows that surround the roads. The pixels in the Input image have here been largely replaced by shaded object forms that human observers can understand.

Slide 59 shows that the filled-in surface representation in Slide 58 is the result of processing the Input image using three different spatial scales: small, medium, and large. The small boundary scale detects local image contrasts best, such as the individual posts on the road. The large boundary scale detects more global features, such as the collinear structure of the road. A separate surface network corresponds to each boundary scale, and fills-in surface brightnesses within the completed boundaries at each of these three boundary scales. The final Surface Filling-in image in Slide 58 is a weighted sum of the three Surface Filling-In images in the bottom row of Slide 59.

4 Neural Models of Boundary Completion by Bipole Cells

The next group of slides explains *how* these processes work in a non-technical way. To this end, Slide 60 asks how our brains compute boundaries inwardly and in an oriented fashion between pairs or greater numbers of approximately collinear inducers with similar orientations?

Slide 61 proposes that the cortical cells which complete boundaries obey a property that I have called the *bipole* property. This name describes the fact that these cells receive signals from nearby cells via receptive fields that have two branches, or poles, on either side of the cell body. Suppose, for example, that a horizontal edge, as in one of the pac men of a Kanizsa square stimulus, activates such a cortical cell (shown in green). It then sends excitatory signals via long-range horizontal connections (in green) to neighboring cells. These signals do not, however, activate these neighboring cells because inhibitory cells (in red) are also activated by the excitatory signals. These inhibitory cells inhibit the cells that the excitatory cells are trying to excite. The excitatory and inhibitory signals are approximately the same size, so the target cell cannot get activated. It is a case of "one-against-one".

Slide 62 shows the case in which an entire Kanizsa square is the stimulus. Now there are two pac men that are like-oriented and collinear on each side of the stimulus. Consider the pair of pac men at the top of the figure. Each of them can activate a cell whose long-range excitatory connections try to activate intervening cells. As before, they also activate inhibitory interneurons that try to inhibit these target cells. Why, then, does not the total inhibition cancel the total excitation, as before?

This does not happen because the inhibitory interneurons also inhibit each other (see red connections). This recurrent inhibition converts the network of inhibitory interneurons into a recurrent, or feedback, competitive network. I proved in Grossberg [23] that such a network tends to normalize its *total* activity. Thus, no matter how many inhibitory interneurons get activated, their total output remains approximately the same. The total inhibition to the target bipole cell thus does not summate like the excitatory signals do as more inhibitory cells are activated. This is thus a case of "two-against-one" so that the bipole cell can get activated if two or more approximately like-oriented and collinear neighboring cells send signals to it. This explains why boundary completion occurs inwardly and in an oriented manner from two or more neighboring cells, as noted in Slide 29. Slide 62 also includes, at its upper right corner, a schematic way to represent the longer-range excitatory (in green) and shorter-range inhibitory (in red) effects on a bipole cell's firing.

Do bipole cells exist in our brains? I predicted that they do in an article that I published in 1984. That same year, a famous article was published in *Science* by von der Heydt et al. [66] that provided experimental support for the prediction in cortical area V2; see Slide 27. Slide 63 summarizes key properties of their neurophysiological data. In particular, either direct excitatory inputs to a bipole cell body, or similarly oriented excitatory inputs to both "poles," or receptive fields, of a bipole cell, are needed to activate it. Moreover, an input to a receptive field is still effective in activating the cell if it is moved around within this receptive field. If, however, only one pole gets activated, then no matter how intensely this is done, the bipole cell does not fire.

Slide 64 shows that additional evidence for this kind of horizontal activation of cells in cortical area V1, which is the cortical area that feeds into V2, and which itself receives inputs from the Lateral Geniculate Nucleus, or LGN; see Slide 27. Both the longer-range excitatory influence (in blue) and the shorter-range inhibitory influence (in red) were found both in psychophysical and neurophysiological experiments by Kapadia et al. [49]. These excitatory effects are, however, of shorter range than they are in V2, and typically modulate, or sensitize, V1 cells to fire more to inputs directly to them, rather than fire them without such direct inputs.

Slide 65 shows some of the anatomical evidence for cells with long-range oriented horizontal connections.

The top left image in Slide 66 shows the oriented bipole cell receptive field that Ennio Mingolla and I used to simulate boundary grouping and completion properties in an article of ours we published in 1985 (Grossberg and Mingolla [39, 40]. The dot at the center of this image represents the position of the bipole cell body. The lines at either side of the cell body represent how strongly the cell body gets activated by inputs to the cell's two receptive fields. In particular, the length of each line at every position and orientation represents the relative strength of the connection to the bipole cell body in response to an input with that position and orientation. Note that inputs can be received by the cell body from both collinear and nearly collinear positions and orientations, with the most collinear positions and orientations delivering the largest inputs, other things being equal. The upper right image represents psychophysical data of Field et al. [17] that support bipole cell properties. The two images in the bottom row represent the bipole receptive fields that were used in modeling studies by two sets of other authors.

5 Boundary Formation by the Laminar Circuits of Visual Cortex

We are now ready to consider some of the main concepts and mechanisms of Laminar Computing which, as Slide 68 notes, is another new paradigm for understanding how our minds work. Laminar Computing tries to clarify why all neocortical circuits are organized into layers of cells, often six characteristic layers in perceptual and cognitive cortices. Said more directly: What do layers have to do with intelligence?

Slide 69 depicts a simplified diagram of the circuits in cortical layer 2/3 that carry out perceptual grouping using long-range, oriented, horizontal excitatory connections, supplemented by short-range disynaptic inhibitory interneurons, in the manner that I already summarized in Slides 61–66. This slide also summarizes some of the article authors and dates that have supported this conception. Slide 70 asks what happens before layer 2/3. In particular, how do inputs reach the grouping layer 2/3?

Slide 71 provides more information about how the oriented local contrast detectors called simple cells, that were mentioned in Slide 31, do their job. Simple cells are the first cortical stage at which cells fire in response to preferred orientations at their preferred positions and spatial scales. Each simple cell can respond to either an oriented dark-to-light contrast or an oriented light-to-dark contrast, but not both. Slide 72 notes that simple cells are not sufficient, as I already noted when discussing Slide 30. As already noted in Slide 31, Slide 73 reminds us that simple cells of like orientation and position, but opposite contrast polarities, add their output signals at complex cells.

Slide 74 notes that complex cells are also not sufficient because they do not respond adequately at line ends or corners. Indeed, as Slide 75 remarks, multiple processing stages are needed to accomplish another hierarchical resolution of uncertainty. This one compensates for weaknesses in the ability of simple cells to detect oriented contrasts.

Slide 76 illustrates what goes wrong if only simple and complex cells process line ends. At a bar end, these oriented cells can respond at each position, as illustrated by the red lines in the left image. However, they cannot respond at a line end, as illustrated by the gap in the red boundary there. This problem occurs for every choice of simple cell scale. One just needs to choose the width of the line accordingly. Slide 77 asks: Who Cares? Why is this a problem in the first place?

Slide 78 shows that it is, in fact, a very serious problem because color could flow out of every line end during the process of surface filling-in, thereby leaving the scenic representation awash in spurious color.

Slide 79 summarizes the problem that needs to be solved: Somehow the brain needs to create a line end, called an *end cut*, after the stage where complex cells act. After the end cut forms, color will be contained within the line end. Slide 80 emphasizes that the process which creates end cuts carries out a context-sensitive pattern-to-pattern map, not a pixel-to-pixel map, since it would be impossible, looking just at a pixel with no boundary, to decide if it needs to be part of an end cut, or just left alone because nothing is happening in the scene at that pixel.

Yet another processing stage is needed to carry out this hierarchical resolution of uncertainty. Slide 81 depicts a circuit that contains, in addition to simple and complex cells, a subsequent stage of hypercomplex (or endstopped complex) cells that are capable of generating end cuts. The hypercomplex cells respond in two stages. The first competitive stage is defined by an on-center off-surround, or spatial competition, network. Using this network, each complex cell excites like-oriented hypercomplex cells at its position while inhibiting like-oriented hypercomplex cells at nearby positions. In addition to receiving these excitatory and inhibitory inputs, these hypercomplex cells are also tonically active; that is, they are activated even in the absence of external inputs, due to an internal source of activation.

In the absence of inputs from the first competitive stage, firing of the hypercomplex cells due to their tonic activation is inhibited by the second competitive stage, which is realized by a competition between hypercomplex cells at the same position that are tuned to different orientations. Maximal inhibition is delivered between hypercomplex cells that are preferentially tuned to perpendicular orientations. When all the hypercomplex cells receive only tonic activation, they can inhibit each other equally using this orientational competition.

Slide 82 explains how end cuts are created at the end of a vertical black line on a white background. Near the end of the vertical line, its vertical edges can activate vertical complex cells which, in turn, can activate vertical hypercomplex cells at its position, and inhibit vertical hypercomplex cells at nearby positions, including positions beyond the end of the line. Inhibition of these vertically oriented hypercomplex cells removes their inhibition from other oriented hypercomplex cells at the same positions. The most inhibition is removed from hypercomplex cells that are tuned to perpendicular orientations. When the activities of these cells are disinhibited, their tonic activation can drive them to fire. An end cut can hereby form.

Slide 83 shows the results of a computer simulation of how complex cells (left image) and hypercomplex cells (right image) respond to a line end. The line end is shown in gray in both images. The lengths of the oriented lines are proportional to the responses of the cells at those positions and orientational preferences. The complex cell responses in the left image exhibit strong vertically, and near vertically, oriented responses along the vertical sides of the line. Despite these strong responses along the sides of the line, there are no responses at the bottom of the line. This is due to the elongated shape of oriented simple and complex cells. In the current simulation, the receptive field size is shown by the dark dashed lines.

The hypercomplex cell responses in the right image of Slide 83 show a strong end cut that is perfectly aligned with the bottom of the line end (hyperacuity!) but also generates responses at multiple nearly horizontal orientations (fuzzy orientations). These near-horizontal hypercomplex cell responses result from the near-vertical complex cell responses.

Slides 84–86 illustrate some of the consequences of these end cut properties. In particular, Slide 84 notes that some kinds of printed fonts, such as Times and Times New Roman fonts, build in their own end cuts, in the form of serifs, which are marked in red. Thus, despite the fact that "our brains try to make their own serifs" using end cuts, adding serifs in fonts can facilitate readability. Slide 85 notes that the fuzzy orientations that occur in end cuts allow lines that are not perfectly parallel to nonetheless generate emergent boundaries by cooperation among their end cuts. Finally, Slide 86 notes that the global grouping that forms through line ends may, or may not, go through their preferred perpendicular orientations. In the upper two images, the emergent boundary is perpendicular to all the line ends. In the lower image, it is not. The boundary that ultimately forms is the one has the most support from all the inducers with which it can group.

Slide 87 reminds us that all of these possibilities are due to the fuzzy receptive fields of individual bipole cells. This state of affairs raises the question: Why are not all the groupings that form using fuzzy bipole cells themselves fuzzy, which would cause a significant loss of acuity if it were true? Why, moreover, do bipole cells have such fuzzy receptive fields in the first place?

Slide 88 suggests that a fuzzy band of possible groupings often does form initially (left image), and that this is a good property: If bipole cell receptive fields were too sharply defined, then there would be a close-to-zero probability that a grouping could ever get started. Keep in mind that our brains are made of meat, not silicon. Initial fuzziness is essential to initiate the grouping process using such an imperfect medium. Having gotten a grouping started, then the challenge is to choose the grouping with the most evidence, while suppressing weaker groupings (right image). This is done using another hierarchical resolution of uncertainty.

Slide 89 notes that sharp boundaries emerge from fuzzy bipole cells due to interactions within the larger network of which bipole cells form a part.

The computer simulations that are summarized in Slide 90 illustrate some of the sharp groupings that bipole cells can create in such a network. Images (a), (c), (e), and (g) represent the inputs to such a network. Each line in these images is proportional to the size of the input to a cell centered at the middle of the line and with the vertical orientational preference of the line. Thus, every input is composed of a "bar" of vertical features. The inputs differ only in whether or not the bars are aligned in rows, columns, or both. In (a), only the columns are aligned. In (c), both columns and rows are aligned. In (e), only the rows are aligned. And in (g), the rows are aligned and closer together.

Images (b), (d), (f), and (h) depict the steady-state responses of the bipole cells in this network. In (b), vertical boundaries are created between the bars. In (d), vertical and horizontal boundaries are created. In (f), horizontal boundaries are created. And in (h), both horizontal and diagonal boundaries are created, even though there are no diagonal orientations in the inputs. These simulations illustrate that the network is sensitive to the colinearity and orientations of input inducers, and that sharp boundaries can be completed using fuzzy bipole cell receptive fields. The simulation in (h) also shows how emergent diagonals can be created if there is enough evidence for them in the input inducers, just as they are in response to the bottom display in Slide 86. The rows needed to be brought closer together for this to happen so that they fell within the span of the diagonally oriented bipole cell receptive fields.

Slide 91 includes images that induce percepts which illustrate the properties of the simulations in Slide 90. In response to the upper left image of an E that is composed of smaller A's, the top horizontal boundary of the E groups diagonal orientations of the A boundaries. The top horizontal boundary of the S emerges from the perpendicular line ends of the H's, whereas the right vertical boundary of the S emerges from collinear grouping of the right sides of the H's.

These properties have inspired works of art. Slide 92 shows a typography portrait of Sean Williams in which all the facial features and the hair exploit these properties of boundary completion.

Slides 93–98 show how the processes that have already been reviewed can explain the percept of neon color spreading. Slide 94 depicts a neon color spreading image that is composed of black crosses abutting red crosses. In this image, the contrast of the red crosses with respect to the white background is smaller than the contrast of the black crosses with respect to the white background. In response to this image, one of several percepts can be perceived. One can either perceive red neon color filling local shapes around the individual red crosses, such as diamonds or circles, or one can perceive diagonal streaks of color passing through a collinear array of red crosses.

Slide 95 depicts how neon color can appear to spread beyond a red cross and be contained by the illusory circle that is induced where the black and red regions touch. Let us now see how the first steps in generating a neon percept are caused in the simple-complex-hypercomplex network of Slide 96.

Slide 97 considers what happens where a pair of collinear black and red line ends touch. Vertically oriented complex cells respond along their vertical boundaries. Because the black-to-white contrast is larger than the red-to-white contrast, the complex cells that are along the black line end become more active than those along the red line end. Because of the first competitive stage, the black vertical complex cells inhibit red vertical hypercomplex cells more than conversely near where the two line ends touch. As a result, these red boundaries are inhibited, or at least significantly weakened, thereby causing a hole, or weakening, in them that is called an *end gap*. Red color can spread outside the red crosses through these end gaps during surface filling-in.

Due to the second competitive stage, the weakening of the red vertical hypercomplex cell activities disinhibits other oriented hypercomplex cells at those positions, especially horizontal hypercomplex cells, thereby creating end cuts, just as in the case of the line end in Slides 82 and 83.

After these end cuts form, the bipole cells that they activate can create an emergent boundary that best interpolates the end cuts, as illustrated by Slide 98. The red color that spreads outside the red crosses is blocked from spreading beyond this circular illusory boundary.

We can now apply these insights to better understand how various paintings look, starting with the paintings of Jo Baer (Slide 99). Slide 100 shows a group of three of Jo Baer's paintings side-by-side. All of them have a black border. Within this border is a less contrastive border with a specific color: red, green, or blue, from left to right. The percepts show reddish, greenish, and bluish hues spread throughout the intervening canvas. How does this percept happen?

The main effect can be explained by the spatial competition of the first competitive stage (Slide 96), followed by surface filling-in. The black-to-white and black-to-red contrasts are larger than the red-to-white contrasts in the leftmost image. As a result, the red-to-white boundary is weakened, so red color can spread through the interior of the canvas. The same holds true for the green and blue contrasts.

A more vivid version of this effect was developed by Baingio Pinna, who calls it the watercolor illusion [56, 57]. In the image in Slide 101, there are four closed regions in which a dark blue wiggly line abuts a light blue wiggly line, which encloses a white interior region. The percepts within these regions is one of light blue color filling their interiors. This happens for the same reason that the Joe Baer effects do, because the dark blue contrast with respect to both the white background and the light blue contrast, is larger than the light blue contrast with respect to the white background. The effect is made stronger by using corrugated, or wiggly lines, whose surface area relative to the surrounded white interiors is much larger than straight lines would allow, thereby creating many more positions at which light blue color can flow within the weakened boundaries to fill the white interiors.

Slide 102 calls attention to the fact that the bluish regions also seem to bulge slightly in front of the white backgrounds that surround them. This may be explained as a special case of how cells with multiple receptive field sizes, or spatial scales, influence how we see objects in depth. Slide 103 shows more examples of this using shaded images that create compelling percepts of objects in depth. These techniques are called chiaroscuro and trompe l'oeil. Slide 104 notes that similar effects make many shaded and textured objects in 2D pictures appear to have a 3D rounded shape. I will now explain how responses of receptive fields with multiple sizes can create form-sensitive webs of boundaries that control filling-in of surfaces at multiple depths, thereby leading to these rounded percepts.

Slide 105 describes one factor that helps to explain how this happens. As an object approaches an observer, it gets bigger on the retina. As a result, other things being equal, a larger retinal image is closer. Slide 106 notes that smaller scales can respond better to small scales, whereas larger scales can respond better to larger scales so that, other things being equal, bigger scales can be associated with nearer depths during years of experience with perception-action cycles.

A big image on the retina is not, however, always due to a nearer object. For example, a very large object far away, and a smaller object nearby, can both generate retinal images of the same size. Both retinal image size and depth from an observer need to work together to disambiguate these different situations. How this "size-disparity correlation" generates more informative depth percepts is explained in Grossberg [27, 28].

Slides 107–113 describe some of the processes that enable an object like a shaded ellipse in a 2D picture to generate a compelling percept of a 3D ellipsoid. Slide 107 notes that, if boundaries were just edge detectors, there would be just a bounding edge of the ellipse (shown in red). Slide 108 shows how the ellipse would then look after filling-in occurs. It would have a uniform gray color after filling-in within the bounding edge, and would look flat. We know, however, from Slide 71 that simple cells are oriented local contrast detectors, not just edge detectors.

Slide 109 notes that, because of the way that simple cells respond to shaded images, different size detectors generate dense form-sensitive boundaries, that I have called "boundary webs" for short, at different positions and depths along the shading gradient. Slides 110–112 show that increasingly large receptive fields are sensitive to broader bands of shading, starting from the bounding edge and working toward the ellipse interior. Other things being equal, the small scales signal "far", larger scales signal "nearer", and the biggest scales signal "nearest", other things being equal.

As noted in Slide 113, the boundary web corresponding to each scale captures the gray shading in the small form-sensitive boundary compartments that it projects to the surface stream, where it regulates how the gray color will fill-in within that scale. We see this pattern of shading as it is distributed across all the scales. Because different scales tend to be associated with different depths, we perceive a shaded percept in depth.

This view of how 3D shape percepts are generated is supported by many computer simulations of human data about visual perception. In particular, it has succeeded in quantitatively simulating psychophysical data about human judgments of depth in shape-from-texture experiments. In Slide 114, although the 2D images of all of the five disks are composed of spatially discrete black shapes on a white disk, the ones to the left appear to have a rounded shape in depth, whereas those to the right appear to be increasingly flat. These percepts were quantitatively simulated using multiple-scale boundary webs and the multiple-scale filled-in surface representations that they induce.

Coming back in Slide 115 to the watercolor illusion, we can now explain its bulge in depth as a consequence of a multiple-scale boundary web, albeit one that is generated by just a few abutting wiggly lines of decreasing contrast. The chiaroscuro and trompe l'oeil images in Slide 116 also generate multiple-scale boundary webs but use gradual changes in contrast to induce them, so that more scales can be involved, leading to more gradual and vivid perceived changes in depth.

Slides 117–120 propose why the famous paintings by Claude Monet of the Rouen cathedral at different times of day lead to different conscious percepts. In Fig. 6 (Slide 118), the cathedral was painted at sunset when lighting was almost equiluminant across most of the pointing. As a result, color, rather than luminance, differences defined most of the boundaries, which were correspondingly weakened. Fine

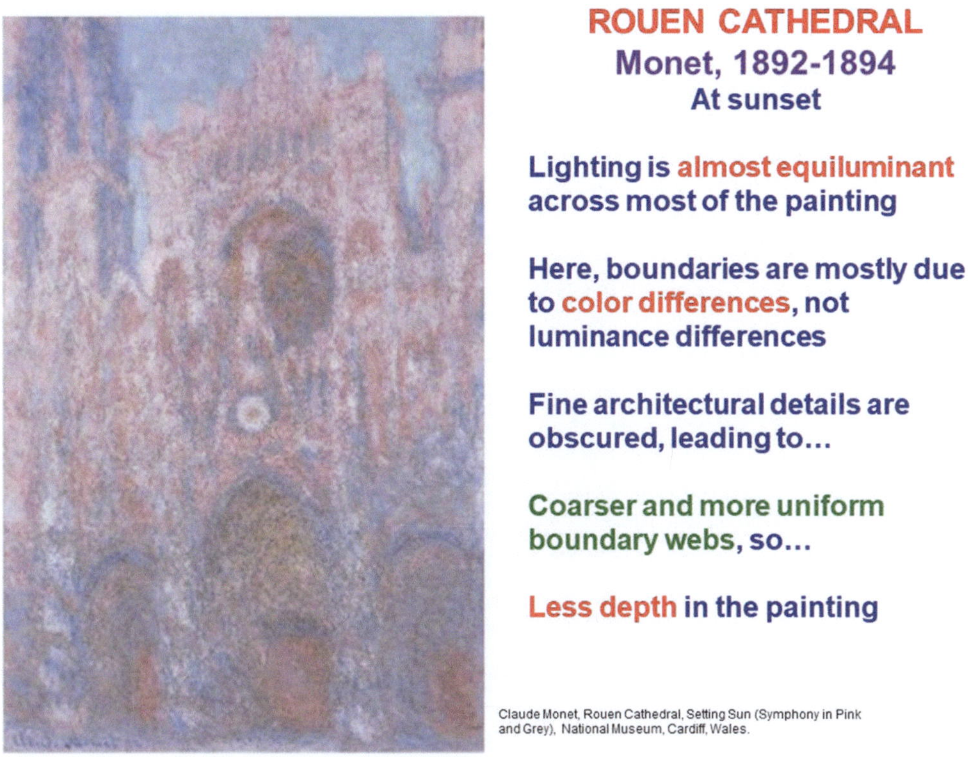

Fig. 6 Equiluminant light creates less depth in the painting (Slide 118)

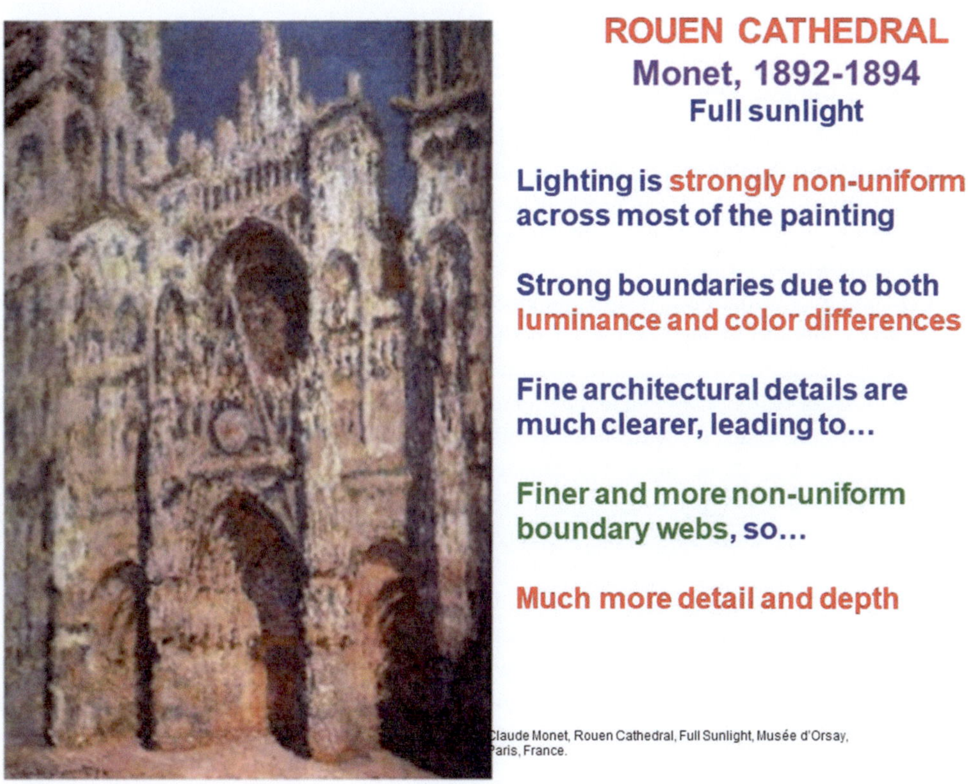

Fig. 7 Strongly non-uniform light creates more depth in the painting (Slide 119)

architectural details were not represented, so that coarser and spatially more uniform boundary webs were created, thereby leading to less perceived depth in the painting.

Figure 7 (Slide 119), in contrast, shows the cathedral in full sunlight that is very non-uniform across the painting, thereby creating strong boundaries due to both luminance

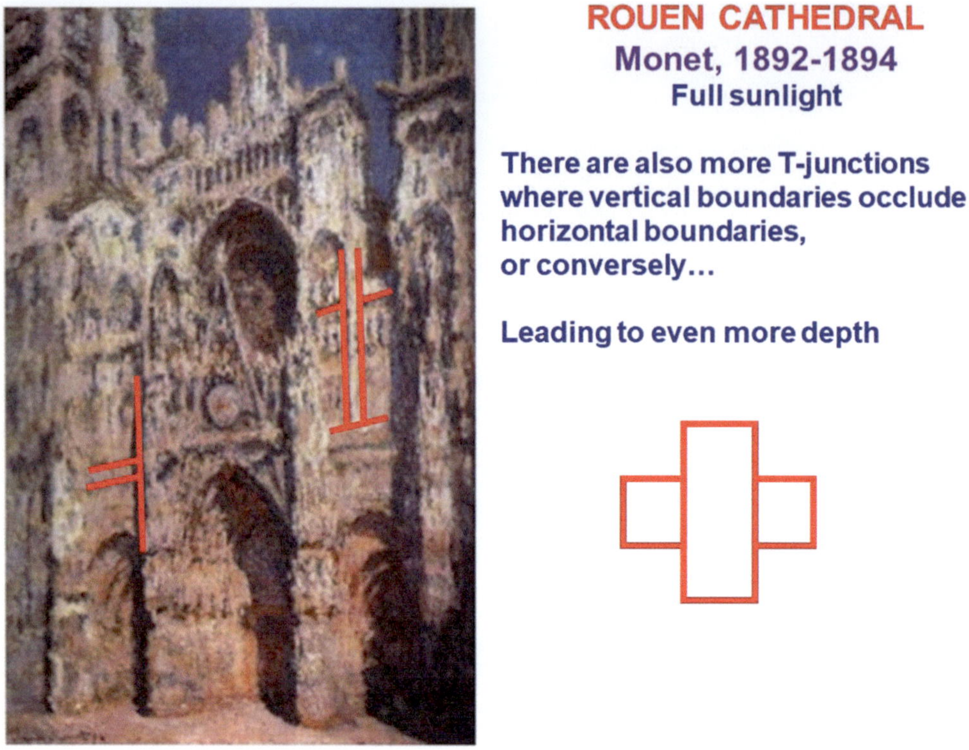

Fig. 8 T-junctions where vertical boundaries occlude horizontal boundaries, or conversely, lead to more depth in the painting (Slide 120)

and color differences. Due to the increased amount of detail, the boundary webs that form are finer and more non-uniform, leading to a more depthful percept.

Figure 8 (Slide 120) emphasizes another consequence of full sunlight by marking some of the T-junctions that are now clearly visible in the painting, leading to additional cues to perceiving relative depth, as in the percept of a partially occluded rectangle shown in red in this slide, and further discussed in Slides 34 and 35.

Let us now consider how these same mechanisms help to explain how quite different combinations of painterly properties are perceived. Let us start with the color field paintings of Jules Olitski (Slide 121). Slide 122 summarizes four of these "spray" paintings, so called because of the method that was used to create them. Slide 123 contrasts the percepts created by these spray paintings with those of Monet and other Impressionists. In the spray paintings, there are no discrete colored units (or at least very few), and no structured color or luminance gradients. Instead, diffuse boundary webs are spread over the entire surface. When they fill in, the resulting surface percepts are of a space filled with a colored fog and a sense of ambiguous depth. The quote of Olitski at the bottom of Slide 123 summaries his intention to create this kind of effect.

Quite different percepts are seen in paintings of Ross Bleckner (Slide 124). Slide 125 refers the reader to some of his paintings that create self-luminous effects. To explain self-luminous percepts requires a deeper analysis of how we see surface color and brightness. Slide 126 claims that at least two different processes can create these effects: Boundary web gradients and lightness anchoring.

Slide 127 presents some examples of how a picture can seem to glow if boundary web gradients exist; that is, if the shading that creates boundary webs varies systematically across space, from darker to lighter. Because the stronger boundaries can inhibit the weaker boundaries more than conversely, brightness can spread out of the inhibited weaker boundaries into regions where it can be trapped. The four images in the upper left corner illustrate how this brightness is trapped within the interior square of the images.

These four images, working from left to right in the top row, and then from left to right in the bottom row, have increasingly steep boundary web gradients. The steepest gradients enable stronger boundaries to more completely inhibit the weaker boundaries near to them, allowing more brightness to flow beyond them. This brightness summates in the interior square, thereby creating an increasing bright result that, in the final square, appears self-luminous.

The right column of Slide 127 shows a similar effect in its top row with the example of the double brilliant illusion. The rows beneath that summarize computer simulations using the Anchored Filling-In Lightness Model (aFILM) that I developed with my Ph.D. student, Simon Hong [35]. More will be said about aFILM in the next few slides, since it can explain the brightening effects due to boundary web gradients, as well as those due to lightness anchoring.

A remarkable percept is shown in the left pair of images in the bottom row, where two vases are shown side by side.

The rightmost vase looked matte, or dull. A highlight was manually attached to this dull vase to create the vase in the left image. Now the entire vase looks glossy! This can be explained by the fact that the highlight includes luminance gradients that match the shape of the surrounding vase. The \boundary web of the highlight can thus be assimilated into the boundary web of the rest of the vase, thereby allowing brightness to spread from the highlight across the vase. Beck and Prazdny [2], who reported this percept, also rotated the highlight and removed its luminance gradients. Both effects prevented the rest of the vase from looking glossy, as would be expected from the above explanation because the brightness could then not flow into other shape-sensitive boundary webs of the vase.

Slide 128 asks what is lightness anchoring, while Slide 129 furthermore notes that we have thus far only considered how discounting the illuminant preserves the *relative* activities of luminance values, without saturating, as they are converted into perceived brightnesses. The phenomenon of lightness anchoring shows that more is going on when we perceive brightness.

Lightness anchoring additionally raises an issue that is summarized in Slide 129; namely, how is the *full dynamic range* of a cell used, not just its *relative activities*? Another way of saying this is to ask: How do our brains compute what is perceived to be white in a scene?

Slide 130 summarizes one hypothesis about how white is perceived. The great American psychologist, Hans Wallach, suggested that the highest luminance in a scene is perceived as white, the so-called HLAW rule. Slide 131 shows that this rule sometimes works, as in the top row of images. However, the bottom row of images shows that, if there is a very intense light source in a scene, renormalizing it to make the light source white can drive the rest of the scene into darkness.

My Ph.D. student, Simon Hong, and I realized that if one, instead, computes the *blurred* highest luminance as white (BHLAW), then that problem can be avoided, as shown by the computer simulations in Slide 132.

Slides 133 and 134 illustrate how the BHLAW rule works. Slide 133 shows a cross-section of a luminance profile in green, and the spatial kernel that defines the BHLAW rule in red. In this situation, the width of the luminance step is considerably narrower than that of the blurring kernel. As a result, when this scene is anchored to make the blurred highest luminance white, the maximal brightness of the step is more intense than white. It therefore appears to be self-luminous.

In contrast, if as shown in Slide 134, if the luminance step in a scene is at least as wide as the blurring kernel, then when the scene is anchored to make the blurred highest luminance white, the entire luminance of the step is seen as white.

Returning now to look at at the two examples of Bleckner's paintings in Slide 135, we can see that the small bright regions look self-luminous because of lightness anchoring, whereas larger spatial luminance gradients look self-luminous due to the escape of brightness from graded boundary webs.

6 How Do We Consciously See a Painting?

None of the above results would make much sense if we could not consciously see objects in the world, including paintings. Fortunately, there has been considerable progress during the past 40 years to incrementally understand both how and why, from a deep computational perspective, we become conscious. Slide 138 summarizes a definition of the Hard Problem of Consciousness that expresses these issues. Readers who want to study more details about the Hard Problem than I will summarize here are invited to read my non-technical article Grossberg [34] about this topic that I published Open Access and also put on my web page sites. bu.edu/steveg. In particular, Slide 138 asks why any physical state is conscious rather than unconscious, and why conscious mental states "light up" in an observer's brain. Slides 139–141 summarize my hypothesis that our brains "light up" to embody a conscious state when they go into a resonant state. Slide 142 additionally proposes that "all conscious states are resonant states". As Slide 143 notes, not all brain dynamics are resonant, so consciousness is not just a "whir of information processing.

Slide 144 provides a non-technical definition of what a resonant state is. Namely, a resonant state is a dynamical state during which neuronal firings across a brain network are amplified and synchronized when they interact via reciprocal excitatory feedback signals during a matching process that occurs between bottom-up and top-down pathways.

Slide 145 summarizes my central claim that conscious states are part of adaptively behavioral capabilities that help us to adapt to a changing world. Conscious seeing, hearing, and reaching help to ensure effective actions of one kind or another. In particular, conscious seeing helps to ensure effective looking and reaching, conscious hearing helps to ensure effective communication and speaking, and conscious feeling helps to ensure effective goal-oriented action. This lecture does not describe the brain machinery that clarifies *why* evolution may have been driven to discover conscious states. Grossberg [34] does attempt to do this.

In brief, that article argues that evolution was driven to discover conscious states in order to use them to mark perceptual and cognitive representations that are complete, context-sensitive, and stable enough to control effective actions. This link between seeing, knowing, consciousness, and action arises from the fact that our brains use design principles such as complementary computing, hierarchical resolution of uncertainty, and adaptive resonance. In particular, hierarchical resolution of uncertainty shows that multiple processing stages are needed to generate a sufficiently complete, context-sensitive, and stable representation upon which to base a successful action. Using earlier stages of processing could trigger actions that lead to disastrous consequences. Conscious states "light up" the processing stages that compute representations that can control effective actions.

Slides 37–39 already illustrated this problem in the case of visual perception. How, for example, can you look at a part of a scene that is occluded by the blind spot? As summarized in Slide 39, processes like boundary completion and surface filling-in at higher processing stages are needed to overcome these occlusions. Boundary completion and surface filling-in are examples of hierarchical resolution of uncertainty. After a sufficiently complete surface representation is generated, a resonance develops that marks this representation as an adequate one upon which to base looking and reaching.

Slide 146 focuses on this question for the case of seeing and reaching. Slide 147 asks: What is this resonance? It proposes that a surface-shroud resonance "lights up" surface representations that are proposed to occur in prestriate visual cortical area V4. Surface-shroud resonances are predicted to occur between V4 and the posterior parietal cortex, or PPC, where a form-fitting distribution of spatial attention occurs in response to an active surface representation, and begins to resonate with it in the manner that I will explain in Slides 154–157.

Slide 148 proposes that, just as a surface-shroud resonance supports conscious seeing of visual qualia, a feature-category resonance supports conscious recognition of, or knowing about, visual objects and scenes.

How are feature-category resonances formed? Slides 149–153 briefly describe how feature-category resonances are generated using mechanisms and circuits of Adaptive Resonance Theory, or ART. As summarized in Slides 149 and 150, ART models how we learn to attend, recognize, and predict objects and events in a changing world, without being forced to forget things that we already know just as quickly. In other words, ART proposes a detailed mechanistic solution of the brain processes whereby our brains solve the *stability-plasticity dilemma* that is summarized in Slide 149; namely, how can we learn quickly without being forced to forget just as quickly? I am glad to be able to write that ART is currently the most advanced cognitive and neural theory, with the broadest explanatory and predictive range, about how our brains learn to attend, recognize, and predict objects and events in a changing world. These predictive successes include psychological and neurobiological experiments that have supported all of the main ART predictions.

ARTs explanatory range has also enabled it to shed mechanistic insight on how brain mechanisms may become imbalanced to generate mental symptoms of mental disorders that afflict millions of individuals, including Alzheimer's disease, autism, Fragile X syndrome, schizophrenia, ADHD, visual and auditory neglect, medial temporal amnesia, and problems with slow wave sleep ([19, 29, 33, 34, 36, 42]).

In addition to applications of ART to clarify properties of mental diseases, it has been used in many large-scale applications to engineering and technology that need these properties. Some of these applications are listed in Slide 151, including the use of ART by the Boeing company in a parts design retrieval system that was used to design the Boeing 777.

ART can be used with confidence because its properties of learning, recognition, and prediction have been mathematically proved and demonstrated through extensive computer simulations on benchmark problems in a series of articles with Gail Carpenter during the 1980s and 1990s (e.g., Carpenter [5, 6]; Carpenter et al. [7–14]), including the property that it solves the stability-plasticity dilemma, which is also often called the problem of catastrophic forgetting. Most learning algorithms do experience catastrophic forgetting, including the currently popular Deep Learning algorithm. During learning by such an algorithm, an unpredictable part of previously learned memories can suddenly collapse. In other words, learning in these algorithms is unreliable.

Their learning is also often inexplicable. One cannot verify that even correct predictions have been made for sensible reasons. This is a serious drawback when considering whether to depend upon them for life and death decisions, such as medical decisions. In contrast, the adaptive weights of ART algorithms such as Fuzzy ARTMAP [9] can, at any stage of learning, be represented as Fuzzy IF-THEN rules which provide a transparent explanation of how the algorithm is making its decisions.

How does ART manage to achieve these useful properties. Intuitively, it is because ART models learn expectations about the world that focus attention upon the combinations of features that it expects to be useful. But *why* do we learn expectations and pay attention? *Why* are we intentional and attentional beings? Slide 152 notes that top-down attentive feedback encodes learned expectations that dynamically stabilize learning and memory. In other words, learned expectations and attention help us to solve the stability-plasticity dilemma! ART models the neural networks that embody how top-down expectations are learned, and how they enable us to focus our attention upon information that is expected from past experience to be informative.

Feature-category resonances are part of this stability-plasticity expectation-attention story. Slide 153 summarizes how a feature-category resonance develops between an attended pattern of features, called a *critical feature pattern* (depicted in light green), and an active recognition category at the next processing stage. The reciprocal bottom-up and top-down excitatory signals synchronize, amplify, and prolong cell activations. During such a resonance, the adaptive weights, or LTM traces, in both the bottom-up adaptive filters and the top-down expectations can learn to selectively fire the active critical feature pattern and category when a similar input pattern is experienced in the future. It is because such a resonance triggers learning that I have called the theory *Adaptive* Resonance Theory.

Feature-category resonances help to support conscious recognition of visual objects and scenes, but they do not directly support conscious "seeing". Slides 154–158 provide some basic information about the surface-shroud resonances that do support conscious seeing. But first, what is an attentional shroud? Slide 155 notes that an attentional shroud is a surface-fitting distribution of spatial attention. Several excellent visual experimentalists had earlier noted that spatial attention tends to fit itself to surfaces that are attended. I predicted, in addition, how such a shroud enables learning of view-invariant object categories [19]. A view-invariant object category is a recognition category that can be activated by any view of an observed familiar object. I showed how shrouds support learning of such invariant categories by controlling how the cells that will become invariant categories can remain active as our eyes explore its various views to drive the category learning process. This insight was later generalized to explain how view-, position-, and size-invariant categories are learned ([14–16, 18, 38]). How this learning process is proposed to happen is reviewed in Grossberg [34]. Some of the archival articles that preceded this review were written with various Ph.D. students, postdoctoral fellows, and other faculty. They are listed in Slide 155. Here I focus on related issues.

Slide 156 illustrates a one-dimensional cross-section of a simple scene in which two luminous bars occur, the left one a little more luminous than the right one. Both bars send topographic bottom-up excitatory signals to the spatial attention region, where they trigger a widespread spatial competition for attention.

In addition, as Slide 157 summarizes, the activated spatial attention cells send topographic top-down excitatory signals back to the surfaces that activated them. The totality of these interactions defines a *recurrent*, or feedback, on-center off-surround network whose cells obey the membrane equations of neurophysiology, also called shunting interactions. I mathematically proved in Grossberg [23]—see also the review in Grossberg [24]—how such a network can contrast-enhance the attentional activities that focus upon the more luminous bar while also inhibiting the attention focused on the less luminous one. Because such a network tends to *normalize* the total activity across the network, increasing attention to one bar automatically diminishes the attention that is paid to the other bar.

The net effect of these recurrent interactions is a surface-shroud resonance. Due to the top-down excitatory signals, the attended surface appears to have greater contrast, a property that has been reported both psychophysically and neurophysiologically.

Slide 158 summarizes the claim that an active surface-shroud resonance means that sustained spatial attention focuses on the object surface. The recurrent interactions sustain the attentional focus.

Slide 159 summarizes the critical claim that, in addition to its role in sustaining spatial attention on an object, a surface-shroud resonance supports conscious seeing of the attended object, in particular, a painting, while our eyes explore it. The talk does not summarize the large amount of psychological and neurobiological data that are consistent with this claim, but my article Grossberg [34] does do this.

Slide 160 summarizes the distinct resonances that support knowing versus seeing. A surface-shroud resonance, with the shroud in posterior parietal cortex (PPC), supports conscious seeing, whereas a feature-category resonance, with the category in inferotemporal cortex (IT), supports knowing. We can know about a familiar object when we see it because both resonances can synchronize their activities via shared circuits in prestriate visual cortical areas such as V2 and V4.

This distinction also enables us to understand various clinical data. For example, Slide 161 notes that, if the knowing resonance is damaged, then patients with visual agnosia can nonetheless accurately reach toward an object even if they cannot describe the orientation or other properties in space of the object that they are reaching. This example dramatizes the claim that seeing supports reaching, even if knowing does not occur.

Slide 162 emphasizes dual, but coordinated, functions of PPC in doing this. First, there is the top-down attention from PPC to V4 that focuses sustained spatial *attention* upon an object as part of a surface-shroud resonance. In addition, there is a bottom-up command from this attentive focus to motor control networks further downstream that carries out an *intention* to move to the attended object. Attention and intention are well-known to both be parietal cortical functions, and some of the articles that have contributed to this insight are listed. The theory clarifies why this so from the perspective of explaining how and why we become conscious of visual qualia.

My final Slide 163 summarizes some of the brain designs that this lecture has used to explain properties of how we consciously see and know things, and how these processes help to guide artists in making visual art. These designs clarify that our brains compute very differently than traditional computers, and from the currently popular algorithm in machine learning and AI called Deep Learning. Adaptive Resonance Theory has also been used in machine learning and AI applications, as Slide 151 has illustrated. ART can thus shed light upon the artistic process as well as provide algorithms for large-scale applications in engineering and technology that require autonomous adaptive intelligence in response to rapidly changing environments that may be filled with unexpected events. As I have already noted above, ART has also been used to provide mechanistic neural explanations of mental disorders that afflict millions of individuals, such as Alzheimer's disease, autism, Fragile X syndrome, schizophrenia, ADHD, visual and auditory neglect, medial temporal amnesia, and problems with slow wave sleep. How ART contributes to such an understanding is explained in a series of articles with several collaborators [19, 29, 33, 34, 36, 42]. Deep Learning cannot do any of these things. I therefore welcome artists, as well as scientists and technologists, to further study ART and to help develop its ability to provide new insights and applications in all of these fields.

References

1. Andersen, R.A., Essick, G.K., Siegel, R.M.: Encoding of spatial location by posterior parietal neurons. Science **230**, 456–458 (1985)
2. Beck, J., Prazdny, S.: Highlights and the perception of glossiness. Percept. Psychophys. **30**, 407–410 (1981)
3. Bosking, W., Zhang, Y., Schofield, B., Fitzpatrick, D.: Orientation selectivity and the arrangement of horizontal connections in tree shrew striate cortex. J. Neurosci. **17**, 2112–2127 (1997)
4. Cao, Y., Grossberg, S., Markowitz, J.: How does the brain rapidly learn and reorganize view- and positionally-invariant object representations in inferior temporal cortex? Neural Netw. **24**, 1050–1061 (2011)
5. Carpenter, G.A.: Distributed learning, recognition, and prediction by ART and ARTMAP neural networks. Neural Netw. **10**, 1473–1494 (1997)
6. Carpenter, G.A.: Neural network models of learning and memory: leading questions and an emerging framework. Trends Cogn. Sci. **5**, 114–118 (2001)
7. Carpenter, G.A., Grossberg, S.: A massively parallel architecture for a self-organizing neural pattern recognition machine. Comput. Vis. Graph. Image Process. **37**, 54–115 (1987)
8. Carpenter, G.A., Grossberg, S.: ART 2: self-organization of stable category recognition codes for analog input patterns. Appl. Opt. **26**, 4919–4930 (1987)
9. Carpenter, G.A., Grossberg, S., Markuzon, N., Reynolds, J.H., Rosen, D.B.: Fuzzy ARTMAP: a neural network architecture for incremental supervised learning of analog multidimensional maps. IEEE Trans. Neural Netw. **3**, 698–713 (1992)
10. Carpenter, G.A., Grossberg, S., Reynolds, J.H.: ARTMAP: Supervised real-time learning and classification of nonstationary data by a self-organizing neural network. Neural Netw. **4**, 565–588 (1991)
11. Carpenter, G.A., Grossberg, S., Rosen, D.B.: Fuzzy ART: fast stable learning and categorization of analog patterns by an adaptive resonance system. Neural Netw. **4**, 759–771 (1991)
12. Carpenter, G.A., Markuzon, N.: ARTMAP-IC and medical diagnosis: instance counting and inconsistent cases. Neural Netw. **11**, 323–336 (1998)
13. Carpenter, G.A., Milenova, B.L., Noeske, B.W.: dARTMAP: a neural network for fast distributed supervised learning. Neural Netw. **11**, 793–813 (1998)
14. Carpenter, G.A., Tan, A.-H.: Rule extraction: from neural architecture to symbolic representation. Connection Sci. **7**, 3–27 (1995)
15. Chang, H.-C., Grossberg, S., Cao, Y.: Where's Waldo? How perceptual cognitive, and emotional brain processes cooperate during learning to categorize and find desired objects in a cluttered scene. Front. Integr. Neurosci. (2014). https://doi.org/10.3389/fnint.2014.0043
16. Fazl, A., Grossberg, S., Mingolla, E.: View-invariant object category learning, recognition, and search: how spatial and object attention are coordinated using surface-based attentional shrouds. Cogn. Psychol. **58**, 1–48 (2009)
17. Field, D.J., Hayes, A., Hess, R.F.: Contour integration by the human visual system: Evidence for a local "association field". Vi. Res. **33**, 173–193 (1993)
18. Foley, N.C., Grossberg, S., Mingolla, E.: Neural dynamics of object-based multifocal visual spatial attention and priming: Object cueing, useful-field-of-view, and crowding. Cogn. Psychol. **65**, 77–117 (2012)
19. Franklin, D.J., Grossberg, S.: A neural model of normal and abnormal learning and memory consolidation: adaptively timed conditioning, hippocampus, amnesia, neurotrophins, and consciousness. Cogn. Affect. Behav. Neurosci. **17**, 24–76 (2017)
20. Gilbert, C.D., Wiesel, T.N.: Columnar specificity of intrinsic horizontal and corticocortical connections in cat visual cortex. J. Neurosci. **9**, 2432–2442 (1989)
21. Gnadt, J.W., Andersen, R.A.: Memory related motor planning activity in posterior parietal cortex of macaque. Exp. Brain Res. **70**, 216–220 (1988)
22. Goodale, M.A., Milner, A.D., Jakobson, L.S., Carey, D.P.: A neurological dissociation between perceiving objects and grasping them. Nature **349**, 154–156 (1991)
23. Grossberg, S.: Contour enhancement, short-term memory, and constancies in reverberating neural networks. Stud. Appl. Math. **52**, 213–257 (1973)
24. Grossberg, S.: How does a brain build a cognitive code? Psychol. Rev. **87**, 1–51 (1980)
25. Grossberg, S.: Outline of a theory of brightness, color, and form perception. In: Degreef, E., van Buggenhaut, J. (eds.) Trends in mathematical psychology, pp. 59–85. North-Holland, Amsterdam (1984)
26. Grossberg, S.: Cortical dynamics of three-dimensional form, color, and brightness perception, II: Binocular theory. Percept. Psychophys. **41**, 117–158 (1987)
27. Grossberg, S.: 3-D vision and figure-ground separation by visual cortex. Percept. Psychophys. **55**, 48–120 (1994)
28. Grossberg, S.: Cortical dynamics of three-dimensional figure-ground perception of two-dimensional figures. Psychol. Rev. **104**, 618–658 (1997)
29. Grossberg, S.: The imbalanced Brain: From normal behavior to schizophrenia. Biol. Psychiat. **48**, 81–98 (2000)
30. Grossberg, S.: Towards a unified theory of neocortex: laminar cortical circuits for vision and cognition. In: Cisek, P., Drew, T., Kalaska, J. (eds.) For Computational Neuroscience: From Neurons to Theory and Back Again, pp. 79–104. Elsevier, Amsterdam (2007)
31. Grossberg, S.: Cortical and subcortical predictive dynamics and learning during perception, cognition, emotion and action. Philos. Trans. Roy. Soc. Lond. B Biol. Sci. **364**, 1223–1234 (2009)

32. Grossberg, S.: Adaptive resonance theory: how a brain learns to consciously attend, learn, and recognize a changing world. Neural Netw. **37**, 1–47 (2013)
33. Grossberg, S.: Acetylcholine neuromodulation in normal and abnormal learning and memory: vigilance control in waking, sleep, autism, amnesia, and Alzheimer's disease. Front. Neural Circ. (2017). https://doi.org/10.3389/fncir.2017.00082
34. Grossberg, S.: Towards solving the hard problem of consciousness: the varieties of brain resonances and the conscious experiences that they support. Neural Netw. **87**, 38–95 (2017)
35. Grossberg, S., Hong, S.: A neural model of surface perception: lightness, anchoring, and filling-in. Spat. Vis. **19**, 263–321 (2006)
36. Grossberg, S., Kishnan, D.: Neural dynamics of autistic repetitive behaviors and Fragile X syndrome: basal ganglia movement gating and mGluR-modulated adaptively timed learning. Front. Psychol. Psychopathol. (2018). https://doi.org/10.3389/fpsyg.2018.00269
37. Grossberg, S., Kuhlmann, L., Mingolla, E.: A neural model of 3D shape-from-texture: multiple-scale filtering, boundary grouping, and surface filling-in. Vis. Res. **47**, 634–672 (2007)
38. Grossberg, S., Markowitz, J., Cao, Y.: On the road to invariant recognition: explaining tradeoff and morph properties of cells in inferotemporal cortex using multiple-scale task-sensitive attentive learning. Neural Netw. **24**, 1036–1049 (2011)
39. Grossberg, S., Mingolla, E.: Neural dynamics of form perception: boundary completion, illusory figures, and neon color spreading. Psychol. Rev. **92**, 173–211 (1985)
40. Grossberg, S., Mingolla, E.: Neural dynamics of perceptual grouping: textures, boundaries, and emergent segmentations. Percept. Psychophys. **38**, 141–171 (1985)
41. Grossberg, S., Mingolla, E., Ross, W.D.: Visual brain and visual perception: How does the cortex do perceptual grouping? Trends Neurosci. **20**, 106–111 (1997)
42. Grossberg, S., Seidman, D.: Neural dynamics of autistic behaviors: cognitive, emotional, and timing substrates. Psychol. Rev. **113**, 483–525 (2006)
43. Grossberg, S., Todorovic, D.: Neural dynamics of 1-D and 2-D brightness perception: A unified model of classical and recent phenomena. Percept. Psychophys. **43**, 241–277 (1988)
44. Grossberg, S., Zajac, L.: How humans consciously see paintings and paintings illuminate how humans see. Art & Percept. **5**, 1–95 (2017)
45. Hawthorne, C.W.: Hawthorne on painting. Dover, Mineola, New York (1938/1939)
46. Heitger, F., von der Heydt, R.: A computational model of neural contour processing: figure-ground segregation and illusory contours. In: Proceedings of the 4th International Conference on Computer Vision, pp. 32–40. Berlin, Germany (1993)
47. Hirsch, J.A., Gilbert, C.D.: Synaptic physiology of horizontal connections in the cat's visual cortex. J. Neurosci. **11**, 1800–1809 (1991)
48. Hubel, D.H., Wiesel, T.N.: Receptive fields and functional architecture of monkey striate cortex. J. Physiol. **195**, 215–243 (1968)
49. Kapadia, M.K., Ito, M., Gilbert, C.D.: Improvement in visual sensitivity by changes in local context: parallel studies in human observers and in V1 of alert monkeys. Neuron **15**, 843–856 (1995)
50. Kellman, P.J., Shipley, T.F.: A theory of visual interpolation in object perception. Cogn. Psychol. **23**, 141–221 (1991)
51. Kennedy, J.M.: Subjective contours, contrast, and assimilation. In: Nodine, C.F., Fisher, D.F. (eds.) Perception and Pictorial Representation. Praeger, New York (1979)
52. Mingolla, E., Ross, W., Grossberg, S.: A neural network for enhancing boundaries and surfaces in synthetic aperture radar images. Neural Netw. **12**, 499–511 (1999)
53. Peterhans, E., von der Heydt, R.: Mechanisms of contour perception in monkey visual cortex. II. Contours bridging gaps. J. Neurosci. **9**, 1749–1763 (1989)
54. Perlovsky, L.: Musical emotions: functions, origins, evolution. Phys. Life Rev. **7**, 2–17 (2010)
55. Perry, L.C.: Reminiscences of Claude Monet from 1889 to 1909. Am. Mag. Art **18**, 119–126 (1927)
56. Pinna, B.: Un effetto di colorazione. In: Majer, V., Maeran, M., Santinello, M. (eds.) Il laboratorio e la città. XXI Congresso degli Psicologi Italiani, pp. 158 (1987)
57. Pinna, B., Grossberg, S.: The watercolor illusion and neon color spreading: a unified analysis of new cases and neural mechanisms. J. Opt. Soc. Am. A **22**, 2207–2221 (2005)
58. Redies, C., Spillmann, L.: The neon color effect in the Ehrenstein illusion. Perception **10**, 667–681 (1981)
59. Robichaux, J.W.: Hensche on painting. Dover, Mineola (1997)
60. Schmidt, K., Goebel, R., Löwel, S., Singer, W.: The perceptual grouping criterion of colinearity is reflected by anisotropies of connections in primary visual cortex. Eur. J. Neurosci. **9**, 1083–1089 (1997)
61. Snyder, L.H., Batista, A.P., Andersen, R.A.: Coding of intention in the posterior parietal cortex. Nature **386**, 167–170 (1997)
62. Snyder, L.H., Batista, A.P., Andersen, R.A.: Change in motor plan, without a change in the spatial locus of attention, modulates activity in posterior parietal cortex. J. Neurophysiol. **79**, 2814–2819 (1998)
63. Thorell, L.G., De Valois, L.G., Albrecht, D.G.: Spatial mapping of monkey V1 cells with pure color and luminance stimuli. Vision. Res. **24**, 751–769 (1984)
64. Thorpe, S., Fize, D., Marlot, C.: Speed of processing in the human visual system. Nature **381**, 520–522 (1996)
65. Todd, J., Akerstrom, R.: Perception of three-dimensional form from patterns of optical texture. J. Exp. Psychol. Hum. Percept. Perform. **13**, 242–255 (1987)
66. Von der Heydt, R., Peterhans, E., Baumgartner, G.: Illusory contours and cortical neuron responses. Science **224**, 1260–1262 (1984)
67. Williams, L.R., Jacobs, D.W.: Stochastic completion fields: a neural model of illusory contour shape and salience. Neural Comput. **9**, 837–858 (1997)

Stephen Grossberg is Wang Professor of Cognitive and Neural Systems; Professor of Mathematics & Statistics, Psychological & Brain Sciences, and Biomedical Engineering; and Director of the Center for Adaptive Systems at Boston University. He is a pioneer in computational neuroscience, connectionist cognitive science, and neuromorphic technology. His work focuses upon how individuals adapt autonomously in real-time to unexpected environmental challenges, and includes models of vision and visual cognition; object, scene, and event learning and recognition; audition, speech, and language learning and recognition; development; cognitive information processing; reinforcement learning and cognitive-emotional interactions; consciousness; visual and path integration navigational learning and performance; social cognition and imitation learning; sensory-motor learning, control, and planning; mental disorders; mathematical analysis of neural networks; experimental design and collaborations; and applications to neuromorphic technology. Grossberg founded the International Neural Network Society and the journal Neural Networks, and has served on the editorial boards of 30 journals. He is a fellow of AERA, APA, APS, IEEE, INNS, MDRS, and SEP. He has published 17 books or journal special issues, over 550 research articles, and has 7 patents. He received the 2015 Norman Anderson Lifetime Achievement Award of the Society of Experimental Psychologists (SEP), the 2017 Frank Rosenblatt award of the Institute for Electrical and Electronics Engineers (IEEE), and the 2019 Donald O. Hebb award of the International Neural Network Society (INNS). Webpage: sites.bu.edu/steveg.

Is Beauty in the Eye of the Beholder or an Objective Truth? A Neuroscientific Answer

Hassan Aleem, Maria Pombo, Ivan Correa-Herran, and Norberto M. Grzywacz

1 Introduction

What makes something beautiful? The enigmatic nature of beauty has preoccupied philosophers and scientists alike since antiquity. For philosophers, short of defining beauty, the principal question has been to discover where it lies. Specifically, is beauty a quality of objects (objectivist view) or does it come from within the beholder (subjectivist view)? From the sixth century BCE until the eighteenth century CE, most philosophers fell in the objectivist camp [25]. For example, both Plato and Aristotle held that things were beautiful if they respected certain mathematical forms. Later, in the Middle Ages, Augustine argued that things gave delight because they were beautiful, not the other way around. The philosophers and artists of the Renaissance extended these classical principles, placing beauty in mathematical properties of objects like proportions, perspective, symmetry, and compositional geometry [1]. It was not until the end of the seventeenth century that philosophers such as Locke, Hume, and Kant started to think of beauty in a more subjective manner [25]. Locke, for instance, pointed out that experiencing color, a major aspect of beauty, was unique to the individual [18]. In turn, Hume, one of the biggest proponents of the subjectivist view, wrote, "beauty of things exists merely in the mind which contemplates them" [12]. The debate over where beauty lies continues to this day and has spilled over beyond the realm of philosophy into the fields of cognitive and neural sciences. For instance, in recent years, the field of neuroaesthetics has seen a significant growth [8]. Increasingly, neuroscientists are beginning to use modern tools to see whether they can give insight into the age-old question of beauty.

The objectivist viewpoint of beauty has considerable support from scientific studies across the globe. These studies explore whether measurable features of stimuli can account for people's preferences. An example of one such feature is symmetry. Research shows that symmetry is highly preferred across cultures, genders, and age groups [5]. Additionally, this symmetry preference exists across many domains, whether it be in faces, foods, buildings, inanimate objects, or technological interfaces [31]. Therefore, symmetry is one of the most prominent examples of an objectively defined characteristic of beauty. Other features such as balance, color, fractality, complexity, and curvature also point toward the existence of objective, universal standards [17]. For example, complexity is known to follow a universal "inverted U" shape in relation to beauty and liking [4]. Hence, individuals prefer moderate amounts of complexity to something very simple or extremely complex. At a higher conceptual level, features such as prototypicality, novelty, and semantic content also show universality of preference. For example, people prefer more prototypical faces, shapes, cars, and paintings, and prefer figurative as compared to abstract art [19, 33]. The reason for such universality can be explained by our shared evolutionary history and consequently because of similar processing mechanisms in our brains (discussed in greater detail in Sect. 2).

H. Aleem (✉) · N. M. Grzywacz
Interdisciplinary Program in Neuroscience, Georgetown University, Washington D.C, USA
e-mail: ha438@georgetown.edu

M. Pombo
Department of Psychology, Georgetown University, Washington D.C, USA

I. Correa-Herran · N. M. Grzywacz
Department of Neuroscience, Georgetown University, Washington D.C, USA

I. Correa-Herran
Facultad de Artes, Universidad Nacional de Colombia, Bogotá, Colombia

N. M. Grzywacz
Department of Physics, Georgetown University, Washington D.C, USA

N. M. Grzywacz
Graduate School of Arts and Sciences, Georgetown University, Washington D.C, USA

J. L. Contreras-Vidal et al. (eds.), *Mobile Brain–Body Imaging and the Neuroscience of Art, Innovation and Creativity*, Springer Series on Bio- and Neurosystems 10, https://doi.org/10.1007/978-3-030-24326-5_11

For every instance of objective standards discussed above, there are equally as many examples of subjectivity [20]. These examples exist at both the socio-cultural and the individual levels. An example of cultural differences comes from a study comparing British and Egyptian students' preferences of graphic stimuli [5]. This study found that the Egyptian group overall gave higher ratings to all types of symmetry (horizontal, vertical, and rotational). However, Egyptians liked less complex versions of symmetry than the British counterparts did. Similarly, a study of fractality preference found that European and African populations picked images with greater complexity than did people from North America or Central Asia [29]. These differences can likely be attributed to different levels of exposure as well as the culturally dependent values of those variables. For example, in Middle Eastern countries the holy sites are often adorned with symmetric patterns, increasing the cultural value of some types of symmetry [5]. Higher cognitive factors such as visual content processing can also be culturally modulated [22]. For instance, when looking at a visual scene Westerners tend to focus on focal objects, while East Asians tend to have a more holistic approach. This difference is evident in eye-tracking studies as well as in functional brain imaging [11]. Unsurprisingly, these differences also influence aesthetic preferences, with Westerners preferring images with central objects and less contextual information as compared to East Asians [20]. These differences also likely stem from cultural beliefs and values as they are evident in cognitive domains outside of visual processing as well [22].

So far, we have seen evidence for universal (objective) as well as cultural (subjective) dependence of visual aesthetic preferences. In the next two sections, we look at these differences from the perspective of neuroscience. We begin with discussing a cognitive psychology theory and show that one of its consequences is the existence of objective aspects of beauty (Sect. 2). We then discuss how subjectivity arises from the networks in the brain responsible for learning and motivation (Sect. 3).

2 The Processing Fluency Theory and Objectivity in Beauty

Certain physical properties of objects in the world are important for survival regardless of one's environment or social setting. For example, as social beings, detecting and recognizing human faces quickly and correctly is valuable to us. Therefore, through evolution, our brains have developed specialized neural structures to process information from faces [14]. Similar neural circuitry also exists for certain visual properties such as symmetry, complexity, and balance [9, 13, 32]. As a result, barring some cultural variability, the neural and cognitive mechanisms underlying the processing of these features are largely similar across individuals. Consequently, our cognitive responses to these features, including liking and disliking, are also largely similar, thereby creating a semblance of objectivity through universality [17]. A prominent theory in Neuroaesthetics, the processing fluency theory, links the evolutionary basis of these universals to aesthetic values. In this section, we discuss this component of processing fluency theory and present evidence from our research showing its applicability to aesthetics.

The processing fluency theory states that the easier it is for a perceiver to process the properties of a stimulus, the greater its aesthetic response will be [24]. Therefore, the theory depends on both the dynamics of the perceiver as well as the object. This theory has four assumptions. However, for our purposes, we will only consider the two primary ones. First, the processing fluency theory assumes that objects differ in their fluency. Specifically, the extent to with what ease one perceives and conceptualizes an object defines how fluent it is. Therefore, this assumption implies that a component of fluency relies on the constituent features of the object. Examples of these features include symmetry, proportion, balance, contrast, and complexity. What mediates the fluent processing of these variables? As discussed above, such variables have dedicated neural circuitry. Consequently, this allows these variables to be processed more efficiently and "fluently". In this way, evolutionarily important variables which have their own real-estate in the brain form a major part of? processing fluency. Second, the processing fluency theory assumes that fluency is hedonically marked, so objects with higher fluency are perceived more positively than are those with lower fluency. Why are these features and their fluency hedonically marked? The answer has to do with evolution and the nature of perception. Our only access to the surrounding world is perceptual estimation through our senses. We use these estimates to make decisions about the world (sometimes life or death). Therefore, it is highly advantageous for evolution to associate rewards with those features in the outside world, improving their estimates and letting us make better decisions. For example, detection of imbalance in visual scenes is necessary for survival, because lack of balance codes for visual outliers, and may thus indicate danger or other features of interest [13]. Therefore, the amount of imbalance in a visual scene indicates its salience and will thereby attract our visual attention. This allows us to immediately spot and direct our attention to, for example, a lion hiding in the bushes. Overall, the assumptions of the processing fluency theory have considerable support from several psychophysiological studies in cognitive psychology, but also in marketing, technology, education, and other fields [24]. It is evident that processing fluency, in part due to its evolutionary roots, can account for a wide variety of psychological phenomena related to preference. However, we wondered if it could explain aspects of aesthetics in art as well. Specifically, we were interested to see if certain universal biases would emerge in artworks.

To understand whether the processing fluency theory could account for certain aspects of visual art, we measured symmetry, balance, and complexity in Early Renaissance Portraits [2]. We chose these variables because of their evolutionary importance, dedicated circuitries in the brain, and prominence in art theory [3, 9, 13, 31]. To give a detailed example, consider the case of symmetry. Symmetry is of high evolutionary importance due to its prominence in important biological structures such as faces, plants, and body plans (Fig. 1a). In biological contexts symmetry is often a signal of good health and disruption of symmetry can signal genetic or natural abnormalities [27]. Apart from its importance in the natural world, symmetry is one of the defining principles in art (Fig. 1b). Additionally, symmetry serves as a "perceptual glue" allowing efficient grouping of visual input to separate objects from backgrounds [31]. Considering how much important visual information symmetry can deliver, it is not surprising that it has dedicated neural structures in the brain. Functional magnetic resonance imaging (fMRI) studies have shown that areas early in the visual processing hierarchy are activated when looking at symmetric stimuli (Fig. 1c—[26]). Not surprisingly, therefore, the processing of symmetry is fast and fluent [32]. Consequently, the processing fluency theory accounts for symmetry being a hallmark of visual aesthetics.

Based on the premises of the processing fluency theory, we predicted that master painters would show biases toward maximizing fluency variables [2]. To test this prediction, we first developed computational measures for symmetry, balance, and complexity. We then measured these fluency variables in three types of images: portrait paintings, carefully posed photographic portraits, and spontaneously snapped photographic portraits (Fig. 2). All portraits included only one subject. The portrait paintings were from master artists from the Early Renaissance, using a variety of mediums. The posed control portraits consisted of carefully framed frontal, angled (45°), and profile (90°) pictures of volunteer participants. With the carefully posed frontal pictures, we could ask whether the master painters achieved optimal amounts of symmetry and balance. In turn, the spontaneously snapped pictures were meant to have no artistic intent. Hence, they allowed us to figure out whether painted portraits showed more balance and symmetry than those obtained spontaneously.

Comparing spontaneous portraits with those by master painters from the Early Renaissance gave support to the processing fluency theory. An example of one such comparison appears in Fig. 3a. Here we measured the amount of vertical bilateral balance in each different type of image. There are many definitions of pictorial balance, for our analyses we defined balance as the difference between the total pixel intensities across the vertical midline of the image. Our results show that the mean index of imbalance for Renaissance portraits is lower than is that for spontaneous portraits. Hence, Renaissance master painters were not making spontaneous portraits, but composing their painting to increase balance. These results stemming from the analysis of balance were similar to those for the index of symmetry [2]. As for complexity, the analysis separated information based on pixel intensities from spatial organization. There are many definitions of complexity, all which essentially measure the amount of information [9]. We defined Complexity of Order 1 as the total amount of variability in pixel brightness, for example, a uniform image compared to static noise, with more variability leading to greater complexity. We then defined Complexity of Order 2 as the spatial organization of those pixels, such as a detailed image versus a uniform shape or an object, where greater detail would lead to greater complexity. To do this analysis, we first converted the images into grayscale. The results showed that the Complexity of Order 1 of canvases was less from those of photographs because of the limited choices of oil pigments and hence less variability in intensities [2]. However, we found that master painters may have consciously or subconsciously compensated by increasing Complexity of Order 2. They did so by making paintings more realistic, thus increasing their level of detail. This gives more information to the viewer, which increases its fluency as predicted by the processing fluency theory.

However, master painters did not make balance, symmetry, and complexity as large as possible. For example, Fig. 3a shows that by carefully posing subjects frontally, one can achieve indices of imbalance that are lower from those seen in Early Renaissance portraits. Careful posing yielded similar results for symmetry [2]. Are these results in violation of the processing fluency theory, which predicts a maximization of fluency variables such as balance and symmetry? Intriguingly, art historians have observed that Early Renaissance master painters tended to avoid frontal portraits, thereby reducing perfect balance and symmetry [23].

A probable reason for why Early Renaissance master painters did not maximize balance, symmetry, and complexity was the competition of these variables against each other. For instance, if one increases the symmetry in an image, it becomes less complex [9]. In a symmetric image, knowing the color of a point on the left side of the canvas automatically tells us the color of the equivalent point on the right side. Therefore, the amount of information or complexity falls as the symmetry (or balance) increases. This reduction of complexity would explain why Early

Renaissance painters tended to avoid frontal poses. These painters might want to increase complexity to give more information about their subject. That variables like complexity compete against symmetry or balance means that each master painter must decide how to equilibrate them. Some may emphasize the complexity, while others may choose to highlight balance and symmetry. Figure 3b shows how three different master painters from the Early Renaissance equilibrated balance and complexity individually. If one thinks of the possible values of balance and complexity as spanning a space, then individual painters exist in different portions of this space. We conceptualize the full space as multidimensional. It would include variables like complexity, balance, and symmetry, but also others that influence aesthetic values, such as color and texture. We call the possible values of these variables the "neuroaesthetic space". We propose that preferences existing in different regions of the neuroaesthetic space are a major component of individuality in artistic production and appreciation.

In conclusion, our work supports the processing fluency theory and thus, the existence of some universal aesthetic variables such as balance, symmetry, and complexity, and therefore, a degree of objectivity in beauty. Importantly, while the processing fluency theory is centered around the perceiver, it applies to visual artists equally as well since they actively perceive and revise their work [7]. While the aforementioned variables may have different meanings for a professional artist and a naïve viewer, the principle remains the same. Overall, the theory emphasizes that different from classical and Renaissance thinking, objectivity does not stem from elegant mathematical relations but from utilitarian evolutionary mechanisms. However, the processing fluency theory is likely to be incomplete. It does not capture the competition between different fluency variables and the resulting individuality. What leads different individuals to exist in distinct portions of the neuroaesthetic space? One reason could be external constraints, such as employer demands or availability of materials like oil versus fresco

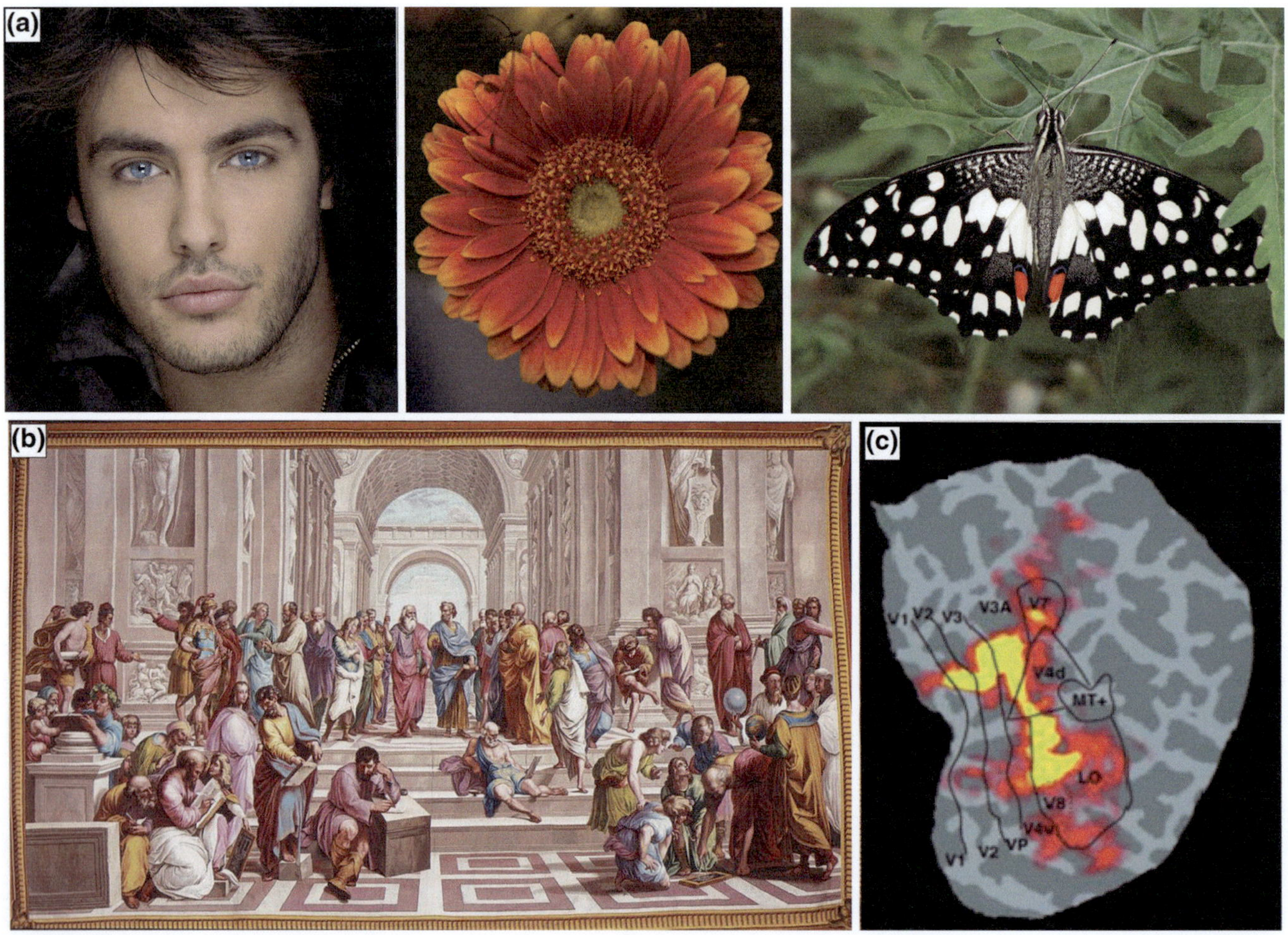

Fig. 1 Fluency of symmetry and its relation to art. **a** Symmetry is prominent in important structures in nature, such as faces [*Image Source* (https://pixabay.com/en/man-singer-musician-portrait-67467/)] plants [*Image Source* (https://pixabay.com/en/flower-flowers-summer-flowers-1431010/)], and body plans [*Image Source* (https://flic.kr/p/526sbH)]. **b** Symmetry is central in art, for example, *School of Athens* by Raphael [*Image Source* (https://library.artstor.org/asset/ARTSTOR10341822001612454)]. **c** The brain has dedicated areas devoted to symmetry (Image Reproduced with the author's permission [26]) which allow for its fluent processing

[2]. Other reasons could be internal, such as differences in perception due to, for instance, eyesight acuity. In addition, differences could emerge in how much individuals value certain aesthetic variables. In the next section, we explore this possibility by investigating how an individual's unique learning and motivation have a role in individuality.

3 Learning and Motivation as Roots of Subjectivity in Beauty

In this section, we focus on the cognitive mechanisms underlying subjectivity in beauty. What is considered beautiful is often largely cultural. Therefore, the roots of subjectivity are likely due to differences in our environment and experience with it. How do these differences manifest in our brain? We know from other fields of neuroscience that our brains can change in both structure as well as function as a result of experience [22]. Therefore, it is not surprising that the same may happen as a result of culture. Specifically, each culture has its own unique beliefs and values and in order for us to survive, we must learn and adopt these values. Therefore, learning is fundamental for differences in subjectivity. In the brain, the learning of such cultural values may largely undergo through a mechanism known as "reinforcement learning" [30]. We will discuss the details of this process further below. Additionally, it is important to note that although populations may learn the same values, no two individuals in the same culture are exactly similar in their likes and dislikes. One reason lies in the internal states of the individual. We further discuss how reinforcement learning can be directly modulated by internal factors such as motivation or drive. Therefore, we would expect that key mechanisms in the brain giving rise to subjectivity might be related to learning and motivation.

To begin, we focus on the neural structures that most likely underlie learning of aesthetic values. We preface this discussion by emphasizing that learning of aesthetic values may not be any different than learning of values in general. Discovering the true underlying aesthetic response in the

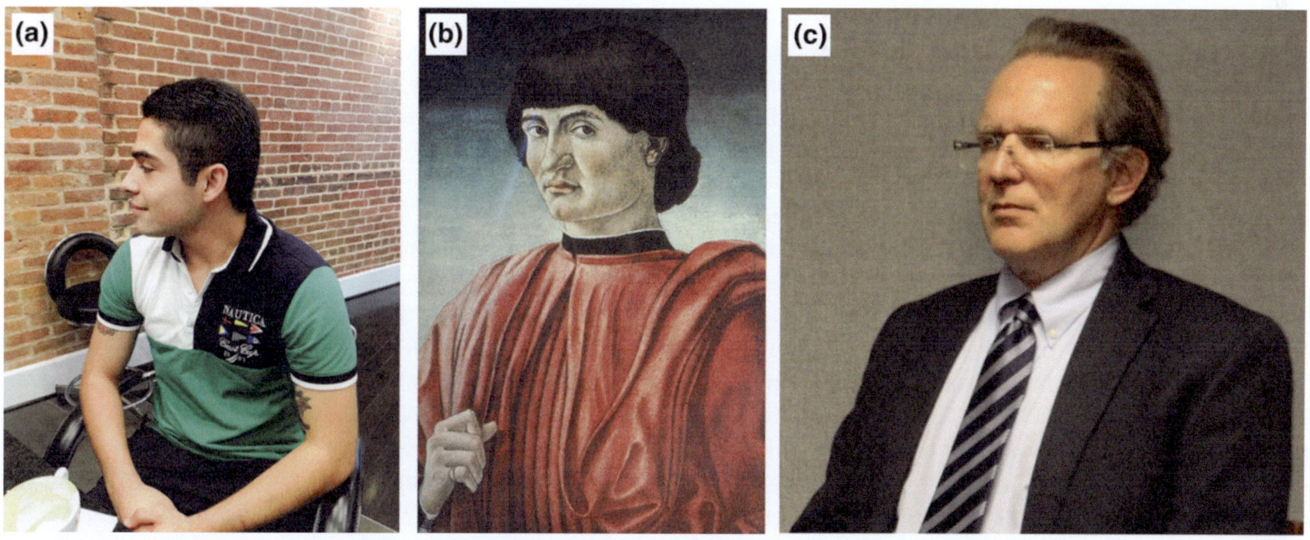

Fig. 2 Examples of images used in study of processing fluency in art **a** Spontaneously taken photograph **b** Early renaissance portrait painting, *Portrait of a Man,* by Andrea del Castagno [*Image Source* (https://www.nga.gov/collection/art-object-page.19.html)]. **c** Posed portrait photograph. *Note* All images were converted to grayscale for actual analysis

Fig. 3 Statistical analysis of early renaissance paintings. **a** Comparing balance across image categories. **b** Painters differed in their composition of complexity and balance

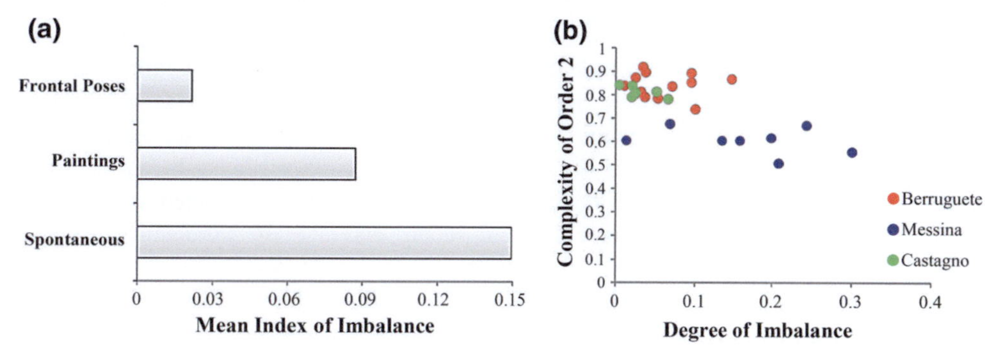

brain has been a challenging task for neuroscience. Part of the challenge stems from the complexity and variety of stimuli that can elicit aesthetic emotions. For example, faces, paintings, food, and music can all have their own respective aesthetic responses making it difficult to tease out what the true aesthetic response to beauty is. To tease this out, the earliest fMRI studies of beauty in art asked subjects whether they liked or disliked certain stimuli (paintings). As expected, these studies found activations in a wide array of visual areas as well as spatial, motor, emotional, and reward structures [15]. Since then, other studies have found a similar and seemingly widespread array of brain activations [8]. How can we reconcile these results? In particular, does a generalized network of brain regions that is responsible for aesthetic judgements irrespective of sensory modality exist? For example, are the brain mechanisms in "I like this painting of food" the same as in "this food is delicious"? To answer these questions, neuroscientists have used meta-analytic approaches. This approach combines the results of a range of neuroimaging studies to find the most concordant brain regions. The result of one such meta-analysis involving 93 fMRI studies of aesthetics in vision, taste, audio, and olfaction revealed a network of appraisal-related brain regions common to all sensory modalities. Specifically, the analysis found that three of the most concordant regions of activation were the orbitofrontal cortex (OFC), anterior insula, and the ventral basal ganglia [6]. This evidence suggests that aesthetic appraisal may be a special case of generalized appraisal mechanisms in the brain. We will now discuss these processes in greater detail.

The brain regions underlying appraisal are closely tied to learning of values from experience. In particular, previous research has shown the importance of these brain areas in processing and learning from rewards [6]. For example, two major functions of the OFC involve multisensory integration and tracking their sensory reward values. Similarly, the anterior insula is largely involved in interoception and assigning valence to objects concerning the motivational state of the organism. Lastly, the parts of the basal ganglia are involved in processes such as making predictions and keeping track of errors in those predictions. Combined, these areas allow the overall process of reward-based learning to occur [21]. Due to their intimate connection with sensory processing, reward, motivation, value, and learning, these regions are ideal candidates for neural circuitry underlying aesthetic learning and appraisal. A key component of this process being reinforcement learning.

To help better understand how reinforcement learning works and may be applied to aesthetics, let's look at an example. Consider the case of an individual seeing and smelling a red apple (Fig. 4a). The visual and olfactory regions would transmit pieces of sensory information to the OFC, which would integrate them into one percept. Based on this percept, parts of the basal ganglia help make a prediction about the reward gained by eating the apple, for example, "This will be sweet." Then, depending on that individual's internal motivational state, for example, "I am hungry," or "I am satisfied" as signaled for example, by the anterior insula, the person would act on the apple to test the initial prediction. Once the individual acts and eats the apple, the outcome (apple was bitter/apple was sweet) will be compared with the initial prediction. This comparison is the crux of the learning process. Here, again at the basal ganglia, the parameters of value models for the sensory inputs will be updated/learned given the reward. Thus, a certain property of apples, for example, "how red they are," is then "reinforced" and given a value. In the future, our brain can use this value as an initial guide for a prediction, allowing the individual to learn from experience and make better decisions. Similarly, the learned value will also influenced that person's preference, i.e. liking more red apples. This framework then largely encapsulates how we navigate and learn from our surroundings, and how that in turn affects our future decisions and preferences. In reality, the neural basis of these processes is much more nuanced, with much overlap. However, we have only considered those areas directly related to reward-based learning and their major roles.

From the example above, it can be seen that reinforcement learning is evolutionarily important and essential to our survival. This form of learning allows us to keep up with our ever-changing surroundings by constantly learning and updating an internal value model. Considering the fundamental nature of this process and the evidence from neuroimaging, we suggest that these same mechanisms apply when learning aesthetic values as well.

Let us now consider how this learning framework would apply when the same individual later looks at a painting of an apple (Fig. 4b). All the initial steps of the framework would be the same up to the prediction point. However, crucially, the individual cannot eat the painted apple to test the value prediction. Why may then the individual still enjoy looking at the painting? We propose that the answer lies in the previously learned "value" of the sensory aspects of the painting (for example, red equals good). This value then becomes the "aesthetic value." Just as how in processing fluency evolutionarily important features are hedonically marked, we propose a similar mechanism for aesthetic value within the reinforcement-learning framework. Thus, subjective aesthetic values may be formed in a similar manner to objective ones, albeit at a much shorter timescale. To further investigate the dynamics of exactly how these values form, we formulated a computational model based on Fig. 4.

By simulating the model, we got predictions for the dynamics of learning, individuality of aesthetic values, and cultural differences. For the purposes of this chapter, we present only an example subset of features of the model and

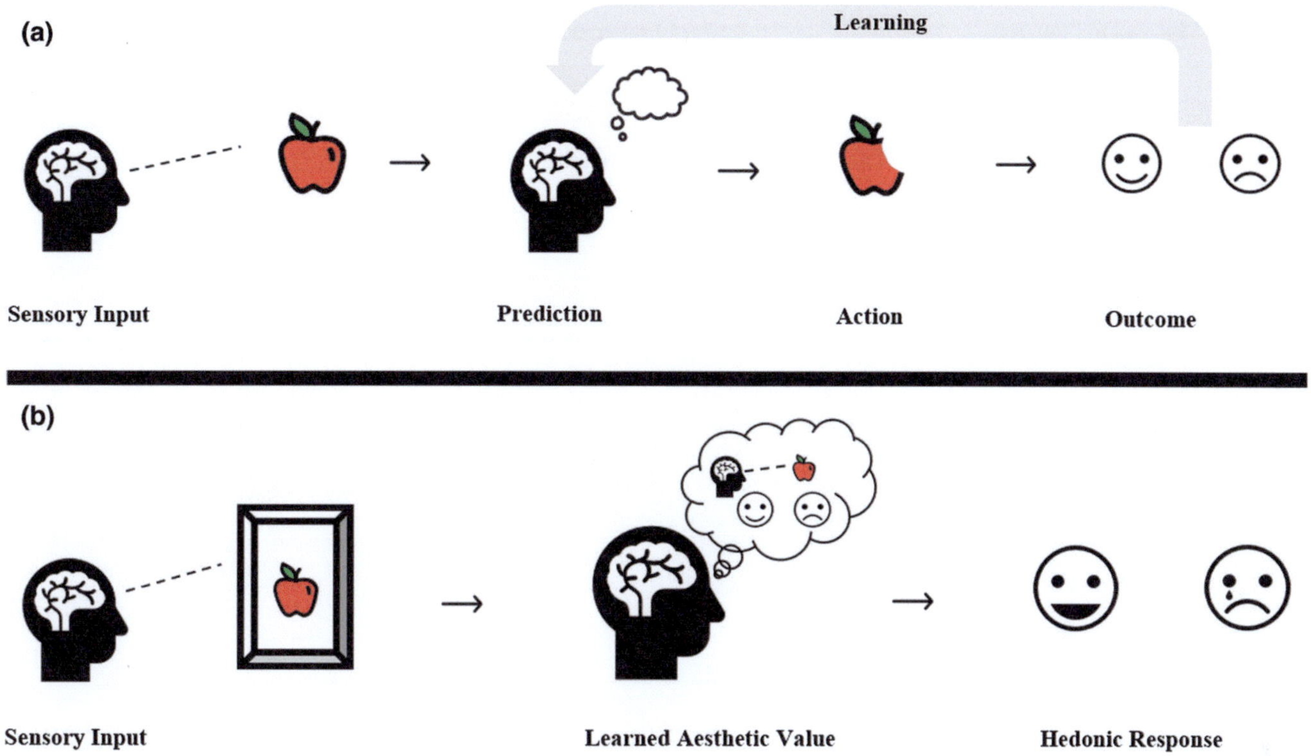

Fig. 4 Simplified illustration of reinforcement learning. **a** Case where individual encounters actual object in the environment. **b** Case where individual encounters a work of art with similar statistics to the object in the environment

in schematic form. In this example, we considered individuals who learn to weigh the aesthetic values of balance and complexity of their sensory inputs. The first feature of the learning model that we illustrate here is the motivation function (Fig. 5a). This function is the conditional probability that the individual will act given a sensory input. Thus, this function is set independently for each individual. In this example, the individual is motivated to act around certain levels of input complexity, while the motivation is independent of input balance. The second feature that we illustrate is the reward function (Fig. 5b). This function is the conditional probability that the individual will receive a "social" reward if the individual acts with the given sensory inputs. Thus, the reward function is set across all individuals of a social group. In this example, the reward increases linearly with the level of input balance. In the example of Fig. 5, we set the initial conditions of the simulations at zero, that is, the individual had no initial bias for balance and complexity.

A schematic representation of the simulated value weights for balance and complexity for an example individual appears in Fig. 5c. The weights began at zero and rose quickly. This fast rise was due to the tendency of high balance and complexity to be rewarding (see for example, Fig. 5b). However, after the rapid rise, the balance and complexity weights began to diverge. The latter went up slowly, while the former went down. Consequently, although balance and complexity had positive aesthetic values, this divergence phase indicated their inter-competition as described in the introduction to Fig. 3b. This competition phase lasted a relatively long time, eventually converging to a steady state. These results suggest an intriguing hypothesis for how we may learn aesthetic values. For example, the bulk of learning may be witnessed either early in development or when there is a dramatic change in environment, such as moving to a foreign country. Many questions about the timescale dynamics of aesthetic learning remain. We are currently performing behavioral experiments to shed more light on this issue.

Next, we investigated how aesthetic preferences would vary for individuals undergoing learning under different motivation functions. In our example, we considered two different individuals with preference for lower and higher levels of complexity (left panel of Fig. 5d). We informally thought of them as risk-averse and risk-taker individuals respectively. The results suggested that difference in risk-taking could lead to drastically different endpoints in aesthetic value (middle panel of Fig. 5d). For the risk-taker, the complexity and balance weights tended to be high and low respectively at steady state. However, for the risk-averse

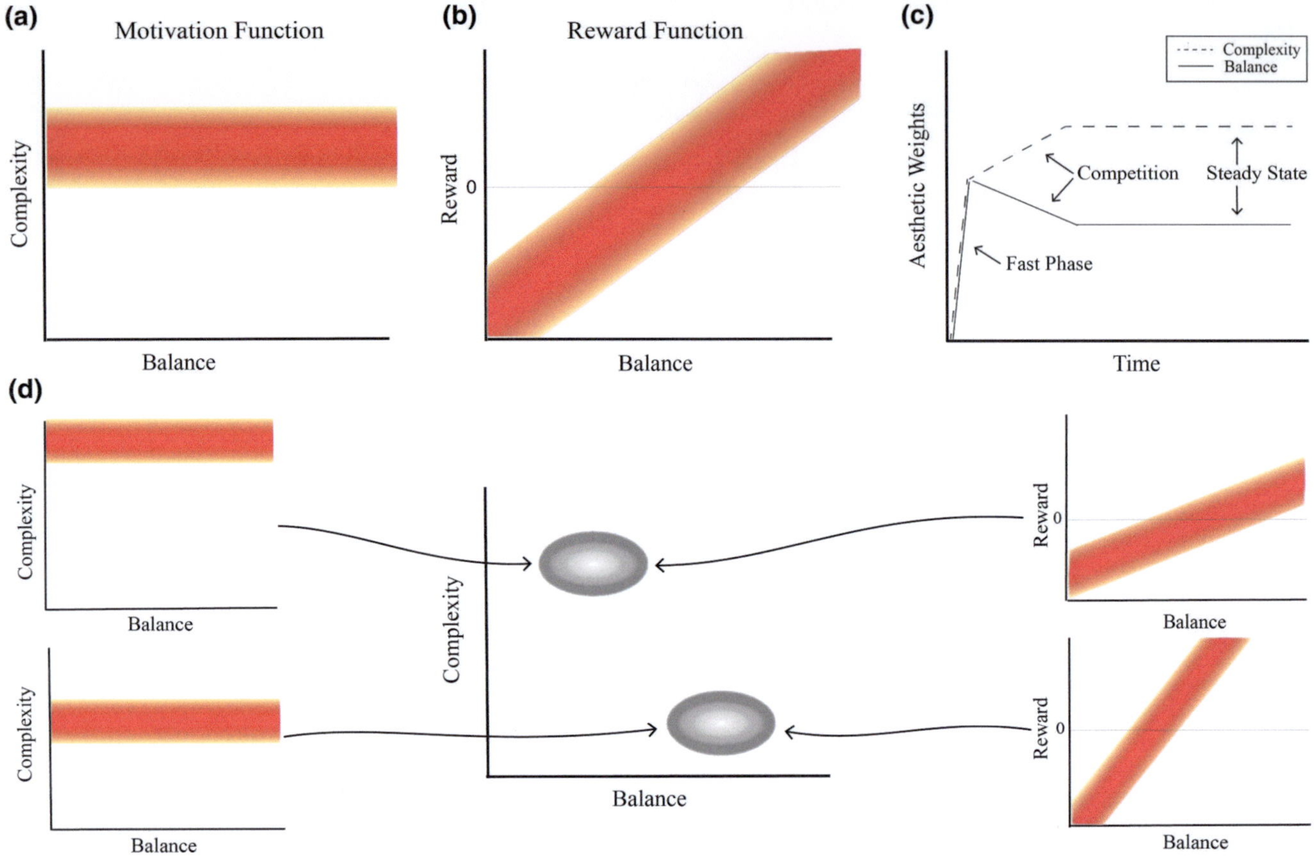

Fig. 5 Schematic overview of some features of the model and its predictions. **a** The conditional probability distribution of motivation given complexity and balance. **b** The conditional probability distribution of reward given balance. **c** An illustration of the dynamics of how the aesthetic weights of balance and complexity are updated. **d** An illustration of how changes in motivational state or social reward affect learning of aesthetic values. The left panel shows examples of two different motivation functions as in A. The right panel shows examples of two different reward functions as in B. The center panel illustrates the distribution of aesthetic weights at steady state (see Panel C) because of changes in motivation and reward functions

individual, the opposite happened. These results suggested that difference in motivation during learning are a factor underlying aesthetic individuality (see the end of Sect. 2 for more factors). In terms of real-world implications, there is convincing behavioral evidence for personality traits being a determining factor in aesthetic preferences. These studies are consistent with our results that show greater preference for complexity with more risk-taking personality traits [10]. Our results can therefore serve as a possible computational basis for these findings.

Lastly, we also investigated how changes in social reward functions could give rise to distinct aesthetic preferences. In our example, we performed simulations with different reward functions. To do this, we varied the slopes of the balance reward function. The results in Fig. 5d again show that just like internal motivational states, external factors such as social rewards can also result in individuals ending up with distinctly different aesthetic values. In particular, the results showed that when the balance reward function had steeper slopes, the complexity and balance weights tended to be low and high respectively at the steady state. However, for shallower slopes of the balance reward function, the opposite happened. Hence, different cultures with distinct reward functions could lead to divergence of aesthetic values. This finding is consistent with previously discussed evidence for the cultural dependence of aesthetics.

Overall, the results from our computational model suggest some possible mechanisms for how aesthetic subjectivity arises. Hence, subjectivity can arise from a multitude of factors, ranging from external differences such as culture, to internal differences such as motivation and learning dynamics. How each one of us arrives at our respective preferences may then be a unique function of the interaction of these two dynamics.

4 Discussion

Our search for the neuroscientific basis of objectivity and subjectivity in beauty ended up revealing something unexpected to us: both are reflections of utilitarian brain mechanisms. Beauty may not be the direct result of objective mathematical properties as once thought by Plato or Renaissance thinkers. Instead, objectivity may have arisen in part to our evolutionary history and principles captured by the processing fluency theory. Here, it is important to note that our definition of objectivity may differ with that of philosophy. Instead of objectivity being purely a priori qualities of the world, we extend it to mean the universality of response in human observers. For example, symmetry may be universally preferred because of its fluent processing as a result of shared dedicated neural circuitry [31]. Similarly, brain circuitry evolved for survival, particularly reward-based learning, may have given rise to subjectivity in beauty. Neuroimaging studies suggest that aesthetic appraisal depends in part on reinforcement-learning and motivational state circuitries in the brain [6]. We extend this framework in a computational model to show how it could be a basis for subjectivity. Additionally, we emphasize the role of the individual in aesthetic learning. While learning is central for social and environmental adaptability, individual motivational states help us choose actions that are best for ourselves. Thus, learning under the constraints of motivational states could give rise to subjective individual experiences of beauty. Lastly, we stress the "naturalistic" viewpoint of aesthetics [6, 28]. We propose that same evolutionary, learning, and motivational mechanisms that are involved in the appraisal of values in everyday decisions are also involved in aesthetic appreciation. It is possible then that same generalized value-computing brain circuitries may be "co-opted" for the appraisal of beauty as well. More specifically, we propose that the estimated value is akin to aesthetic value.

Taken together with evidence from neuroimaging studies, our results suggest that the processes underlying objective and subjective aesthetics are no different from the mechanisms of appraisal. Therefore, our hypotheses imply that both objective and subjective aspects of beauty lie within the perceiver's brain. This contrasts with the early philosophical perspectives that subjectivity is internal, while objectivity is external. We argue that this is not the case, that is, both are internal, with objective beauty also depending on underlying brain mechanisms. Thus, objectivity and subjectivity may represent two different ways of building values. Objectivity may be at the scale of evolution, thus more rigid and universal. In contrast, subjectivity may be at the scale of reinforcement learning, being more flexible and individualized. In turn, the interaction of these two mechanisms can account for both the universality as well as the individuality in human preferences across the globe. While all of us may be born with similar aesthetic biases, over time these biases are shaped by our experience through learning.

What are the implications for neuroscientists, artists, or anyone who appreciates beauty given our assertion that both objectivity and subjectivity may be internal to the brain? That beauty may manifest from the same fundamental evolutionary mechanisms as learning and survival should not diminish its importance. At the same time, maintaining an esoteric viewpoint of beauty will not further its understanding. The field of neuroaesthetics could benefit greatly by investigating aesthetic phenomena in the context of other fundamental brain processes such as memory and emotion. As for artists, knowing the neuroscientific basis of aesthetics may allow them to better understand the reasoning behind their academic principles, as well as allowing them to innovate and improve their art. For example, better understanding the interplay between internal and external dynamics of the viewer's brain may allow artists to create a better, more individualized museum experience [16]. While we are far away from uncovering the true nature of beauty and aesthetic appreciation, we must persist with knowing that the knowledge gained can improve our understanding both in the lab as well as in the studio.

References

1. Alberti, L.B., Leoni, G.: The Architecture of Leon Batista Alberti in Ten Books. E. Owen (1755)
2. Aleem, H., Correa-Herran, I., Grzywacz, N.M.: Inferring master painters' esthetic biases from the statistics of portraits. Front. Hum. Neurosci. **11**, 94 (2017)
3. Arnheim, R.: Art and Visual Perception: A Psychology of the Creative Eye. Univ of California Press (1956)
4. Berlyne, D.E.: Aesthetics and psychobiology. JSTOR (1971)
5. Bode, C., Helmy, M., Bertamini, M.: A cross-cultural comparison for preference for symmetry: comparing British and Egyptians non-experts. Psihologija **50**(3), 383–402 (2017)
6. Brown, S., Gao, X., Tisdelle, L., Eickhoff, S.B., Liotti, M.: Naturalizing aesthetics: brain areas for aesthetic appraisal across sensory modalities. Neuroimage **58**(1), 250–258 (2011)
7. Bryson, N.: Vision and Painting: The Logic of the Gaze. Springer (1983)
8. Chatterjee, A.: Neuroaesthetics: a coming of age story. J. Cogn. Neurosci. **23**(1), 53–62 (2011)
9. Donderi, D.C.: Visual complexity: a review. Psychol. Bull. **132**(1), 73 (2006)
10. Furnham, A., Walker, J.: Personality and judgements of abstract, pop art, and representational paintings. Eur. J. Pers. **15**(1), 57–72 (2001)
11. Han, S., Northoff, G.: Culture-sensitive neural substrates of human cognition: a transcultural neuroimaging approach. Nat. Rev. Neurosci. **9**(8), 646 (2008)
12. Hume, D.: 1985 of the standard of taste. In: Essays: Moral, Political and Literary (1757)

13. Itti, L., Koch, C., Niebur, E.: A model of saliency-based visual attention for rapid scene analysis. IEEE Trans. Pattern Anal. Mach. Intell. **20**(11), 1254–1259 (1998)
14. Kanwisher, N., McDermott, J., Chun, M.M.: The fusiform face area: a module in human extrastriate cortex specialized for face perception. J. Neurosci. **17**(11), 4302–4311 (1997)
15. Kawabata, H., Zeki, S.: Neural correlates of beauty. J. Neurophysiol. **91**(4), 1699–1705 (2004)
16. Kontson, K., Megjhani, M., Brantley, J.A., Cruz-Garza, J.G., Nakagome, S., Robleto, D., et al.: 'Your brain on art': emergent cortical dynamics during aesthetic experiences. Front. Hum. Neurosci. **9**, 626 (2015)
17. Lindell, A.K., Mueller, J.: Can science account for taste? Psychological insights into art appreciation. J. Cog. Psychol. **23** (4), 453–475 (2011)
18. Locke, J.: An Essay Concerning Human Understanding (1689)
19. Martindale, C., Moore, K.: Priming, prototypicality, and preference. J. Exp. Psychol. Hum. Percept. Perform. **14**(4), 661 (1988)
20. Masuda, T., Gonzalez, R., Kwan, L., Nisbett, R.E.: Culture and aesthetic preference: comparing the attention to context of East Asians and Americans. Pers. Soc. Psychol. Bull. **34**(9), 1260–1275 (2008)
21. O'doherty, J., Dayan, P., Schultz, J., Deichmann, R., Friston, K., and Dolan, R.J.: Dissociable roles of ventral and dorsal striatum in instrumental conditioning. Science **304**(5669), 452–454 (2004)
22. Park, D.C., Huang, C.-M.: Culture wires the brain: a cognitive neuroscience perspective. Perspect. Psychol. Sci. **5**(4), 391–400 (2010)
23. Pope-Hennessy, J.W.: The portrait in the renaissance. Princeton University Press (1966)
24. Reber, R., Schwarz, N., Winkielman, P.: Processing fluency and aesthetic pleasure: is beauty in the perceiver's processing experience? Pers. Soc. Psychol. Rev. **8**(4), 364–382 (2004)
25. Sartwell, C.: "Beauty". In: Zalta, E.N. (ed.) The Stanford Encyclopedia of Philosophy (Winter 2017). Metaphysics Research Lab, Stanford University (2017)
26. Sasaki, Y., Vanduffel, W., Knutsen, T., Tyler, C., Tootell, R.: Symmetry activates extrastriate visual cortex in human and nonhuman primates. Proc. Natl. Acad. Sci. **102**(8), 3159–3163 (2005)
27. Scheib, J.E., Gangestad, S.W., Thornhill, R.: Facial attractiveness, symmetry and cues of good genes. Proc. Roy. Soc. London B: Biol. Sci. **266**(1431), 1913–1917 (1999)
28. Skov, M.: The pleasure of art. Pleasures of the brain 270–283 (2010)
29. Street, N., Forsythe, A.M., Reilly, R., Taylor, R., and Helmy, M.S. (2016). A complex story: Universal preference vs. individual differences shaping aesthetic response to fractals patterns. *Frontiers in human neuroscience* 10, 213
30. Sutton, R.S., Barto, A.G.: Introduction to Reinforcement Learning. MIT press Cambridge (1998)
31. Treder, M.S.: Behind the looking-glass: a review on human symmetry perception. Symmetry **2**(3), 1510–1543 (2010)
32. Tyler, C.W.: The symmetry magnification function varies with detection task. J. Vis. **1**(2), 7–7 (2001)
33. Vessel, E.A., Rubin, N.: Beauty and the beholder: highly individual taste for abstract, but not real-world images. J. Vis. **10** (2), 18–18 (2010)

Hassan Aleem is a Ph.D. candidate in the Interdisciplinary Program for Neuroscience at Georgetown University. He previously obtained his bachelor's degrees in Molecular and Cellular Biology and Physiology with a minor in Psychology. His research interests include the neural representation of aesthetic preferences, reward, pleasure, and how aesthetic preferences form. He is additionally interested in topics related to computational neuroscience and artificial intelligence.

Maria Pombo is an undergraduate at Georgetown University's College of Arts and Sciences majoring in Psychology and minoring in Cognitive Science. She is interested in how reward and value are represented in the brain and how these guide human behavior. She recently began research in Neuroaesthetics.

Ivan Correa-Herran is an Associate Professor of Architecture at Facultad de Artes, Universidad Nacional de Colombia, UNC. He got his masters in Architecture from Graduate School of Design, Harvard University. Professional degree in Architecture from Universidad de los Andes. Doctoral Studies, in the History and Theory of Art and Architecture at UNC. Research Scholar, Department of Neuroscience, Georgetown University.

Norberto M. Grzywacz is Dean of the Graduate School of Arts and Sciences, and Professor of Physics and Neuroscience at Georgetown University. Previously, he was Chair and Professor of Biomedical Engineering at the University of Southern California. He received his Ph.D. from the Hebrew University of Jerusalem and was a postdoctoral fellow and research faculty at the Massachusetts Institute of Technology.

Part IV

Cognitive and Medical Applications: How Can Arts and Neuroscience Research Improve Physical and Mental Health and Promote Wellbeing?

Introduction

Juliet King
Department of Art Therapy, The George Washington University, Washington D.C., USA

The art of healing comes from nature, not from the physician. Therefore, the physician must start from nature, with an open mind.
—Paracelsus

Paracelsus was a physician, alchemist, and pioneer of the German Renaissance who embraced a range of perspectives in the treatment of the human condition and its maladies. Also a toxicologist, he studied the composition and structure of things and the changes that take place when their properties interact. Paracelsus tested bold ideas with the scientific method and in this process helped to discover how deadly poisons can become useful medications. The themes of transformation are embedded in the human body and soul and can be explored through a variety of contexts. While not discussing botulinum toxin, penicillin, or warfarin, the dynamic and visionary group that participates in the annual *Brain on Art* (BOA) and *Mobile brain/body imaging* (MoBI) meetings investigate something similar: How do the arts and neurosciences teach us about the intersecting and transformational nature of our brains and bodies, and how does this information contribute to the improvement of physical and mental health? These questions clarify how to apply the scientific method to understand an inherently subjective process, knowing that individuality and variance is central to this understanding. Here it is useful to acknowledge that what might be considered rigor in a scientific experiment that cultivates data inclusive of generalizability is just as important as arts-based research that calls upon intuition and phenomenological evidence that contributes to what it is we are seeking to understand.

Healing through the arts is one of the oldest practices in the world and takes on innumerable applications. Art helps to clarify scientific questions just as applying science to artistic processes illuminates the potentials for art to provide evidence that helps us understand the human condition. To conceptualize the therapeutic benefits of art, it is necessary to consider a range of applications, from the recreational and relaxing engagement of creating to the assessment and intervention of the clinically trained art therapist. The profession of art therapy is grounded in specific developmental, psychological, and psychotherapeutic processes that require a masters-level training to understand and apply. Ethical obligations that accompany the work of the psychotherapist, the intrapsychic change that is considered throughout the treatment, and the identification and implementation of specific goals to facilitate symptom reduction are essential when making distinctions between what can be understood as a continuum of therapeutic arts and art therapy interventions. These crucial distinctions help to define, connect with, and support the therapeutic arts while at the same time maintaining the integrity of the training and interventions of clinical art therapy practitioners. Including the arts on equal grounds with more empirically-based measures that define science is enhanced through the merging with the neurosciences, and these partnerships cultivate opportunities to service the physical and mental health needs of patients and conduct interdisciplinary research. The ways of approaching collaboration require an integrative philosophy of treatment, an open-minded process of engagement, a shared language, and common context within which to conceptualize potentials.

We weave our worlds together throughout a series of systems and functions that are mostly based on the integration of sensory stimuli, information, experiences, and behavior, all of which is conducted through muscle contraction or secretion and memories. Our senses are the portals for experiencing the world and the data is processed within corresponding association networks in our brain. Information from the external world combines with the internal and a result is mental imagery and sense perceptions, which might be considered symbolic content. Humans tend to think in images and this imagery is an important way to communicate phenomenological and lived experiences; often times it is essential as language does not always capture the essence of what it is we are trying to say. The use of only words to formulate and express our thoughts and feelings in therapy is a limiting experience. This is true in general, and especially for those that suffer from overwhelming stress, neurological problems, trauma, depression, abuse, and mental illness. As Dr. Girija Kaimal notes in her chapter, "art therapy helps contain and externalize positive and negative emotions, thereby offering the patient or client an alternative visual perspective of his or her condition: breaking the cycle of rumination and providing hope for a possibly fulfilling future." Sometimes visual expression becomes a key factor in the process of healing, especially when a person has endured trauma or brain injury, as the memories experienced are housed in less conscious areas of the brain. Further, recollection of traumatic memories might elicit a deactivation in the Broca's area of the brain, which significantly limits the ability to recall and effectively express the memories with language. In her chapter, art therapist Melissa Walker articulates the value of the mask-making process and resultant products by military service members to communicate multidimensional information that these objects carry with them and the value that dialogue within the context of the therapeutic relationship holds for client progress.

Where there is art, there is power, and art therapy helps to describe a "new symbiosis between art and the experience of war (https://vimeo.com/77617525). Engaging in the creative process helps to synthesize complex conscious/explicit and unconscious/implicit processes and is indicated as a treatment of choice for those who have endured trauma. In the documentary *Veterans Coming Home*, art therapy participant Andrew explained that making objects on paper helped him "… realize that there is still beauty left in the world," while Dusty said that when making art "You can concentrate and focus on what you're doing….your body and your mind relax and give you a sense of well-being for that period of time." (https://video.wfyi.org/video/wfyi-local-productions-veterans-coming-home-healing-arts/). These veterans offer convincing, yet anecdotal, evidence of the value of symbolic communication and healing capacities of creative expression.

Dr. Jose Luis Contreras-Vidal, BOA conference lead, engineering professor and University of Houston Brain–Machine Interface Systems lab director explained how the "arts provide a window" to study our individually unique brains. He articulates that "the more we understand the way the brain responds to the arts, the better we can understand ourselves." Nobel Prize winner Eric Kandel [1] writes prolifically on how in studying the disease state of the brain we also learn about healthy functioning, and this might be seen as a parallel to the use of the arts in illuminating our knowledge of the complex brain processes that make us who we are. New questions and broader capacities of the arts and sciences are possible through intersecting lenses produced by the BOA and MoBI conferences. Neuropsychologisy Dr. Sebastian Crutch et. al are exemplars of these capacities, as described in their chapter "Created Out of Mind: Shaping Perceptions of Dementia Through Art and Science", where the vision for transdisciplinary and person-centered research with the dementias is applied through community engagement and pragmatic program implementation. With years of experience and great sensitivity, this group describes how although the "overwhelming majority of funded research studies in the field of dementia are quite rightly located within objective science and clinical medicine…everyone has different journeys when living with chronic disease and we believe that creative activity has more flexibility to address that complexity than generic therapies or drugs." Crutch et al. go on to say that the subjective experience ultimately adds immense value to the objective learning, and here we are reminded how the arts become an important way of maintaining the integrity of the subjective process and emphasizing the value of $n = 1$.

At the BOA conference in Valencia, Spain in 2017, Crutch described the case study of his patient William Utermohlen, whose self-portraits are distributed worldwide and have likely become the most visible and compelling example of the devastation of Alzheimer's Disease. In this presentation, Crutch remarked that Utermohlen's artwork and its disintegration of formal elements throughout the course of the disease was "much more powerful than any data set." While compelling, and true, integrating the arts and sciences carries as many challenges as potentials in terms of research. That what makes Utermohlen's case so special, the art part, is what makes it more difficult to define. This is similar to the entire profession of art therapy, which despite emerging in the 1940s, continues to be societally challenged as an effective intervention likely due to the dearth of randomized control trials which are notoriously difficult to conduct when there is an emphasis on the $n = 1$. To address this, I have developed a set of precise tenets that help to define the profession and the capacities for research, all of which can be underscored with neuroscience theory and tested with the application of contemporary imaging technologies: (1) The art-making process and the artwork itself are integral components of treatment that help to understand and elicit verbal and nonverbal communication within an attuned therapeutic relationship. (2) Creative expression is healing and life-enhancing. (3) The materials and methods utilized effect self-expression, assist in emotional self-regulation, and are applied in specialized ways [2]. These tenets provide a framework within which to contextualize the many ways of approaching art therapy practice and research.

The clinical implications of merging arts and neurosciences are substantial and becoming increasingly recognized, exemplified by the chapters in this section. We often need an objective presence to help us see ourselves and it is important to provide opportunities to engage in different methods for healing. The field of neuroaesthetics provides a lens within which to explore visual information processing systems and connections with emotional and effort-based reward systems inherent in artistic expression. Expanding into the exciting realm of contemporary neuroimaging and disruptive technologies offers opportunities to generate new evidence and contribute to a commonly shared and accessible language that serves to de-silo respective disciplines. Mobile brain/body imaging utilizes bold methods to capture cortical activity while moving in natural environments and holds exciting possibilities to push forward advances in the arts, neurosciences, and related therapeutics. When we approach the intersections of the brain and art, and the changes that take place when their properties interact, we are better equipped to understand, transform and heal beyond what we might have ever seen to be possible.

References

1. Kandel, E.: *The disordered mind: What unusual brains tell us about ourselves*. Farrar, Straus, and Giroux, New York (2018)
2. King, J.L. (ed.): *Art therapy, trauma, and neuroscience: Theoretical and practical perspectives*. Routledge, New York, NY (2016)

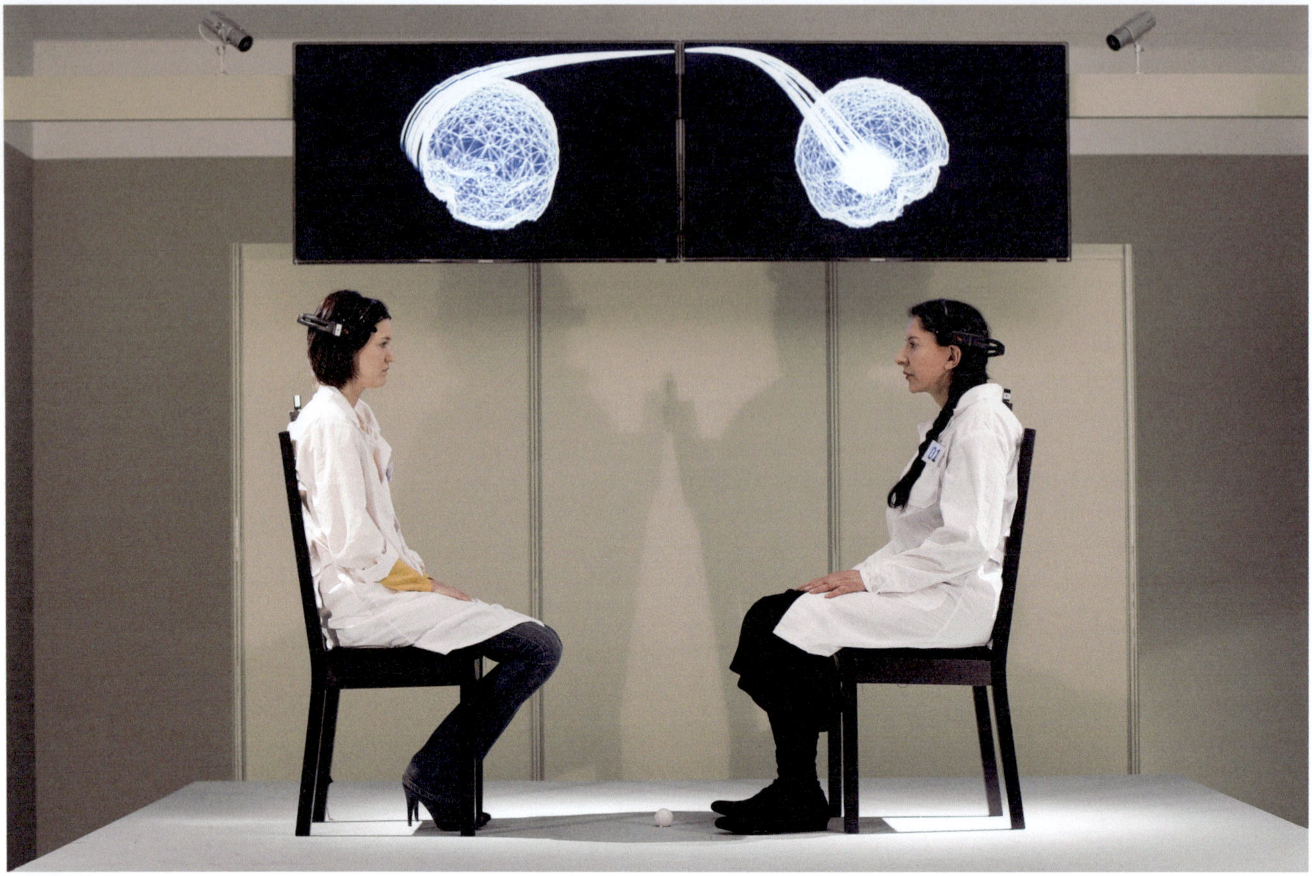

Suzanne Dikker & Matthias Oostrik. In collaboration with Marina Abramovic and participants of the Art & Science: Insights into Consciousness workshop, Watermill Center NY

Measuring the Magic of Mutual Gaze

Neurofeedback video display | EMOTIV EPOC headsets

Measuring the Magic of Mutual Gaze restages Marina Abramovic' The Artist is Present (MoMA, 2010) as an interactive art installation/neuroscience experiment, investigating the relationship between human connectedness and brainwave synchrony between people. Pairs of audience members engage in mutual gaze for 30 minutes while wearing EMOTIV EPOC EEG headsets. Their dominant brain frequencies and moments of brain-to-brain synchrony are visualized in real time on two rotating brains (video: https://www.youtube.com/watch?v=Ut9oPo8sLJw). Results from ~150 visitors of the Garage Museum of Contemporary Art in 2011 showed a significant increase in brain-to-brain synchrony during mutual gaze

Venues:

Marina Abramović, The Artist is Present, Garage Museum for Contemporary Art, Moscow 2011

Soft Control: Art, Science and the Technologically Unconscious, Maribor 2012

Outcomes of Art Therapy Treatment for Military Service Members with Traumatic Brain Injury and Post-traumatic Stress at the National Intrepid Center of Excellence

Melissa S. Walker

The identification of specific products, scientific instrumentation, or organization is considered an integral part of the scientific endeavor and does not constitute endorsement or implied endorsement on the part of the author, DoD, or any component agency. The views expressed in this chapter are those of the author and do not reflect the official policy of the Department of Army/Navy/Air Force, Department of Defense, or U.S. Government.

Between the September 2001 (9/11) attacks and September 2015, 2.77 million US military service members (SMs) served on more than 5.4 million deployments. Notably, over 20% of those SMs deployed three or more times [1]. Factors such as frequency and duration of deployments lead to compounding injuries, specifically traumatic brain injury (TBI) and post-traumatic stress disorder (PTSD) which are referred to as the signature and invisible wounds of the post-911, Global War on Terror conflicts. TBI has been found to be a significant predictor in the development of PTSD [2, 3] often with an overlap of symptoms such as anxiety, depression, cognitive deficits, irritability, sleep disruptions, and embodied memory experiences [4, 5]. A recent study also found that the co-occurrence of PTSD with mild TBI (mTBI) worsens post-concussive symptoms in post-9/11 veterans, including greater pain catastrophizing and intensity, worse recall, and greater illness-focused coping than in veterans with mTBI alone [6]. It is estimated that 19.5–22.8% of Operation Enduring Freedom (OEF) and Operation Iraqi Freedom (OIF) SMs and veterans have sustained a TBI, with 383, 941 diagnoses as of the first quarter of 2018 [7], and 7–20% of SMs and veterans who served in OEF and OIF are living with or at some point met the criteria for PTSD diagnosis [8]. While experiencing symptoms associated with post-traumatic stress is natural following an event and often resolve on their own, the criteria for the diagnosis of a disorder include re-experiencing, avoidance, thought/mood disturbance, and hyperarousal which impair life functioning for longer than a month [9, 10]. Chronic PTSD in our veterans can span a lifetime if left untreated.

Due to the complex and unique nature of SMs overcoming comorbidity of TBI and underlying psychological health (PH) conditions including PTSD, specialized treatment facilities have begun operating within both the Departments of Defense (DoD) and Veterans Affairs (VA). The National Intrepid Center of Excellence (NICoE), a directorate of Walter Reed National Military Medical Center (WRNMMC) in Bethesda, MD, US, utilizes a holistic, interdisciplinary approach to clinical care for military SMs whose comorbid TBI and PH conditions have not responded to traditional treatment. At the NICoE (Fig. 1), a facility designed to be a healing environment, SMs receive behavioral health and rehabilitation treatments which target the mind, body, and spirit, and foster and encourage resilience, well-being, and self-management through active engagement in their care [11]. Since 2010, the NICoE has developed and implemented a four-week intensive outpatient program (IOP) which employs 17 medical and integrative health disciplines, such as neurology, neuropsychology, psychiatry, family therapy, physical therapy, and speech-language pathology, which offer an array of recovery techniques and tools for active-duty SMs with mTBI. The NICoE also offers long-term outpatient programming in which SMs whose TBIs range from mild to severe are referred to various treatments based on their individual recovery goals. As part of interdisciplinary programming under the behavioral health umbrella, each SM is scheduled to partake in creative arts therapies treatment delivered by trained and certified art, music, and dance/movement therapists.

M. S. Walker (✉)
National Intrepid Center of Excellence, Walter Reed National Military Medical Center, Bethesda, MD, USA
e-mail: melissa.s.walker12.civ@mail.mil

M. S. Walker
Creative Forces: NEA Military Healing Arts Network, National Endowment for the Arts, Washington, D.C., USA

J. L. Contreras-Vidal et al. (eds.), *Mobile Brain–Body Imaging and the Neuroscience of Art, Innovation and Creativity*, Springer Series on Bio- and Neurosystems 10, https://doi.org/10.1007/978-3-030-24326-5_12

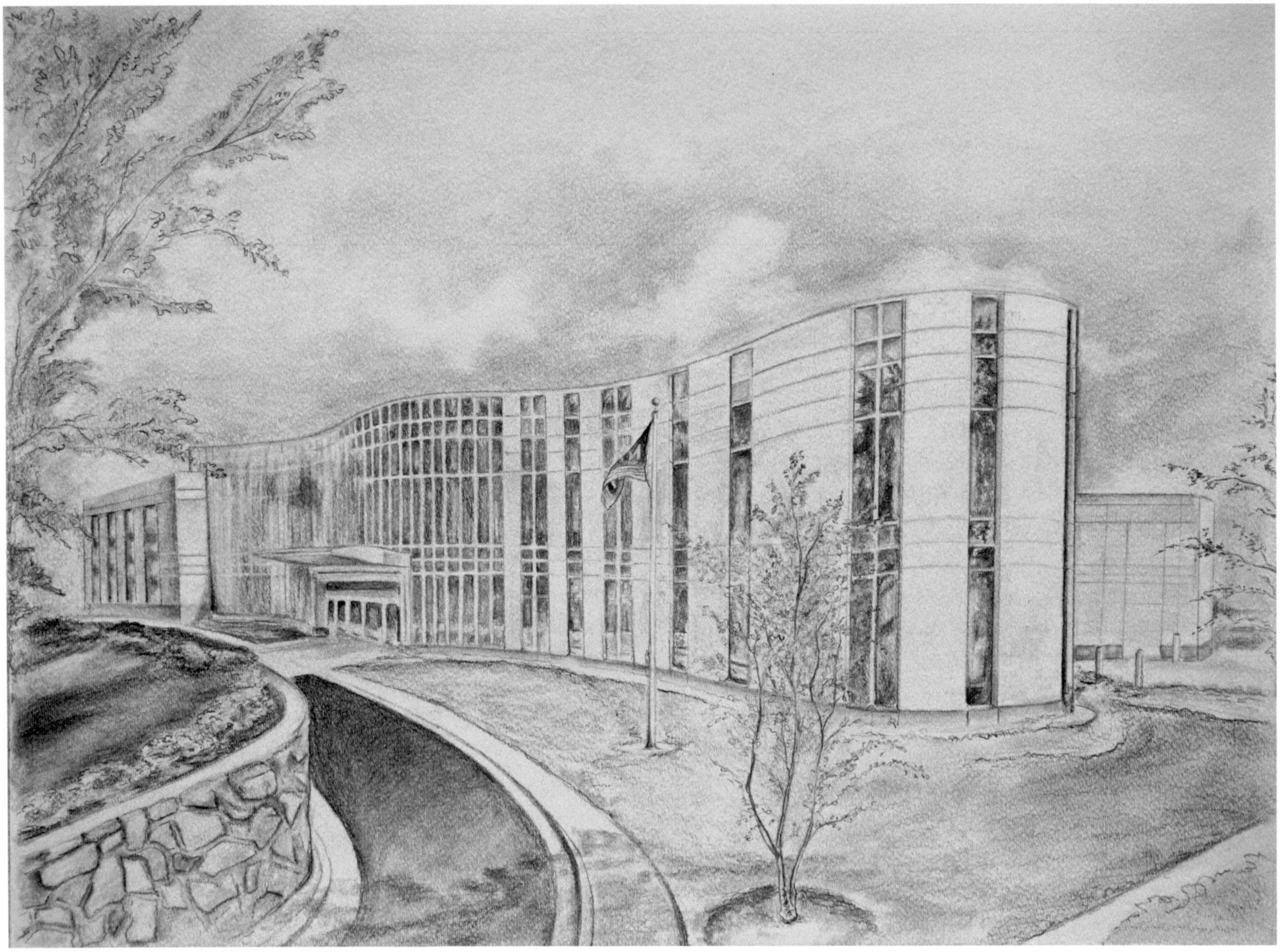

Fig. 1 Drawing of the NICoE by art therapist Melissa S. Walker

At the NICoE, a facility designed to be a healing environment, SMs receive behavioral health and rehabilitation treatments which target the mind, body, and spirit, and foster and encourage resilience, well-being, and self-management through active engagement in their care [11].

This chapter focuses solely on art therapy programming at the NICoE, and includes content and research outcomes presented at the *2017 International Conference on Mobile-Brain Body Imaging and the Neuroscience of Art, Innovation, and Creativity* conference held in Valencia, Spain. The art therapy program at the NICoE began in 2010 and has since evolved and expanded into various military treatment facilities across the US. The expansion is in large part due to the Creative Forces®: NEA Military Healing Arts Network, an initiative of the National Endowment for the Arts (NEA). The NEA Creative Forces initiative is a partnership between the NEA, the DoD and VA, and the state and local arts agencies, with administrative support provided by Americans for the Arts. The NEA Creative Forces initiative has supported all NICoE art therapy research published to date, which is briefly summarized. Although only art therapy is discussed, all Creative Forces publications through 2018 which include art and music therapy, and therapeutic writing, can be found open access through the NEA Creative Forces initiative's Research and Scholarly Manuscripts Inventory [12].

1 Art Therapy with Military Service Members

Art therapy is a regulated, integrative mental health and human services profession that enriches the lives of individuals, families, and communities through active art-making, creative process, applied psychological theory, and human experience within a psychotherapeutic relationship [13], and is delivered by registered art therapists (ATRs) certified by the Art Therapy Credentials Board (ATCB). All art therapists hired by the DoD, VA, and the NEA Creative Forces initiative are master's level trained clinicians who have either received or are actively working toward ATR certification. The employment of certified art therapists is imperative, as the utilization of psychotherapy through art with vulnerable populations such as injured military SMs and veterans can conjure traumatic content, feelings, and emotions which are difficult to process and contain safely if an individual is not adequately trained [4, 14]. Art therapy encourages SMs to be "active engagers rather than passive recipients, elevate the personal nature of treatment, and motivate the SMs to be more invested in their care" [15]. It is through art therapy that service members find a "visual voice", using imagery, symbolism, and metaphor to externalize that for which they may not be able to find the words [4, 16]. Walker et al. [15] add that while the objective findings of conventional imaging technologies such as fMRI offer useful information on brain structure and function, they do not capture the unique, personalized injury experience of individuals with TBI—noting that no two TBIs and the associated psychological occurrences are the same. The authors elaborate on the benefit of individuality in art therapy, stating "the creative and psychotherapeutic processes provide patients with the opportunity to freely express symptoms and associated trauma, which are not merely physical in nature. The ability to express specific experiences, emotions or memories and the associated granularity/richness of those subjective psychological elements cannot fully be expressed with standard clinical testing… The "open canvas" concept of art therapy, therefore, offers a more robust and holistic option for describing trauma and disability in a respectful and dignified manner [2, 15, p. 185]". The art therapy products (artwork) help healthcare providers identify the nature of the comorbidity of TBI/PTSD, as well as common themes SMs are most focused on during treatment. Art at its core exists to communicate, to help bridge gaps and further our understanding of each other, in turn strengthening community. These themes seem to surface time and time again and in our findings are those of the importance of community and sense of purpose [2].

When SMs come together as a team to achieve a common goal, whether to destroy or to create, they reportedly feel stronger and healthier. In *Tribe: On Homecoming and Belonging*, Junger [17] quotes Lyons [18], stating "When people are actively engaged in a cause their lives have more purpose… with a resulting improvement in mental health. It would be irresponsible to suggest violence as a means of improving mental health, but the Belfast findings suggest people will feel better psychologically if they have more involvement with their community." Group art therapy is observed to be particularly effective with SMs because it integrates the strength of the community-based military culture to benefit the individual participant [19, 20]. A mask currently hung in the art therapy studio at the NICoE includes the *Young Guns* movie quote, "See, if you got three or four good pals, why, you got yourself a tribe. There ain't nothin' stronger than that." Indeed, recent research has shown that isolation is a significant indicator for the development of depression in veterans, with perceived stress a mediator which feeds both [21]. Likewise, blast-induced TBI in active duty led to isolation and sad feelings, mixed with the challenge of accepting one's injuries. At the NICoE, art therapy products are displayed proudly by the service members, creating a visual community which encourages dialogue between the SMs and their families, providers, and peers—and when highlighted by the media, society [22, 23]. Art therapy serves a unique purpose in providing a means to help the internal emerge into the external—bringing to light the lonely psyche of individuals so that they, and others, can make meaning of it all. Ultimately, the art therapy products make the invisible wounds visible and may provide visual evidence and assessment of brain functioning, which will be discussed in the forthcoming text.

Art therapy serves a unique purpose in providing a means to help the internal emerge into the external—bringing to light the lonely psyche of individuals so that they, and others, can make meaning of it all. Ultimately, the art therapy products make the invisible wounds visible and may provide visual evidence and assessment of brain functioning…

It has been observed that art therapy helps SMs safely express and work through identity-related and emotional struggles from psychological and physical wounds of war, including the ability to integrate and process fragmented and sensory memories resulting in trauma via symbolic meaning-making in their artwork [4, 14, 24–26]. Individual and group art therapy have been found effective in helping TBI and PTSD patients with emotional expression, socialization, emotional adaptation to mental and physical disabilities, communication in a creative and nonthreatening way [27–29], and an increase in positive mood via reward perception through the creation of a meaningful product [30, 31]. Kline [32] points to the value of art therapy in supporting brain plasticity and highlights the art therapist's ability to foster a safe and supportive environment while stressing the need for flexibility in TBI treatment. King [33] further emphasizes this need and the role of art-making in TBI, stating "Creative Arts Therapists observe how creative expression in the context of the therapeutic relationship promotes the capacity for the brain to balance itself into a homeostasis… We might observe the neuroplastic pathways of creative expression more closely by looking at the compensatory functions found through artistic expression following a brain injury" (p. 1428). Art therapy cannot be defined without mention of the importance of the therapist–patient relationship, which is present in both Kline and King's observations. Related fields working closely with the military TBI population in a therapeutic context are studying the impact of the relational aspects of treatment. A recent publication presents findings which suggest SMs' evaluations of occupational therapy were based on the overall experience of the clinician/patient encounter, centered by the therapeutic relationship. The relationship proved more important to the SMs than the interventions or technology used [34].

2 Art Therapy Programming at the NICoE

At the NICoE, art therapy treatment is delivered in both group and individual sessions throughout the four-week IOP in a designated clinical, studio space [4, 14, 26]. A maximum of six SMs are admitted weekly and participate in standardized and templated art therapy programming which is woven throughout the SMs' interdisciplinary care. All SMs partake in two-hour group art therapy sessions to include mask-making in the first week [2, 4, 15, 26], and montage painting in the fourth week [19, 4, 26], as well as optional open studio art therapy sessions in their third and fourth weeks. The SMs each receive at least one individual art therapy session in their second week, which are tailored to the SMs' individual care needs and often involve the continuation of the creation and processing of the mask-making products began in group therapy. Follow-up sessions may be requested for scheduling by the SMs, art therapists, and/or treatment team. The group mask-making sessions encourage group cohesion, empathy, and mutual support as the SMs bear witness and affirm each other's feelings and experiences, re-creating a sense of belonging [4, 20], while the one-on-one sessions are more individualized and help SMs process personal grief, loss, trauma, and identity issues which they might not feel comfortable sharing in a group setting [4, 26]. It is in these sessions the art therapists meet the SMs where they are in care and allow the SMs to help navigate what they would like to explore in treatment, whether it be through choosing the directive or having the freedom to express whatever content they'd like through art-making or in the art therapy product itself.

In Howie's book, *Art Therapy with Military Populations: History, Innovation, and Applications*, Walker describes the rationale for the integration of these art therapy directives (2017). Likewise, Jones et al. [4], further describe the successful implementation and adaptation of the NICoE IOP art therapy model into the Intrepid Spirit One (ISO), a NICoE satellite center, at the Fort Belvoir Community Hospital beginning in 2013. Jones shares a structure in which SMs move from IOP to longitudinal care "levels" in art therapy, allowing for deeper work surrounding trauma-processing and also encouraging SMs to integrate the arts into their everyday lives, as well as seek out arts opportunities in the community. Both facilities incorporated program evaluation during their creation in order to effectively tweak directives and timing to suit the needs of the SM population. In a NICoE Walter Reed survey collected 2012 November–2014 June, SMs (n = 358) were asked to indicate which treatment techniques or tools they found most helpful in improving their recovery, SMs ranked art therapy amongst their top five out of 30+ techniques/tools. Likewise, at the Fort Belvoir ISO, the art therapist collected surveys at the end of each level of art therapy care in order to help refine clinical practice and provide helpful information to other healthcare providers. SMs' perceptions of the contribution of art therapy, particularly its role in identity integration and sense of self, indicated change in clinical symptoms including reduced flashbacks and nightmares, awareness of the impact of PTSD and TBI on the self, and the ability to experience positive emotions and find meaning in life [4]. These outcomes encouraged the NICoE to begin to explore how and why SMs were reporting these perceived benefits of art therapy treatment.

In 2013, the NICoE Healing Arts Program and the NICoE Research Department began a collaboration with Drexel University's Ph.D. in Creative Arts Therapies Department in order to explore and publish outcomes of the creative arts therapies treatments for the comorbid TBI and PH population. The importance and power of clinician–researcher partnerships will be evident as the discovered outcomes are discussed. A Walter Reed IRB-approved umbrella protocol allowed for the retrospective analysis of any standardized treatment data collected at the NICoE. This included all art therapist clinician notes documented in the Armed Forces Health Longitudinal Technology Application (AHLTA) which summarize clinician-SM encounters and incorporate images of the art therapy products. The protocol also made it possible for clinicians to correlate art therapy outcomes with other incoming and outgoing assessment data, such as the PTSD checklist-military and civilian (PCL-M/C), Patient Health Questionnaire-9 (PHQ-9), and Generalized Anxiety Disorder 7 (GAD-7) self-report measures, while also taking into consideration service member demographics such as branch, rank, age, ethnicity, and sex.

3 Art Therapy Research Outcomes

The first publication to come out of the collaboration is a detailed case study of a senior-ranking service member who self-reported relief from PTSD symptoms as a result of integrative care treatment including but not limited to psychiatry, neurology, acupuncture, and art therapy [14]. Mask-making, in particular, seemed to benefit the SM as it offered a means for the externalization of an intrusive image of a bloody face which had been "haunting" him (Fig. 2). Walker and team [14, 22] describe the shutdown in the Broca's (speech/language) area of the brain in PTSD patients, which was also evident in the SMs' magnetoencephalography (MEG) neuroimaging scans upon admission, seen on the MEG as reduced alpha activity (see Fig. 3). The deactivation of an area of the brain said to be responsible for semantic representation of personal experience [35–37], and block of the neurologic pathways in speech production, leads to what is often referred to as "speechless terror" [38]. Sensory areas of the SMs' brain saw improvement during the duration of the treatment, and the SM continued to use art therapy to psychotherapeutically work through traumatic content with the therapist. These findings support Gantt and Tinnin's [39] assertion that art therapy can bypass the

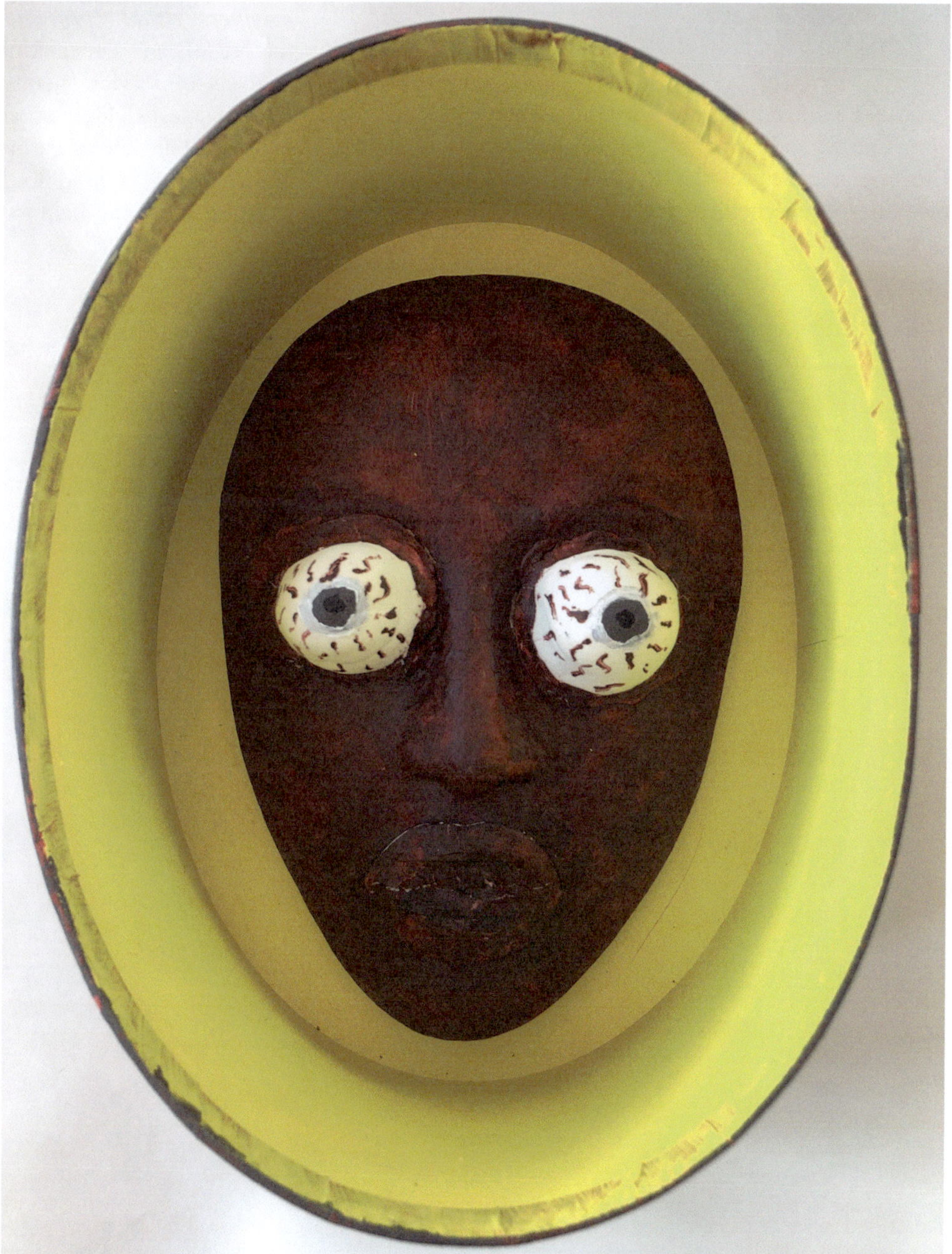

Fig. 2 Mask depicted by SM of intrusive psychological image [14]. License number: 4566560796650

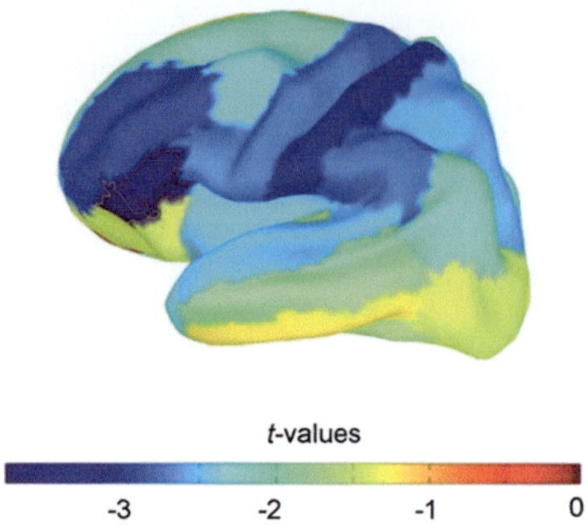

Fig. 3 MEG scan of case study subject upon admission, with darker blue regions indicating reduction in brain wave activity in frontal and temporal lobes [14]. License number: 4566560796650

Broca's block and provide a way to communicate via nonverbal visual and sensory memories associated with trauma. Creating a cohesive narrative out of fragmented trauma memories in psychotherapy has been found to be beneficial for the TBI/PTSD population [40, 41], and the VA currently reports that 53 out of 100 individuals who receive trauma-focused psychotherapy will no longer meet the criteria for PTSD [42]. After receiving treatment at the NICoE the SM stated, "I would continue to make paintings of my hauntings, and each time I see them less, or not at all. In my opinion, I am bringing some compartmentalized fear into the open… [14]." This case study also emphasized the importance of the therapeutic relationship, as well as benefits of interdisciplinary care and collaboration across healthcare and scientific disciplines.

Throughout the past eight years, close to 2000 masks have been created at the NICoE by both the SMs and their families. In the empirical study *Active-duty military service members' visual representations of PTSD and TBI in masks,* the NICoE and Drexel University researchers published an analysis of service members' experiences (n = 370) with making masks, looking closely at the mask products as behavioral health assessments themselves. In art therapy, mask-making is used to improve personal orientation in psychotherapy and help to process identity [43, 44].

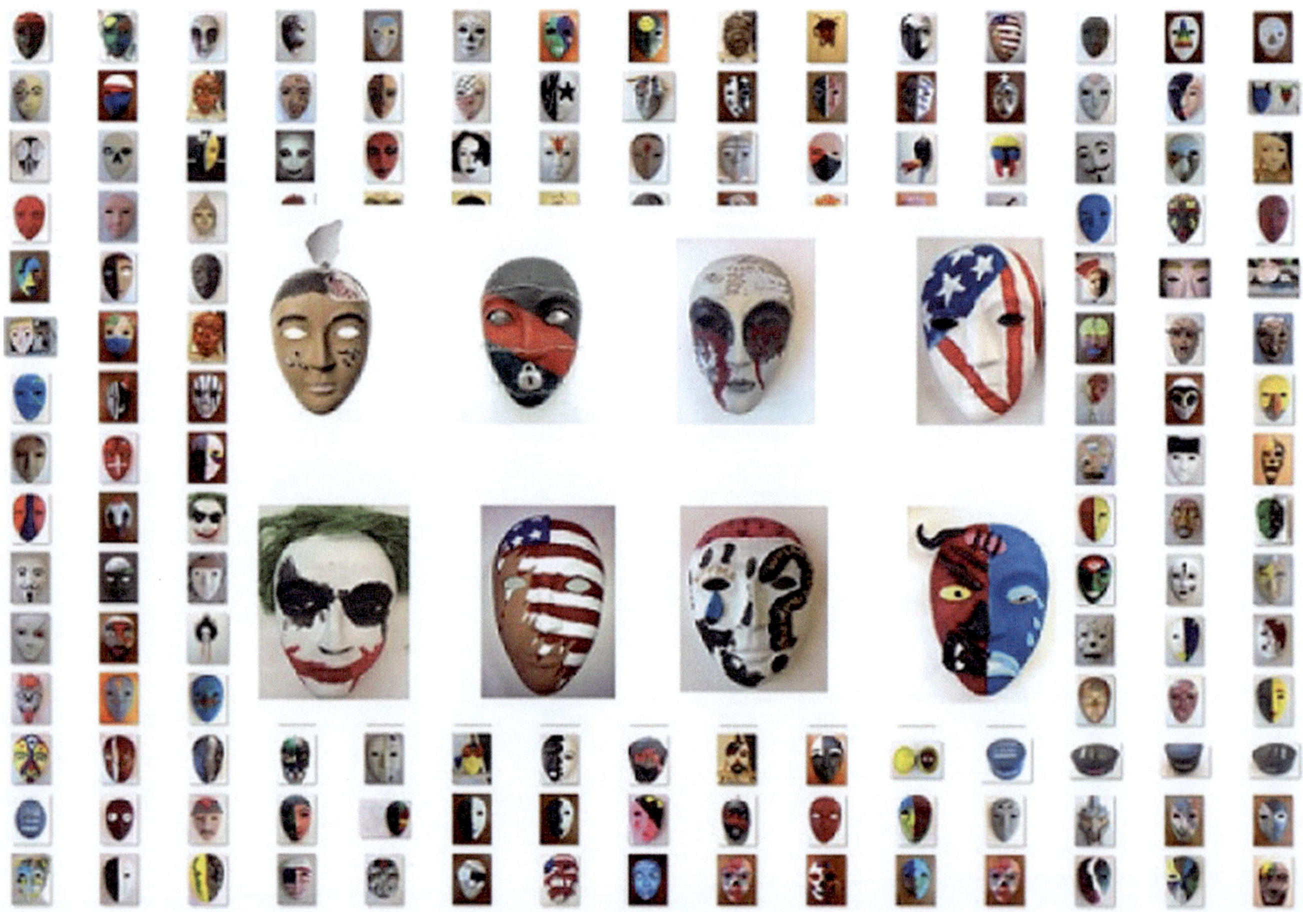

Fig. 4 Examples of themes found in the mask analysis (top left to right: physical injury, psychological injury, mourning and loss, military community; Bottom left to right: cultural references, moral injury, transitions, divided sense of self). Creative commons license CC BY-NC 4.0

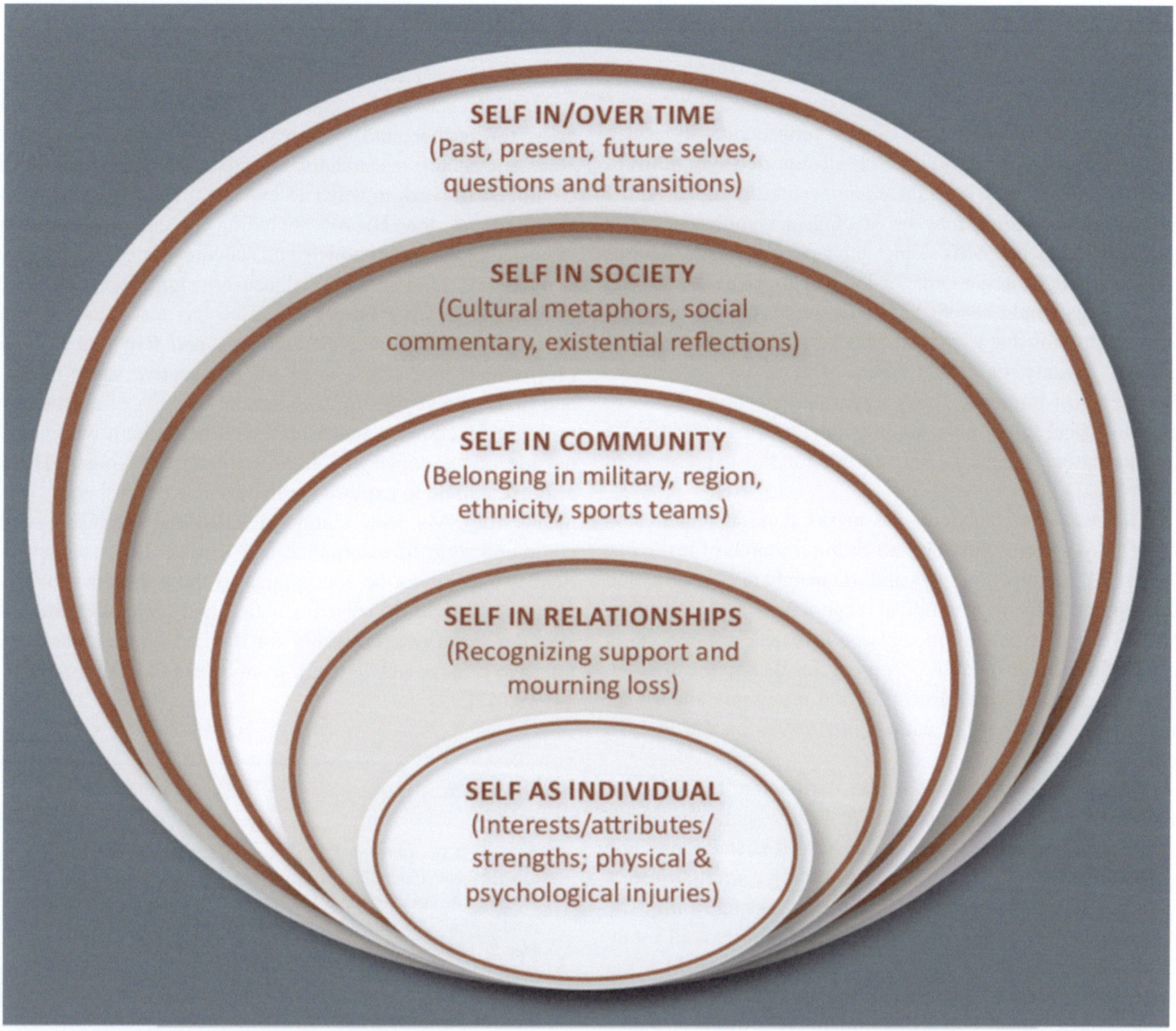

Fig. 5 Framework of representations of self [2] as adapted for the Creative Forces Clinical Research Agenda and Framework (NEA [46]. Creative commons license CC BY-NC 4.0

Kroch [5] identifies several themes in military-related PTSD which also seem to surface in the masks, including trauma remembering; encountering death; hypervigilance in an unsafe world; dualistic psychological ideas (inner vs. outer worlds; public vs. private life; night vs. day); and feeling alien to oneself and others [2]. Using a grounded theory approach [45], the researchers coded and categorized the masks using the mask images as well as clinician notes. Recurring themes were discovered, including: references to physical, psychological, and moral injuries; grief for lost abilities and lost comrades; struggles with transitions and questions about future; and a divided sense of self and disillusionment with their role in and outcomes of war (Fig. 4) [2]. Figure 5 provides examples of themes found in the masks, integrated into a theoretical framework of how participants visually represented aspects of the self within the mask products. The findings of this study highlighted the "unseen" struggles of SMs with comorbid TBI and PTSD, offering an avenue to understand service members' experiences outside of narrative description alone [2].

Of great interest to the researchers was whether or not the thematic patterns of the masks correlate to SMs' standardized clinical self-report measures. In the *Observational study of associations between visual imagery and measures of depression, anxiety, and post-traumatic stress among active-duty military service members with traumatic brain injury at the Walter Reed National Military Medical Center* [47], researchers examined participants' experiences in art therapy and associations between the visual imagery in the masks and clinical data from standardized measures of post-traumatic stress (PCL-M), depression (PHQ-9) and anxiety (GAD-7).

A strong correlation was found between masks which depicted psychological injury and self-reported symptoms of post-traumatic stress, with the creators scoring higher on the PCL-M. This finding may be of clinical significance, as it suggests that when SMs depict their psychological injuries clinicians might target care for the SM to address potential PTSD. This would benefit SMs who are less likely to report their symptoms due to mental health stigma, or the threat of change in duty or career status. The strong association between post-traumatic stress scores and visual depiction of psychological injury also suggests that art therapy might be an alternative avenue or safe forum for the expression of trauma [47].

The use of metaphor in the masks correlated with less anxiety symptoms, possibly indicating a source of resilience when SMs are able to reflect and use insight and imagination to explore their psychological experiences and identities. However, subtypes of metaphors revealed differences in correlations in other self-report measures than in the GAD-7. The use of color symbolism (i.e., red represents anger), for instance, was associated with higher scores on the PCL-M and PHQ-9 [47].

Perhaps most telling was the correlation between the representation of belonging via cohesive symbols of the military unit with lower post-traumatic stress and depression scores, supporting previous findings stating a strong sense of community is a protective factor for SMs' mental health. The development of group identity in the military is well established as a means to ensure trust and effectiveness in a war zone through shared commitment and social cohesion, and studies have shown that social support is a strong negative predictor of post-traumatic stress and depression symptoms [47, 48]. Conversely, the representation of fragmented military symbols in the masks (i.e., tattered/torn flags, pieces of camo fabric) correlated with higher anxiety scores, indicating an uneasiness regarding the impact of their injuries on their lives—feeling "broken"—as well as the discomfort caused by moral injury. Other findings included strong a correlation between the use of nature metaphors and lower PCL-M scores, and historical and cultural character/symbol references correlated with lower PHQ-9 and GAD-7 scores. The findings from this study establish the foundation of a framework for how psychological states might be represented through creative self-expression, and also assist clinicians in identifying sources of strength/protective and of risk factors for SMs with TBI and PTSD [47].

4 Conclusion

Ultimately, each study described built upon findings from the last and recognized common threads as well as several areas for further study. Since conferences proceedings in Valencia, Spain, researchers, scientists, and clinicians have continued to work together to explore art therapy treatment outcomes at the NICoE including potential correlation between art therapy products and specific clinical trajectories of recovery, as well as SMs' neuroimaging data [15, 19]. The NEA Creative Forces initiative has also developed a clinical research strategic framework and five-year agenda [46], which indicates a need for prospective, quantitative, and multi-site studies to examine the benefits of art therapy on brain functioning and overall well-being both within and outside of the context of interdisciplinary treatment. These efforts continue to explore the importance of art therapy as a means for SMs with comorbid TBI/PTSD to use visual communication to externalize and process traumatic experiences that cannot be communicated through verbal means alone, leading to decreased isolation and perhaps creating a stronger sense of community for the SM—which has been found to be a protective factor from PTSD and depression [4, 14, 19, 47].

References

1. Wenger, J.W., O'Connell, C., Cottrell, L.: Examination of Recent Deployment Experience Across the Services and Components. RAND Corporation, RR-1928-A, Santa Monica, CA. https://www.rand.org/pubs/research_reports/RR1928.html (2018). Accessed 9 Dec 2018
2. Walker, M.S., Kaimal, G., Gonzaga, A.M.L., Myers-Coffman, K. A., DeGraba, T.J.: Active-duty military service members' visual representations of PTSD and TBI in masks. Int. J. Qual. Stud. Health Well-Being **12**(1), 1267317 (2017)
3. Yurgil, K., Barkauskas, D., Vasterling, J.J., Nivergelt, C., Larsan, G.E., Schork, N., Baker, D.G.: Association between traumatic brain injury and risk of posttraumatic stress disorder in active-duty marines. JAMA **71**(2):149–157 (2014)
4. Jones, J.P., Walker, M.S., Drass, J.M., Kaimal, G.: Art therapy interventions for active duty military service members with post-traumatic stress disorder and traumatic brain injury. Int. J. Art Therapy (2017). https://doi.org/10.1016/j.psychres.2018.08.019. Epub 2018 Aug 11
5. Kroch, R.: Living with military-related posttraumatic stress disorder (PTSD)-a hermeneutic phenomenological study. Dissertation, University of Calgary (2009)
6. Aase, D.M., Babione, J.M., Proescher, E., Greenstein, J.E., DiGangi, J.A., Schroth, C., Kennedy, A.E., Feeley, S., Tan, M., Cosio, D., Phan, K.L.: Impact of PTSD on post-concussive symptoms, neuropsychological functioning, and pain in post-9/11 veterans with mild traumatic brain injury. Psychiatry Res. **268**, 460–466 (2018). https://doi.org/10.1016/j.psychres.2018.08.019

7. Defense and Veterans Brain Injury Center (DVBIC): DoD Worldwide Numbers for TBI. https://dvbic.dcoe.mil/files/tbi-numbers/worldwide-totals-2000-2018Q1-total_jun-21-2018_v1.0_2018-07-26_0.pdf (2017). Accessed 9 Dec 2018
8. Committee on the Assessment of Ongoing Efforts in the Treatment of Posttraumatic Stress Disorder, Board on the Health of Select Populations: Treatment for Posttraumatic Stress Disorder in Military and Veteran Populations: Final Assessment. Institute of Medicine, National Academies Press (US), Washington, DC. https://www.ncbi.nlm.nih.gov/books/NBK224874/#sec_00047 (2014). Accessed 9 Dec 2018
9. American Psychological Association (APA): Clinical Practice Guideline for the Treatment of PTSD, pp. 7–8. https://www.apa.org/ptsd-guideline/ptsd.pdf (2017). Accessed 9 Dec 2018
10. Bahraini, N.H., Breshears, R.E., Hernández, T.D., Schneider, A.L., Forster, J.E., Brenner, L.A.: Traumatic brain injury and posttraumatic stress disorder. Psychiatr. Clin. North Am. **37**(1), 55–75 (2014)
11. National Intrepid Center of Excellence (NICoE): Intensive Outpatient Program. http://www.wrnmmc.capmed.mil/NICoE/healthcare-services/SitePages/intensive-outreach-program.aspx (n.d.). Accessed 9 Dec 2018
12. National Endowment for the Arts (NEA): Creative Forces Research and Scholarly Manuscripts Inventory. https://www.arts.gov/sites/default/files/Creative-Forces-Research-Inventory-Version-8.13.18.pdf (2018). Accessed 9 Dec 2018
13. American Art Therapy Association: About Art Therapy. https://arttherapy.org/about-art-therapy/ (n.d.). Accessed 9 Dec 2018
14. Walker, M.S., Kaimal, G., Koffman, R., Degraba, T.J.: Art therapy for PTSD and TBI: a senior active duty military service member's therapeutic journey. Arts Psychother. **49**, 10–18 (2016)
15. Walker, M.S., Stamper, A., Nathan, D.E., Riedy, G.: Art therapy and underlying fMRI brain patterns in military TBI: a case series. Int. J. Art Therapy **23**(4), 180–187 (2018). https://doi.org/10.1080/17454832.2018.1473453
16. Stone, A.: How Art Heals the Wounds of War. National Geographic. https://news.nationalgeographic.com/news/2015/02/150213-art-therapy-mask-blast-force-trauma-psychology-war/ (2015). Accessed 9 Dec 2018
17. Junger, S.: Tribe: [on homecoming and belonging]. HarperCollins Publishers, New York (2016)
18. Lyons, H.A.: Civil violence—the psychological aspects. J. Psychosom. Res. **23**(6), 373–393 (1979)
19. Berberian, M., Walker, M.S., Kaimal, G.: 'Master My Demons': art therapy montage paintings by active-duty military service members with traumatic brain injury and post-traumatic stress. Med. Humanit. **0**:1–8 (2018). https://doi.org/10.1136/medhum-2018-011493
20. Herman, J.: Trauma and recovery: the aftermath of violence- from domestic abuse to political terror, p. 214. Basic Books, New York, NY (1992)
21. Martin, J.A., Hartley, S.L.: Lonely, stressed, and depressed: the impact of isolation on U.S. Veterans. Mil. Behav. Health **5**(4):384–392 (2017). https://doi.org/10.1080/21635781.2017.1333064
22. TED: Art can Heal PTSD's Invisible Wounds. TEDMED 2015. https://www.ted.com/speakers/melissa_walker (2016). Accessed 9 Dec 2018
23. Alexander, C.: Healing our soldiers. National Geographic Magazine, Single issue February 2015 (2015)
24. Collie, K., Backos, A., Malchiodi, C., Spiegel, D.: Art therapy for combat-related PTSD: recommendations for research and practice. Art Therapy J. Am. Art Therapy Assoc. **23**(4), 157–164 (2006)
25. Lobban, J.: Factors that influence engagement in an inpatient art therapy group for veterans with post traumatic stress disorder. Int. J. Art Therapy Formerly Inscape **21**(1), 15–22 (2016). https://doi.org/10.1080/17454832.2015.1124899
26. Walker, M.: Art therapy approaches within the National Intrepid Center of Excellence at Walter Reed National Military Medical Center. In: Howie, P. (ed.) Art Therapy with Military Populations: History, Innovation, and Applications, pp. 111–123. Routledge, London, UK (2017)
27. Barker, V.L., Brunk, B.: The role of a creative arts group in the treatment of traumatic brain injury. Music Therapy Perspect. **9**(1), 26–31 (1991)
28. Dodd, F.G.: Art therapy with a brain-injured man. Am. J. Art Therapy **14**(3), 83–89 (1975)
29. Lazarus-Leff, B.: Art therapy and the aesthetic environment as agents for change. J. Am. Art Therapy Assoc. **15**(2), 120–126 (1998)
30. Chilton, G., Gerber, N., Bechtel, A., Councill, T., Dreyer, M., Yingling, E.: The art of positive emotions: expressing positive emotions within the intersubjective art making process. Can. Art Therapy Assoc. J. **28**(1), 12–25 (2015)
31. Kimport, E.R., Robbins, S.J.: Efficacy of creative clay work for reducing negative mood: a randomized controlled trial. Art Therapy J. Am. Art Therapy Assoc. **29**(2), 74–79 (2012)
32. Kline, T.: Art therapy for individuals with traumatic brain injury: a comprehensive neurorehabilitation informed approach to treatment. Art Therapy J. Am. Art Therapy Assoc. **33**(2), 67–73 (2016)
33. King, J.L.: Summary of twenty-first century great conversations in art, neuroscience and related therapeutics. Front. Psychol. **9**, 1428 (2018). https://doi.org/10.3389/fpsyg.2018.01428
34. Cogan, A.M., Huang, J., Philip, J.: Military service member perspectives about occupational therapy treatment in a military concussion clinic. OTJR (2018). https://doi.org/10.1177/1539449218813849
35. Peres, J.F., McFarlane, A., Nasello, A.G., Moores, K.A.: Traumatic memories: bridging the gap between functional neuroimaging and psychotherapy. Aust. N. Z. J. Psychiatry **42**(6), 478–488 (2008)
36. Rauch, S.L., Savage, C.R., Alpert, N.M., Fischman, A.J., Jenike, M.A., et al.: The functional neuroanatomy of anxiety: a study of three disorders using positron emission tomography and symptom provocation. Biol. Psychiat. **42**, 446–452 (1997)
37. Shin, L.M., McNally, R.J., Kosslyn, S.M., Thompson, W.L., Rauch, S.L., Alpert, N.M., et al.: A positron emission tomographic study of symptom provocation in PTSD. Ann. N. Y. Acad. Sci. **821**, 521–523 (1997)
38. van der Kolk, B.A.,van der Hart, O., Marmar, C.R.: Dissociation and information processing in post traumatic stress disorder. In: van der Kolk, B.A., McFarlane, A.C., Weisaeth, L. (eds.) Traumatic stress: the effects of overwhelming experience on mind, body and society, pp. 303–327. New York, NY: Guilford Press (1996)
39. Gantt, L., Tinnin, L.W.: Support for a neurobiological view of trauma with implications for art therapy. Arts Psychother. **36**, 148–153 (2009)
40. Howie, P., Burch, B., Conrad, S., Shambaugh, S.: Releasing trapped images: children grapple with the reality of the September 11 attacks. Art Therapy J. Am. Art Therapy Assoc. **19**, 100–105 (2002)
41. Lobban, J.: The invisible wound: Veterans' art therapy. Int. J. Art Therapy Formerly Inscape **19**(1), 3–18 (2014)
42. US Department of Veterans Affairs (VA): PTSD: National Center for PTSD. https://www.ptsd.va.gov/ (n.d.). Retrieved 9 Dec 2018

43. Hinz, L.D., Ragsdell, V.: Using masks and video in group psychotherapy with bulimics. Arts Psychother. **17**, 259–261 (1990). https://doi.org/10.1016/0197-4556(90)90009-F
44. Fryrear, J.L., Stephens, B.C.: Group psychotherapy using masks and video to facilitate intrapersonal communication. Arts Psychother. **15**, 227–234 (1988). https://doi.org/10.1016/0197-4556(88)90007-X
45. Strauss A., Corbin J.: Basics of qualitative research: techniques and procedures for developing grounded theory, 2nd edn. Thousand Oaks, CA: Sage (1998)
46. National Endowment for the Arts (NEA): Creative Forces Clinical Research: A Strategic Framework and Five-year Agenda. https://www.arts.gov/sites/default/files/CF-Clinical-Research-Framework-and-Agenda-10.23.18.pdf (2018). Accessed 9 Dec 2018
47. Kaimal, G., Walker, M.S., Herres, J., French, L., DeGraba, T.J.: Observational study of associations between visual imagery and measures of depression, anxiety and posttraumatic stress among active duty military service members with traumatic brain injury at the Walter Reed National Military Medical Center. BMJ Open **8**, e021448 (2018). https://doi.org/10.1136/bmjopen-2017-021448
48. James, L.M., Van Kampen, E., Miller, R.D., Engdahl, B.E.: Risk and protective factors associated with symptoms of post-traumatic stress, depression, and alcohol misuse in OEF/OIF veterans. Mil. Med. **178**(2), 159–165 (2013)

Melissa S. Walker ATR, is the Healing Arts Program Coordinator at the National Intrepid Center of Excellence, US, where she implemented and researched creative arts therapies treatment programming for military members with TBI and associated psychological health conditions. Melissa is also Lead Art Therapist for the NEA Creative Forces initiative. http://orcid.org/0000-0001-9375-567X.

Brain on Art Therapy-Understanding the Connections Between Facilitated Visual Self-expression, Health, and Well-Being

Girija Kaimal

1 Artistic Influences: Nature, Nurture, and Heritage

When I was a child, we lived in India and my mother, as was typical for her generation and upbringing, was a stay-at-home wife and mother. It was not her choice, but life circumstances precluded her from pursuing many professional dreams. Despite (or perhaps because of) these limitations, my mother found outlets for creative self-expression around the home. She often wrote little narrative notes and stashed them away in her jewelry case, embroidered pillow cases, sewed clothes, and knitted sweaters for my brothers and me. In addition to these pragmatic creations, she did one thing that has always stayed in my mind: She would take the tops of carrots that were being prepared for cooking and place them in a shallow bowl with just enough water to cover the tops. Slowly the carrot tops would sprout the most beautiful leafy structures, and these would grow to be a few inches tall: fine, bright green leaflets that brought a contrast to the orange carrot top and looked particularly stunning when filtered through the sunlight on the kitchen window sill. This art installation although aesthetically beautiful, served no practical purpose. It was simply an expression of creativity: an up cycling through creative re-creation of kitchen vegetable waste. These early exposures to creative acts taught me that art could be a part of everyday life, an outlet for self-expression, and that aesthetic beauty generated from the simplest of natural sources could bring joy.

I went on to learn many of the art forms practiced by my mother at home, and other traditional forms of expression including dance and art-making were a part of my life for as long as I can remember. When I was a child, I frequently missed school because of a series of illnesses, but my artwork was there in my stead. Working with crayons and paper at that time was my way to communicate with adults around me. Art went on to play a dominant role in my life when I won recognition in art shows and design school. Art then took on a different role when I went on to get my master's degree in art therapy. It further expanded in scope from artistic practice and personal wellness to professional research as I got my doctorate and examined how art and visual story-telling narratives relate to all aspects of human development.

In my present work in the arts, health, human development, and well-being, I am intentionally trying to systematically understand what I might have implicitly sought and experienced in earlier years. The questions I explore include the following: Is art related to beauty and to that which gives us joy? Is beauty essential like food, drink, and social connection? Is it the pursuit of happiness manifested in different forms in all parts of the world? Could the desire for beauty, to be surrounded by elements of nature, be an innate force, a way to be? Is beauty essential for a good life? Do we instinctively create, seek, and replicate what is beautiful to us?

2 Receptive and Expressive Art-Making Experiences

The experience of art-making is ubiquitous in human society and can broadly be divided into two categories: (1) receptive experiences, such as those that involve viewing or experiencing art and (2) expressive experiences such as those that relate to creating or making an art product. But what is the purpose of either aspect? Why do we choose to view or make art? These questions remain largely unanswered and continue to intrigue researchers.

Several scholars have examined the purpose and role of art in the human life. Some evolutionary theorists [13] argue that art is simply a by-product of the human brain's expanded processing capacity, a spandrel that is created simply as a result of improved cortical abilities. This

G. Kaimal (✉)
Creative Arts Therapies Department, Drexel University, Philadelphia, PA, USA
e-mail: gk27@drexel.edu

J. L. Contreras-Vidal et al. (eds.), *Mobile Brain–Body Imaging and the Neuroscience of Art, Innovation and Creativity*, Springer Series on Bio- and Neurosystems 10, https://doi.org/10.1007/978-3-030-24326-5_13

hypothesis is rejected by most scholars who argue that art has several specific roles in society. Dissanayake [9] has referred to arts engagement and art-making as a process of *making special* the everyday and the ordinary by commemorating special events, milestones, and landmarks in the life of an individual or community. Based on observations of traditional communities and viewing art from an anthropological perspective, she argued that art-making is part of human development and history and is not separate and distant, as we have considered the role of the modern artist as the individual expert with works placed in museums. In examining the historical context and role of art-making, Dutton [11] further argued that art-making is not a cultural artifact; rather, it is part of human evolution and a means to capture the imaginative qualities inherent in human functioning. He referred to art not as a technical concept confined to a cultural context but rather as a universal phenomenon of human evolution like language, tool making, and kinship systems. He defined a piece of art as having 12 essential qualities. These include (1) direct pleasure, (2) skill and virtuosity, (3) style, (4) novelty and creativity, (5) criticism, (6) representation, (7) special focus, (8) expressive individuality, (9) emotional saturation, (10) intellectual challenge, (11) art traditions and institutions, and (12) imaginative experience. He argued that the arts are different from activities like sports where there is typically no imaginative experience because the end result of a win or loss is what guides interest in the game. It can be argued though that this is not necessarily a valid critique because games and plays can also be demonstrative of imaginative variations within the core construct of a winning score or loss. Dutton [11] further asserted that art is embedded in a context but is not necessarily always reflective of that context; rather it is the creation of individuals or communities who happen to live in that moment. Recent scholarship in human development and education [16] suggests that working with the visual arts helps us develop a craft (learn to use and care for tools); engage and persist (to stay with a task and persevere to complete it); envision (imagine possibilities not yet seen); express (convey ideas nonverbally); observe (learn to see effectively); reflect (learn to think through with self and others); stretch and explore (learn from mistakes); and learn the artistic practice and professions.

Child developmental theorists like Viktor Lowenfeld have identified stages in artistic development similar to those in cognitive and psychosocial development. In examining millions of drawings, a fairly universal trend was found [28]. Children started with scribbling (up to 2–3 years of age), moved on to create simple images of faces with arms and legs coming out of the head (2–4 years), then added additional details of the neck and formed body parts (4–6 years), added the environment using baselines and skies (6–8) years, added scenes with people and places (age 8 onwards). Art therapists have noted from clinical experience that many children move away from drawing around the age of puberty because they become increasingly critical of their artistic skills and choose not to continue drawing if they do not perceive themselves as being skilled in the visual arts.

3 Art-Making and the Predictive Brain

A common popular perception of the brain is that it is analogous to the computer: accepting and processing information received from the five senses. It is increasingly accepted now that the human brain is not analogous to the computer but is rather a prediction machine [5]. To maximize survival options, the brain is inherently wired to imagine possibilities for the future that enhance safety and resources and minimize risk and danger. We also know that the brain is wired to understand and create stories [12, 36], which is possibly a mechanism to problem solve, learn vicariously, and retain relevant lessons and information. Stories typically follow a chronological sequence where preceding events lead to culminating events and there is a resolution of meaning generated at the end. As human beings, we have an ongoing script for our own stories with new incidents adding to, refining, or defining our stories. Depending on our developmental history and life experiences, we might generate a story that aligns with our interaction with the world. For individuals who are facing stress and adversity, both acute and chronic, the story has the potential to share recovery and resilience or trap the individual in nonrestorative storying.

In brain imaging studies, investigators have demonstrated the activation of the prefrontal cortex during visual arts activities. For example, Chamberlain et al. [6] used magnetic resonance imaging (MRI) scans to study the brain regions associated with drawing skills and artistic training. Their findings suggested that being able to draw from observation was associated with an increase in gray matter density in the left anterior cerebellum and the right medial frontal gyrus in the prefrontal cortex. Schlegel et al. [32] showed that 3 months of art training resulted in changes in prefrontal white matter. In this study of youth who were art students, art-making was associated with plasticity in neural pathways, increases in creative cognition, and to mediate perceptuomotor integration. Bolwerk et al. [3] found a clear difference between producing art compared to viewing art. Visual art production was shown to improve the functional connectivity in several brain areas, particularly between the parietal and frontal cortices, as well as to cause psychological resistance to change [3]. Although these findings suggest that visual art production results in stronger brain connectivity than cognitive art evaluation or viewing art, evidence shows that even passive engagement in art affects the prefrontal cortex [3]. For example, when a person is viewing art, a reward circuitry is engaged that activates the ventral striatum, including the nucleus accumbens, along with the interconnected medial prefrontal cortex (mPFC) and the orbitofrontal cortex and amygdala [26]. Using functional MRI technology, Lacey et al. [26] found that art imagery alone activated the reward circuitry whereas matched non-art images did not. Likewise, activation of the mPFC, along with the rest of the reward circuitry, occurred while the individual was viewing beautiful visual images or architectural spaces [7].

In addition, it is well established that the visual and aural systems are the most developed of our senses (more so than smell, taste, and touch) [11]. Visual processing has been estimated to take up tremendous resources, and is considered our dominant sense including a dedicated area of the brain (occipital lobe) and specialized cells and pathways that track and process visual information [31]. Visual systems are hypothesized to have become a dominant sense especially since human beings evolved to be upright and could see considerable distances. As a result, visual expression, processing, and data could be a tremendous source of information about human experiences and mental states. Dutton asserted that "the greatest works of art are not necessarily the most novel or unusual. They do tend to be somehow the most personal... a strong sense of individual personality (p. 247). To me, this statement illustrates the implications of authentic self-expression, which is what art therapy seeks to do: encourage creative self-expression that represents the authenticity of the individual in the visual representation of the artwork and encourages sharing and re-storying the personal narrative towards health through the facilitative therapeutic relationship.

4 Art-Making in the Context of Art Therapy

Art therapy as a field of study originated in the twentieth century, simultaneously in many parts of Europe and America in response to the needs of clinical populations who were not being served effectively with traditional approaches to medicine and education. Art therapy developed most powerfully with military service members affected by post-traumatic stress syndrome (formerly referred to as "shell shock") and with children with developmental and behavioral challenges. Over the past several decades, art therapists have gone on to work with a range of populations including the elderly, those affected by adversity and violence, individuals facing discrimination and marginalization and relational, developmental, or psychosocial challenges.

The American Art Therapy Association defines art therapy as an integrative mental health and human services profession that enriches the lives of individuals, families, and communities through active art-making, creative processes, applied psychological theory, and human experience within a psychotherapeutic relationship [1]. Art therapists are masters-level trained clinicians who are proficient in art-making and in facilitating expression with a deep understanding of human psychology and psychopathology. There are now more than 6000 art therapists in the United States and 38 credentialed programs that prepare masters-level therapists. The majority of art therapists are clinicians, and the field now offers more than five doctoral programs to further deepen the research base in the field.

As such, we can argue that art therapy taps into the brain's innate predictive and narrative capabilities by enabling individuals to imagine alternate and potentially adaptive and healthy personal narratives. An art therapy session can include groups or individuals and can be short term (on average 1–2 sessions) or longer term (spanning few weeks to several years). Art therapy treatment, depending on patient/client goals, can last a few weeks to several years.

Art therapists serve a range of populations including those with mental and physical illnesses, developmental challenges, and differential abilities and those who have experienced adversity and trauma. A typical art therapy session involves art-making, review of the art product, and verbal processing (as applicable and as the patient/client is able) with a masters-level trained art therapy clinician. The sessions can include individual clients/patients or a group. Figure 1 highlights the components of art therapy including the therapist, the patient/client, the art-making process, and the art product. The components are intentionally shown as intersecting to highlight the interlinking of all four aspects in a session.

Art therapists are often asked how they differ from artists in residence, art educators, or arts facilitators. A core difference lies in the focus on the expressive process and facilitation of the session in a way that supports the development of the individual. Art therapy is distinguished from therapeutic art-making, which can be facilitated by non-clinicians to promote overall health and well-being. Note that a core assumption in art therapy is that everyone is an artist, and art is defined as visual self-expression. Art-making does not judge the aesthetic qualities in the traditional sense; rather art-making is treated as a form of self-expression that allows for communication, learning, and awareness. Art therapists are attuned to the psychosocial needs of patients and clients in the sessions and have a specific skill in facilitating artistic expression that promotes individual strengths. Although patients/clients are not expected to have artistic skills, artistic/expressive knowledge is essential to art therapists' clinical practice. This is one of the key differences between art therapists and artists/art educators who might be more focused on the artistry or quality of the artistic product. The art therapist typically is

Fig. 1 Components of art therapy

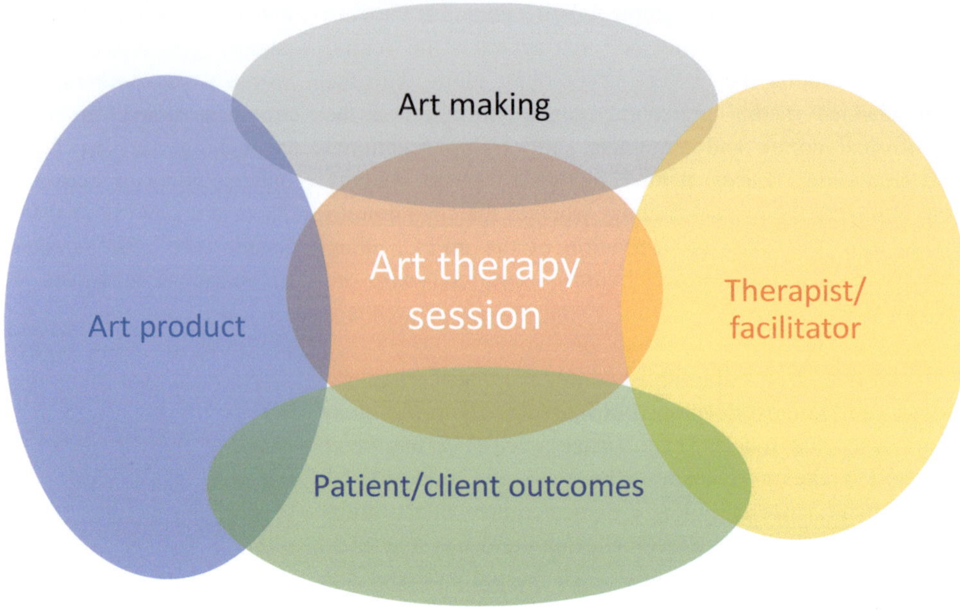

less concerned with the quality of the product than with the process and reflections on the process. The art therapy session is set up as a space that is nonjudgmental about the artistic product; it is a place to express oneself visually in order to learn about one's self and move towards adaptive choices and behaviors and reduced psychopathology in a safe space. The art therapist might work with the patients/clients in individual sessions or in groups based on treatment goals that might include interpersonal functioning, emotional regulation and awareness, physical functioning, or cognitive functioning (e.g., focus and memory). We know that when individuals go through disruptive abusive or traumatic experiences their ability to process sensory information effectively is hampered. We also know that the arts can evoke intense emotion and self-expression for highly stressed or psychologically vulnerable individuals; a lack of adequate support and facilitation can lead to harmful outcomes. Art therapists therefore work with individuals struggling with physical and psychological difficulties to imagine, explore, try out, and finally live out healthier and more adaptive lives.

A question then arises: Is art therapy only for individuals with mental illness conditions or can it be relevant to healthy or normal populations seeking options for well-being. Here I suggest a home-improvement analogy. Many things in the home, for example, can be fixed by the residents of that home; e.g., changing a lightbulb, cleaning and scrubbing dirty floors, drilling nails into walls. Some people might be skilled in doing things like replacing floors and painting walls. However, when we move into the realm of home improvement projects that require more experience and expertise, we are better off hiring an expert. For things like electrical work, roofing, and plumbing, unless we know what we are skilled in, we might do more harm than good by trying to fix these things ourselves. Applying this analogy to art therapy, I argue that some human beings are resourceful, resilient, and capable of taking care of aspects of themselves that need healing and restoration. Most of us can respond effectively to the everyday challenges and adversities of living. However, when the challenges are overwhelming, a trained expert, a clinician, can help address the challenges and problems more effectively than we can by trying to diagnose the problem and fix it properly. Even for things that we might be able to do adequately, working with an expert gives us exposure to the ways in which even simple tasks can be accomplished with beauty, proficiency, and effectiveness. Thus by analogy, if you have artistic skills, practice them by all means, but note that an art therapist can offer you a perspective and facilitate development in a way that cannot be accomplished without the expertise that he or she brings to the situation.

5 What Makes Art Therapy Therapeutic?

The therapeutic power of creative expression in an art therapy session lies in the multidimensionality of the arts: the nonlinearity and timelessness that allow the process to hold several metaphors, associations and meaning concurrently. This creative expression and the unlocking of the imagination lead to a sense of agency and possibility that might previously not have been available to a patient/client. Each condition or cluster of symptoms might result in different mechanisms and pathways of change.

The pathways and mechanisms of change vary by individual, their clinical needs, and the context of care. For example in the case of an individual struggling with feelings of incompetence and inefficacy, the process of art-making involves problem-solving and serves as a trial run for practicing the ability to gain mastery. Thus, for example, for individuals who have experienced trauma, verbal expression is often not an option [34, 35] since the verbal expressive part of the brain tends to be impacted [30]. Thus, being able to say the unsayable becomes really critical in such situations. In other cases with elderly individuals, social isolation might be an issue; thus the emphasis might be on social integration and expression through the art therapy process. Similarly, for an individual struggling with the ability to manage and contain emotional reactions and the process of effectively channeling these struggles through sublimating the emotions in the artwork (instead of harming self or others) in the art therapy session could be the therapeutic element. The therapeutic interaction and opportunity to create also offer opportunities to imagine new possibilities, learn new things about oneself, and experience the rewards of effort-based behaviors [27]. Effort-based reward systems are dopaminergic pathways that connect reward centers with human actions and choices to "make" and "do" things in their environment. Lambert [27] argued that humans evolved to be active and to make things, and just these acts can be ways to release dopamine in the brain and experience positive emotions.

To share an example, a patient came in recently for a session of art therapy as part of one of our research studies on arts and health. He had had surgery for a brain tumor that left him with limited control of his dominant hand. He would hold his right hand with his left hand at the beginning of the session to guide it. In his case, art therapy was a form of relaxation and social reconnection. Once he felt comfortable, relaxed, and safe, he also gained a sense of mastery and self-efficacy. At the end of the session, he was able to use his right without support and to write freely. I asked him if he noticed this change and why he thought it had occurred. His response was that he felt less anxious, he felt good, and he was relaxed at the end of the session; all of these factors helped him with his ability to write and to use his right hand. We could argue that this was the result of feeling comfortable in the session, which could occur in any therapeutic context. However, given that art therapy involves verbal interaction and some form of "making," we might conjecture that multiple activities, processes, outcomes, and systems are at play in a session. The outcomes of individual sessions might be focused interpersonally whereas the outcomes in group art therapy sessions might result in more interpersonal and group transformations. In my workshops and presentations, I find often that group art-making breaks through the proverbial ice really quickly and catalyzes the activation of interpersonal interactions and socialization. The art product offers an externalized object for discussion and mutual engagement as well as a rich resource for learning about each of the group members.

A question that is often asked of art therapists is which patients/clients are best served by the unique contributions of art therapy. The evidence from clinical practice indicates that art therapy is particularly suited for patients who have experienced trauma, identity struggles, physical and psychological stressors, and developmental challenges. Trauma can be overwhelming, which affects how it is integrated into long-term memory and in turn into the personal life story narrative. For individuals who struggle to articulate their lived experiences, challenges, and struggles, art therapy can help initiate expression that leads to reflection, articulation, and a better understanding of their experiences, which allows them to better integrate their experiences into a life narrative that feels empowering and manageable. For example, in working with pediatric oncology patients, Council [8] argued that art therapy promotes self-discovery and emotional and sensory integration that allow young people a safe arena in which to practice skills that can help them confront and transcend life's challenges.

In my own clinical interactions, I remember a young woman who started the session feeling a deep sense of loss of hope and covered the page in black ink. She viewed the image for a while and then added cherry blossoms that were reminiscent of the time of the year. Through engaging in an authentic representation of her emotional state at the start of the session, she was also able to create a layer of new imagery that brought her to the physical present, contrasting the pink and white of the blossoms with her underlying darker state. Art therapy helps contain and externalize positive and negative emotions, thereby offering the patient or client an alternative visual perspective of his or her condition: breaking the cycle of rumination and providing hope for a possibly fulfilling future. See Fig. 2 for her artwork.

6 Brain-Based Research in Art Therapy: What Can We Track?

Several art therapists have developed frameworks for research in art therapy, including the expressive therapies continuum or ETC [17] and the CREATE mode [14]. The theoretical concept of the Expressive Therapies Continuum (ETC), proposed by Kagin and Lusebrink [17], incorporated the approaches to art therapy of several American art therapy pioneers. The ETC comprises three stepwise levels—kinesthetic/sensory, perceptual/affective, and cognitive/symbolic—interconnected by the creative level. The stepwise three-tiered structure of the ETC incorporates concepts from cognitive psychology and art education, namely, perception and imagery, visual information processing, stages of graphic development, and different expressive styles. It

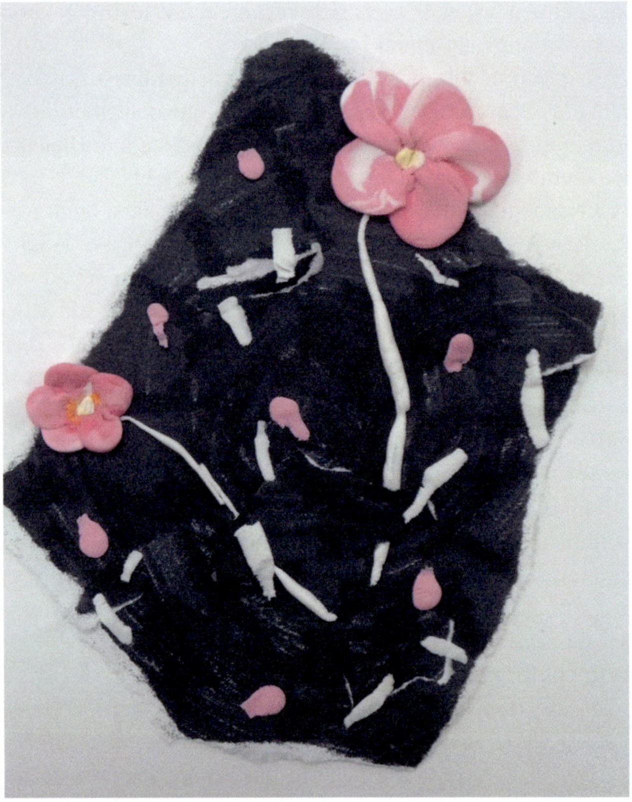

Fig. 2 Artwork illuminating the process of shift in affect through the course of a session

has been hypothesized that the three levels of the ETC reflect three different areas of the brain in processing visual information [4, 29].

A more recent framework in art therapy is the CREATE framework, which demonstrates how the Art Therapy Relational Neuroscience (ATR-N) approach can support resilience in human beings [15]. The framework comprises of six principles included in the acronym CREATE, namely, Creative Embodiment, Relational Resonating, Expressive Communicating, Adaptive Responding, Transformative Integrating, and Empathizing and Compassion [14]. The framework integrates current knowledge of neurobiology with principles of art therapy that emphasize relational development through creative expression and embodiment. The CREATE framework is more grounded in neuroscience compared with the ETC; however, these are both theoretical frameworks developed from clinical experiences and from art therapists' knowledge of neuroscience but have not been empirically tested. Empirical research has been limited to a few empirical studies that have examined outcomes related to artistic skill, drawing tasks, and responses to clay manipulation.

Some findings to date with quantitative electroencephalography (qEEG) indicate that different art media result in different levels of brain activity and that these differences are also associated with whether or not an individual is an artist. Belkofer et al. [2] investigated the differences in patterns of brain activity among artists and non-artists during the process of drawing. Results indicated that there was more activity in the left hemisphere of the brains of artists, whereas more activity was reflected in the frontal lobe of non-artists. This result may have been based on the fact that drawing was a new task for them and that stimulation in this area of the brain is a sign of learning. There was an increased presence of alpha waves for both the artists and the non-artists, indicating potentially relaxed creative opportunities generated by drawing tasks. Similarly, in a quantitative electroencephalographic comparison of working with clay and drawing, activation was noted in regions of memory processes, meditative states, and spatiotemporal processing [25]. King et al. [19] found that art-making resulted in overall increased EEG power compared with a rote motor task, highlighting that there are differences in brain activation in creative versus a pure sensorimotor-based activity.

Art therapy researchers have also focused on the relationship between art and mood states. For instance, art-making has been found to reduce cortisol levels [23, 22] as well as improve mood and self-efficacy [21]. Kaimal et al. [18] examined the outcomes of three different drawing tasks on reward perception as measured using functional near infrared spectroscopy, a technique that examines blood flow using infrared light and that can detect blood flow within up to 3 mm depth of the cortical surface. The underlying assumption in this study was that blood in the mPFC would indicate activation of a reward pathway in the brain. Participants were given three drawing tasks (coloring, doodling, and free drawing) spanning 3 min, each with intermittent rest periods of 2 min each. The findings indicate that the drawing tasks all activated the reward pathway of the brain compared with the no-activity rest conditions, with the doodling condition resulting in maximum activation. These findings are speculated to also mirror the theory of effort-based reward pathways [27] wherein making/creating are related to feelings of reward.

These studies highlight some preliminary work in examining the art-making aspect of art therapy. The relational component, the synchrony between the therapist and the patient/client, and the functional and structural changes that occur in the patient/client him- or herself remain to be better studied and understood. Research on interactional, existential, developmental, neuroscientific and creative processes remains limited because of the incomplete understanding of the processes involved and the complexity involved in measuring these attributes.

Opportunities for future research: The time is optimal for research in art therapy given its increasing visibility and the fact that many funding agencies in the United States, including the National Institutes of Health, the Department of Defense, and the National Endowment for the Arts have issued calls for proposals. Many of the calls are not related to whether art therapy is helpful; rather, the calls are for studies to identify the mechanisms that make it effective. The clinical anecdotes and impressions have been well documented as have initial observational and evaluation studies indicating positive outcomes of art therapy [18–20, 24]. Further research is needed to isolate and identify the short- and long-term functional and structural contributions of art therapy in brain functioning.

Two potentially valuable research areas to pursue are positive emotions and reward perception. A consistent finding in art therapy research has been that art therapy enables individuals to experience positive emotions, often in unexpected ways, including through possibly effort-based rewards pathways. Some of the hormones released by the neuroendocrine system that have been associated with positive emotions include serotonin, endorphin, oxytocin, and dopamine. Serotonin is associated with feeling valued and with self-esteem. Serotonin is the basis of a class of antidepressants and possibly affects self-esteem and perceptions of value. Art therapists work to facilitate this sense of belonging and an inter- and intrapersonal sense of self. Dopamine has been implicated in a range of functions including movement coordination, reward perception, and, most recently [33], in responses to threatening stimuli. Together, these data highlight how dopamine in the mPFC can selectively route sensory information to specific downstream circuits, representing a potential circuit mechanism for valence processing. Dopaminergic reward pathways are overridden by the release of dopamine related to a perceived threat, highlighting the brain's focus on survival and anticipation of potential threats. This process relates the function of dopamine in anticipatory rewards as well since anticipating danger is possibly a way to respond effectively to that threat. Art therapy sessions could help patients regulate this response by better understanding potential triggers in the modern social environment and offsetting the threat with appropriate health-promoting choices. Endorphins are the human body's natural pain-killers. They are released often when an individual cries or after intense physical activity. Participants in art therapy sessions often report feeling temporarily pain-free when they have been deeply engaged in a session.

Oxytocin is understood to be a hormone that encourages relational bonding, primarily in mothers and infants. However De Dreu et al. [10] found that oxytocin can promote bonding sometimes in adversarial contexts including among warriors or tribal groups at war with an adversary. The relational bonding supports survival by making the individuals with heightened oxytocin levels care for and support each other (in-group love) even if possibly hating or wanting to kill an enemy (out-group aggression). Storytelling has also been found to release oxytocin as the narrator and listener [36] engage in an empathic mutuality that engages several parts of the brain. Art therapists help patients and clients create narratives with and through the art-making process, often helping generate verbalizations that might previously have been absent or inaccessible due to histories of trauma and adversity that are known to inhibit narrative production [36].

Challenges and roadblocks in the field: Art therapy evolved as a clinical profession, and much of the knowledge in the field resides within the clinical impressions of experienced clinicians. Given this clinical focus, the profession has predominantly focused on developing masters-level trained clinicians rather than a sound evidence-based research base. Research in art therapy has been constrained by limited funding resources and the capacity of researchers in the field [19]. Given that there are about 6000 credentialed art therapists in the United States and only a few hundred art therapists with doctoral level training, there is a real limitation in capacity to conduct comprehensive, systematic research studies that capture the unique mechanisms and outcomes of the profession. In addition, given the lack of funding in the past, research has tended to be small in scale and done by individual therapists often at their own expense. Thus sample sizes have been small and the majority of the studies have been case-based descriptive summaries. This situation has begun to change because of the increasing recognition and new funding opportunities that have emerged in the United States through the National Endowment for the Arts, the National Institutes of Health, and the Department of Defense. The funding agencies recognize that the creative arts therapies have been effective in alleviating patient symptoms, especially in cases of individuals with long-standing chronic symptoms [24] but understanding of the mechanisms of change and of the generalizable outcomes based on larger population-based studies is limited. As art therapy clinicians and researchers learn to work in collaborative interdisciplinary groups, the evidence base is beginning to expand beyond case studies and small outcome studies to large cohort-based observational studies and randomized controlled trials.

In addition to systemic challenges, a specific problem in art therapy and brain-based research has been the difficulty in capturing the complexity of the session, which includes the artwork, art-making, verbal and nonverbal interactions between the therapist and patient/client, and all the unseen psychological and physiological changes happening internally among the participants in a session. Current brain imaging technologies can focus only on elements of the interaction and/or outcome, not on the all the multiple components that in combination lead to an effective session. Mobile brain/body imaging technologies are best suited for art therapy research because they allow for measurements in natural environments [18, 20]. As these technologies for measuring the response of multiple individual and multilevel changes in human responses develop, including physiological and psychological measures, we will be better positioned to assess the processes and outcomes of an art therapy intervention.

7 Conclusions

In returning to the questions raised at the beginning of this chapter, scholarship to date asserts that art-making is a complex venture with many dimensions and levels of meaning to the human mind. In many ways, our lived experiences of the joys and rewards of art-making are far ahead of our abilities to understand why and how Art therapy integrates the relational facilitative interaction to art-making and thus adds an additional layer of complexity to this experience. Art therapy as a profession has a long-standing clinical history and, based on the foundation of clinical insights, is well positioned to conduct more empirical studies on how an art therapist-facilitated session can impact human physiological and psychological functioning. Clinical impressions of session duration, format, and dosage are key to determining how art therapy sessions can impact human functioning for optimal health and well-being. Brain imaging technologies are now beginning to capture functional and structural changes in patients/clients as a result of art therapy sessions. As imaging technologies develop in capacity and sensitivity, we will be better able to capture the multidimensionality of art and its role in understanding the human experience.

References

1. AATA.: American Art Therapy Association. [Association Web site]. Available at https://arttherapy.org/ (2018)
2. Belkofer, C.M., Van Hecke, A.V., Konopka, L.M.: Effects of drawing on alpha activity: a quantitative EEG study with implications for art therapy. Art Therapy **31**(2), 61–68 (2014). https://doi.org/10.1080/07421656.2014.903821
3. Bolwerk, A., Mack-Andrick, J., Lang, F.R., Dörfler, A., Maihöfner, C.: How art changes your brain: differential effects of visual art production and cognitive art evaluation on functional brain connectivity. PLoS ONE, **9**(7), e101035 (2014). https://doi.org/10.1371/journal.pone.0101035
4. Botvinick, M.: Multilevel structure in behavior and in the brain: a model of Fuster's hierarchy. Philos. Trans. R. Soc. Lond. B Biol. Sci. **362**(1485), 1615–1626 (2007)
5. Bubic, A., von Cramon, D.Y., Schubotz, R.I.: Prediction, cognition and the brain. Fronti. Hum. Neurosci. **22**(4), 25 (2010). https://doi.org/10.3389/fnhum.2010.00025
6. Chamberlain, R., McManus, I.C., Brunswick, N., Rankin, Q., Riley, H., Kanai, R.: Drawing on the right side of the brain: A voxel-based morphometry analysis of observational drawing. Neuroimage **96**, 167–173 (2014). https://doi.org/10.1016/j.neuroimage.2014.03.062
7. Chatterjee, A., Vartanian, O.: Neuroaesthetics. Trends Cogn Sci **18** (7), 370–375 (2014). https://doi.org/10.1016/j.tics.2014.03.003
8. Councill, T.: Art therapy with children. In: Gussak, D.E., Rosal, M.L. (eds.) The Wiley Blackwell handbook of art therapy. Wiley Blackwell, West Sussex, UK (2015)
9. Dissanayake, E.: Art and intimacy: how the arts began. University of Washington Press, Seattle and London (2000)
10. De Dreu, C.K.W., Greer, L.L., Van Cleef, G.A., Shalvi, S., Handgraaf, M.J.J.: Oxytocin promotes human ethnocentrism. PNAS **108**(4), 1262–1266 (2011). https://doi.org/10.1073/pnas.1015316108
11. Dutton, D.: The art instinct: beauty, pleasure, and human evolution. Bloomsbury Press, New York (2010)
12. Gottschall, J.: The storytelling animal: how stories make us human. Mariner Books/ Houghton Mifflin Harcourt, New York (2013)
13. Gould, S.J.: The exaptive excellence of spandrels as a term and prototype. In: Proc. Natl. Acad. Sci. USA **94**(20), 10750–10755. https://doi.org/10.1073/pnas.94.20.10750. pmc2 3474
14. Hass-Cohen, N.: Create: art therapy relational neuroscience principles (ATR-N). In: Hass-Cohen, N., Carr, R. (eds.) Art therapy and clinical neuroscience, pp. 283–309. United Kingdom, Jessica Kingsley, London (2008)
15. Hass-Cohen, N., Findlay, J.C., Carr, R., Vanderlan, J.: Check, change what you need to change and/or keep what you want: an art therapy neurobiological-based trauma protocol. Art Ther. **31**(2), 69–78 (2014). https://doi.org/10.1080/07421656.2014.903825
16. Hetland, L., Winner, E., Veenema, S., Sheridan, K.: Studio thinking: the real benefits of visual arts education. Columbia University Press, New York NY (2007)
17. Kagin, S., Lusebrink, V.: The expressive therapies continuum. Art Psychotherapy **5**, 171–181 (1978)
18. Kaimal, G., Ayaz, H., Herres, J.M., Dieterich-Hartwell, R.M., Makwana, B., Kaiser, D.H., Nasser, J.A.: Functional near-infrared spectroscopy assessment of reward perception based on visual self-expression: coloring, doodling and free drawing. Arts Psychotherapy **55**, 85–92 (2017). https://doi.org/10.1016/j.aip.2017.05.004
19. Kaimal, G.: The road ahead: preparing for the future of art therapy research. In: Carolan, R., Backos, A. (eds.) Emerging perspectives in art therapy: trends, movements and developments, pp. 58–74. Routledge, New York, NY (2017)
20. Kaimal, G., Mensinger, J.L., Drass, J.M., Dieterich-Hartwell, R.: Art therapist-facilitated open studio versus coloring: differences in outcomes of affect, stress, creative agency, and self-efficacy. Canadian Art Ther. Association J. **30**(2), 56–68 (2017). https://doi.org/10.1080/08322473.2017.1375827
21. Kaimal, G., Ray, K.: Free art making in an art therapy open studio: changes in affect and self-efficacy. Arts Health **9**(2), 154–166 (2016). https://doi.org/10.1080/17533015.2016.1217248
22. Kaimal, G., Ray, K., Muniz, J.M.: Reduction of cortisol levels and participants' responses following art making. Art Ther (Alex) **33**(2), 74–80 (2016). https://doi.org/10.1080/07421656.2016.1166832
23. Kaimal, G., Drescher, J., Fairbank, H., Gonzaga, A., Junkin, J., White, G.P.: Learning about leadership from a visit to the art museum. Int. J. Educ. Arts **17**(6) (2016a). Retrieved from http://www.ijea.org/v17n6/index.html
24. Kaimal, G., Jones, J.P., Dieterich-Hartwell, R., Acharya, B., Wang, X.: Evaluation of art therapy programs for Active Duty Military service with TBI and post-traumatic stress. Arts Psychotherapy (2018). https://doi.org/10.1016/j.aip.2018
25. Kruk, K.A., Aravich, P.F., Deaver, S.P., deBeus, R.: Comparison of brain activity during drawing and clay sculpting: a preliminary qEEG study. Art Ther. J. Am. Art Ther. Association **31**(2), 52–60 (2014). https://doi.org/10.1080/07421656.2014.903826

26. Lacey, S., Hagtvedt, H., Patrick, V.M., Anderson, A., Stilla, R., Deshpande, G., Hu, X., Sato, J.R., Reddy, S., Sathian, K.: Art for reward's sake: visual art recruits the ventral striatum. Neuroimage **55**(1), 420–433 (2011). https://doi.org/10.1016/j.neuroimage.2010.11.027
27. Lambert, K.: Lifting depression: a neuroscientist's hands-on approach to activating your brain's healing power. Basic books, New York, NY (2008)
28. Lowenfeld, V., Brittain, W.L.: Creative and mental growth, 8th edn. Prentice Hall, New York (1987)
29. Lusebrink, V.B., Mārtinsone, K., Dzilna-Šilova, I.: The expressive therapies continuum (ETC): interdisciplinary bases of the etc. Int. J. Art Ther. **18**(2),75–85 (2013). https://doi.org/10.1080/17454832.2012.713370van
30. Rauch, S.L., Savage, C.R., Alpert, N.M., et al.: The functional neuroanatomy of anxiety: a study of three disorders using positron emission tomography and symptom provocation. Biol. Psychiatry **42**, 446–452 (1997)
31. Ripley, D., Politzer, T.: Visual disturbance after TBI. Neuro Rehabilitation **27**, 215–216 (2010). https://doi.org/10.3233/NRE-2010-0599
32. Schlegel, A., Alexander, P., Fogelson, S.V., Li, X., Lu, Z., Kohler, P.J., Riley, E., Tse, P.U., Meng, M.: The artist emerges: visual art learning alters neural structure and function. Neuroimage (2015). https://doi.org/10.1016/j.neuroimage.2014.11.014
33. Vander Weele, C.M., Siciliano, C.A., Matthews, G.A., Namburi, P., Izadmehr, E.M., Espinel, I.C., Nieh, E.H., Schut, E.H.S., Padilla-Coreano, N., Burgos-Robles, A., Chang, C.J., Kimchi, E. Y., Beyeler, A., Wichmann, R., Wildes, C.P., Tye, K.M.: Dopamine enhances signal-to-noise ratio in cortical-brainstem encoding of aversive stimuli. Nature **563**(7731), 397–401 (2018). https://doi.org/10.1038/s41586-018-0682-1
34. Van der Kolk, B.A., van der Hart, O., Marmar, C.R.: Dissociation and information processing in posttraumatic stress disorder. In: van der Kolk B.A., McFarlane A.C., Weisaeth L. (eds.) Traumatic stress: the effects of overwhelming experience on mind, body and society, pp. 303–327. Guilford Press, New York, NY (1996)
35. Walker, M., Kaimal, G., Koffman, R., DeGraba, T.J.: Art therapy for PTSD and TBI: a senior active duty military service member's therapeutic journey. Arts Psychotherapy **49**(2), 10–16 (2016). https://doi.org/10.1016/j.aip.2016.05.015
36. Zak, P. Why inspiring stories make us react: the neuroscience of narrative. Cerebrum February. Available at http://www.dana.org/Cerebrum/2015/Why_Inspiring_Stories_Make_Us_React_The_Neuroscience_of_Narrative/ (2015)

Girija Kaimal (Ed.D, MA, ATR-BC) is an Associate Professsor in the Ph. D. Program in Creative Arts Therapies at Drexel University. Her research (currently funded by the US Department of Defense and the National Endowment for the Arts) examines the psychological and physiological outcomes of visual self expression. She works with a range of populations including military service members, cancer patients, and their caregivers. More details on her work are available at www.girijakaimal.com and http://drexel.edu/cnhp/research/research-areas/creative-arts-therapies/KaimalGirija/research/.

Created Out of Mind: Shaping Perceptions of Dementia Through Art and Science

Sebastian J. Crutch, Charles R. Harrison, Emilie V. Brotherhood, Paul M. Camic, Brian Day, and Anthony J. Woods

1 Background to the Hub at Wellcome Collection

In 2014, Wellcome opened The Hub in Wellcome Collection. It was envisaged as an experimental research space where interdisciplinary research could be nurtured and, more importantly, the resident research team would be encouraged to produce a variety of research outputs—not just the traditional fare of academic papers and book chapters. Over the last four years, however, the Hub has evolved into a highly effective transdisciplinary and co-creative research unit that utilises ground-breaking research methodology that is scalable and informs and identifies potential health interventions. That evolutionary process is outlined here as an example of why allowing researchers true academic freedom is highly beneficial.

The idea for a Wellcome Hub originated in 2011. Fundamentally, it grew out of a desire to capitalise upon the rich tradition of research (initially history of medicine and then more broadly medical humanities) associated with Wellcome. The already apparent success of the Wellcome Collection at this time provided the perfect context, as well as further cultural impetus, to the idea of adding a research component that interpreted and exploited the rich collections as well as making the most of the public presence of this increasingly vibrant cultural venue and library. It was regarded as a golden opportunity to fund and facilitate interdisciplinary, often subjective, research that added value to Wellcome's science funding portfolio, alongside the evolving medical humanities/social science grant programmes—but crucially to enable research that could not be carried out within universities (or funded through existing grant programmes). The key to its uniqueness lay in its location, embedded within Wellcome Collection, a free science-based museum and library that aims to challenge how people feel and think about health (https://wellcomecollection.org/pages/Wuw2MSIAACtd3Stq).

This was a bold, experimental idea and it is true to say that nobody had any idea whether it would work or what it would look like 5 years down the line. There was no roadmap or even a destination in mind—this was 'see how it goes research'. Of course, research teams bidding for the space (winners of the biennial competition are awarded £1 million and the space for 22 months) were given some guidance by the funder but this was really designed to give them some comfort rather than be prescriptive or restrictive. The overarching brief was for teams to "bring multiple perspectives to bear on key health challenges and help deliver the Trust's vision of extraordinary improvements in health." Teams were encouraged to do this by:

a. Progressing an interdisciplinary area of work around a theme linked to the vision of the Wellcome Trust;
b. Identifying a group of associates to join the Hub Core team (3–5 academics) for short and long-term residencies to create an environment of knowledge and cultural exchange linked to the theme;
c. Developing and hosting a lively and active programme to foster exchange and development of ideas among the core group, associates, Trust staff and the wider creative

S. J. Crutch (✉) · E. V. Brotherhood
Dementia Research Centre, UCL Queen Square Institute of Neurology, University College London, London, UK
e-mail: s.crutch@ucl.ac.uk

S. J. Crutch · C. R. Harrison · E. V. Brotherhood · P. M. Camic · B. Day
Created Out of Mind, The Hub, Wellcome Collection, London, UK

P. M. Camic
Salomons Centre for Applied Psychology, Canterbury Christ Church University, Tunbridge Wells, Kent, UK

B. Day
Whole-Body Sensorimotor Lab, Department of Clinical and Motor Neurosciences, UCL Queen Square Institute of Neurology, University College London, London, UK

A. J. Woods
Wellcome, Euston Road, London, UK

J. L. Contreras-Vidal et al. (eds.), *Mobile Brain–Body Imaging and the Neuroscience of Art, Innovation and Creativity*, Springer Series on Bio- and Neurosystems 10, https://doi.org/10.1007/978-3-030-24326-5_14

and academic communities associated with Wellcome Collection.

d. The theme must be interdisciplinary and bring together relevant experts from a wide community of disciplines, such as scientists, artists, historians, ethicists, anthropologists, social scientists, philosophers, lawyers and literary scholars.

A call for proposals was released with a start date for the successful bidders of October 2014. Although the rhetoric around the Hub was very much focused on improvements in health, it was acknowledged that any health outcomes would be indirect (i.e. no direct benefits for those with health conditions). What Wellcome was really creating was a "laboratory for the medical humanities" where residents were to be encouraged to focus on literary, bioethical, historical and philosophical analyses of wellbeing, health and healthcare. Thus, a space was created where subjectivity could be taken seriously from a research perspective. However, the space was intended for use by professional researchers; nobody envisaged that people with lived experience of ill health would ever enter the space as part of the research programme. This was certainly true of the first residents—an interdisciplinary team from the University of Durham (their project was entitled Hubbub) who tackled the hitherto neglected topic of Rest (https://wellcomecollection.org/pages/Wuw2MSIAACtd3SsS). This team took their interdisciplinary approach very seriously and published two volumes on their research methodology [4, 5]. They were highly successful in producing the desired variety of academic and non-academic outputs (including a large survey in collaboration with the BBC—The Rest Test—involving 23,000 across the globe, launched on Radio 4) but within Wellcome the Hub was regarded as an experiment and not yet an essential element of its funding portfolio.

2 Created Out of Mind

Created Out of Mind (http://www.createdoutofmind.org/) took up residency in The Hub in 2016 and comprised what we would now call a transdisciplinary team of scientists, artists, clinicians, practitioners and people living with a dementia. The aims of the residency team at the outset were to:

- Reaffirm the value of people with dementia and their experiences.
- Explore, challenge and shape perceptions of dementias.
- Demonstrate the value of the creative process.
- Extend our understanding of the brain.

However, in addressing these aims a shift in methodology was required. An interdisciplinary model—whereby collaborators came from different disciplines and contributed to a part of the project from within their own discipline—was not appropriate to fully take on the challenge the team had set themselves. Transdisciplinary research, however, is, essentially, holistic team enquiry. In a transdisciplinary research endeavour, researchers and practitioners, including non-academic participants (consumers of research, patients, community members), come together to co-design (or co-create) both the research process and its key outputs. It cuts across the traditional divisions of knowledge with a view to making new discoveries via unexpected connections. Individual research team members strive to understand the complexities of the whole project, rather than one part of it. This transdisciplinary way of working was the approach Created Out of Mind came upon by design, purely because it made practical sense.

A number of key principles and issues emerged early in the residency work [3]. One of these was a commitment to '*search before research*', holding the tension between the common scientific practice of executing a pre-determined set of aims, procedures and analyses (research), and an approach arguably more common in artistic practice based on discovery through exploration in which the process may be as important as—if not more important than—the end product (search). Another was to maintain a balance between '*experience and experiment*': introducing elements of evaluation and monitoring into arts-based activities risks fundamentally altering the situation and people's responses to it, so the team had to vary the visibility and extent of evaluation across projects. A further driver was to adopt mixed-methods approaches blending qualitative and quantitative research methods, particularly to explore '*in the moment*' experience, by enriching observational and phenomenological occurrence-type data by capturing important factors such as the duration, intensity, frequency and variability of behavioral, psychological and physiological responses (see Thomas et al. [10]).

But arguably the most important decision was to place the questions, statements and uncertainties of people with a lived experience of rarer dementias at the centre of the residency. Rare dementias include those conditions that have a young age of onset (before 65 years), affect non-memory domains initially (e.g. vision, language, social behavior), may be directly inheritable, and/or are caused by non-Alzheimer or vascular pathology. Through Rare Dementia Support meetings for people with rare, atypical and young-onset dementias (http://www.raredementiasupport.org/), and participatory multi-arts sessions and events, the Wellcome Hub was transformed into a space that hosted and valued people living with a long-term condition and their careers and families. All had observations to share and they were actively encouraged to become involved in every aspect of the research process—from generating the ideas and research questions all the way through to producing the outputs. In this way, The Hub became a space whereby boundaries between research producer and research consumer were blurred.

3 The Importance of Lived Experience: Do I See What You See?

This transdisciplinary way of working can be described by reference to one of the Created Out of Mind outputs, an animated film called *Do I See What You See?* by Simon Ball (http://www.createdoutofmind.org/stories-and-reflections/do-i-see-what-you-see). This film explores the experiences of people living with Posterior Cortical Atrophy (PCA), a visual form of dementia [2, 7]. People living with PCA have difficulty seeing what and where things are owing to degeneration of the visual cortex—an issue of 'brainsight' not eyesight—typically caused by Alzheimer's disease. The film capitalised on the ongoing relationship between people with PCA, their care partners, family and friends, and a wider team of clinicians, researchers and artists through the PCA Support Group (http://www.raredementiasupport.org/posterior-cortical-atrophy-pca/).

The central role of those with a lived experience is illustrated in Fig. 1. The Hub residency placed people with a lived experience of a health condition at the centre (small circle) and drew them into interaction with scientists (small triangle), artists (small square) and others. The value in these connections (double headed black arrows) came not only from working together directly, but also from observing how the other parties involved interacted. Place was critical—much of our previous work occurred in places familiar to scientists (large triangle, e.g. clinical settings, experimental laboratories) or artists (large square, e.g. galleries, museums), whilst The Hub provided a neutral space in which those with a lived experience felt comfortable and valued (large circle).

Outputs of the residency were many and varied (black icons). The exemplar *Do I See What You See?* film has had a variety of impacts. Importantly it resonated with the experiences of others living with PCA, helping people to acknowledge and anticipate their own symptoms, sensations and responses (blue arrows). At the same time, the film exists as an art object in its own right, and has appeared in a number of international film festivals (orange arrow). But perhaps most exciting has been unplanned uses by other members of the PCA Support Group (red arrows). In one case, the wife of someone with PCA showed the film to the ward sister during her husband's hospital admission for a physical health problem; the lead nurse in turn used it to

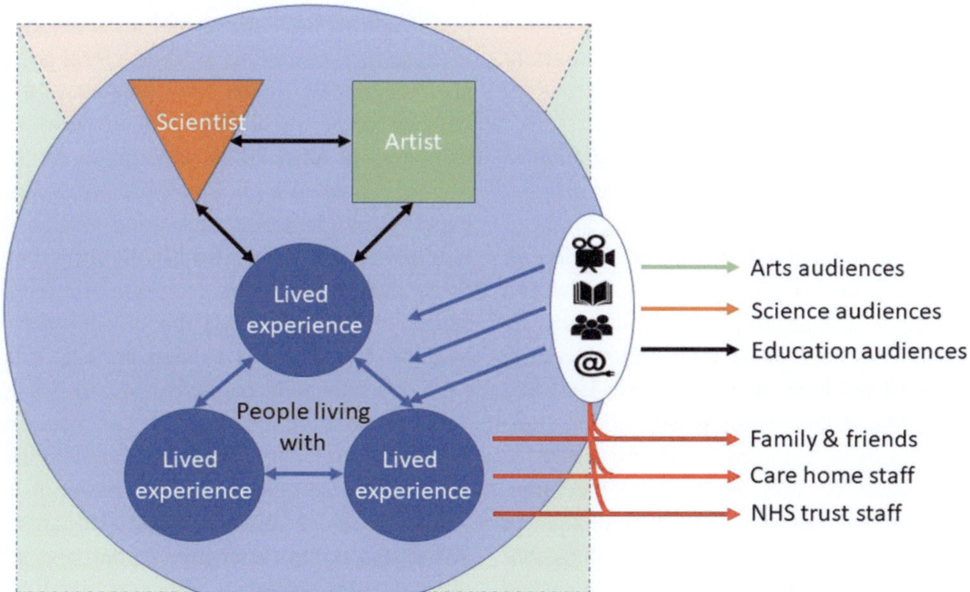

Fig. 1 People with a lived experience of a health condition (small circle) are at the centre of a research process that draws them into interaction (black double-headed arrows) with scientists (small triangle), artists (small square) and others. The research takes place not in a typical scientific (large triangle) or arts environment (large square), but in a space suited to those with a lived experience (large circle). Outputs (black icons) are varied, but address varied audiences including others with a lived experience of a similar condition (blue arrows), and traditional arts, science and education audiences (green, orange and black arrows). Outputs may also be used by those with a lived experience and others for a variety of unplanned and unexpected purposes (red arrows; see text for examples)

educate other ward staff about the condition which facilitated improved clinical care (e.g. the ward staff had interpreted his leaving food as a lack of hunger, when in fact he could not perceive the location of the tray or what was on it). Another support group member describes the film as a tool he can use in his self-chosen new role as an 'advocate ambassador'; he has shared the film with local care home, health and social care professionals, after his wife's care home manager recognised he knew more about the condition than they did.

4 Recognising the Contributions of Those with a Lived Experience: Am I the Right Way Up?

One important lesson from the residency has been the need to better recognise and acknowledge the considerable contributions of those with a lived experience to research. These contributions may take many forms (Crutch et al., in process), but particularly worthy of note are:

1. Inspiration: generating new ideas and hypotheses, and challenging existing assumptions.
2. Context setting and training: opening up the lived experience to other researchers.
3. Project design and development: co-design, revision, feasibility assessment and piloting as an active patient/citizen researcher.
4. Motivation and insights through participation in established projects.
5. Review, reflection and rethinking: scrutinising current work and improving future research.

Contributions by inspiration are particularly important to recognise as too often in science, hypotheses are formed and claimed without full thought being given to the genesis of those ideas [1]. One example that led directly to a variety of artistic, scientific and transdisciplinary responses within Created Out of Mind was a striking statement made by someone living with PCA. At a PCA support group meeting, two family members related how their mother (in-law) had recently asked them, "Am I the right way up?". This comment was remarkable as PCA had been known primarily for its characteristic progressive loss of visual processing. Though not the first hint of non-visual sensory challenges in this condition, this subjective experience strongly suggested a broader disruption of the way in which the balance system integrates different types of sensations.

This comment sparked a number of avenues of research enquiry, each of which were shaped by the contributions of individuals with or caring for someone with PCA [8]. To understand whether this experience was shared by others living with the condition, comments and descriptions of symptoms were collated from a number of support group discussions and question-and-answer sessions. These yielded examples of people walking whilst tilted to one side (physically leaning, but perceiving themselves to be upright) and others feeling as if they might "fall off the edge of the world" (remaining physically well-aligned to the world, but perceiving themselves not to be so).

To understand the neurological underpinnings, those with a lived experience of balance problems were brought into conversation with experts in the neuroscience of balance at UCL's Whole Body Sensorimotor Laboratory in Queen Square. This laboratory is designed to study the balance system and the way in which it relies upon sensory information to relate body position to the gravitational vertical and to report how the body is moving to stop you falling over. Visual, vestibular (from inner ear) and proprioceptive (from muscles and joints) information all play a role but no one sensory system on its own provides the brain with the complete story. Instead, the information from all three has to be combined. Combining these types of information is known to require the parietal lobes of the brain which are particularly vulnerable in PCA. However, the laboratory had never been used before to study PCA, so the feedback given by people living with PCA during a pilot study was critical to the development of the stimuli and apparatus used in a subsequent series of neurological experiments investigating perceived verticality and the impact of visual information on perception of the vertical and on standing balance.

In turn, Created Out of Mind visual artist Charlie Murphy also responded to the environment and processes of the balance study, participating in the tests and observing the responses of other study participants to and within that setting. She created a photographic series aiming to illuminate experiences of disorientation and vertigo which this condition can evoke, while also highlighting the vulnerability that these testing situations can sometimes elicit (see Fig. 2). Murphy also supervised the development of a Central St Martins' BA Ceramic Design project 'What Can Ceramics Do?' working with collective *Studio Senses*, who created a series of vessels exploring how clay could be used to express some of the physical impacts of dementias on the brain. In response to the question "Am I the right way up?", student Rachel Wilcock digitally modelled an 'illusion vase' to visualise a complete change in the vase's orientation when viewed from different angles. Collectively, these artistic responses to and extension of the 'Am I the right way up?' project have proven to be a powerful catalyst for a wide range of artworks, which have been exhibited extensively in the UK.

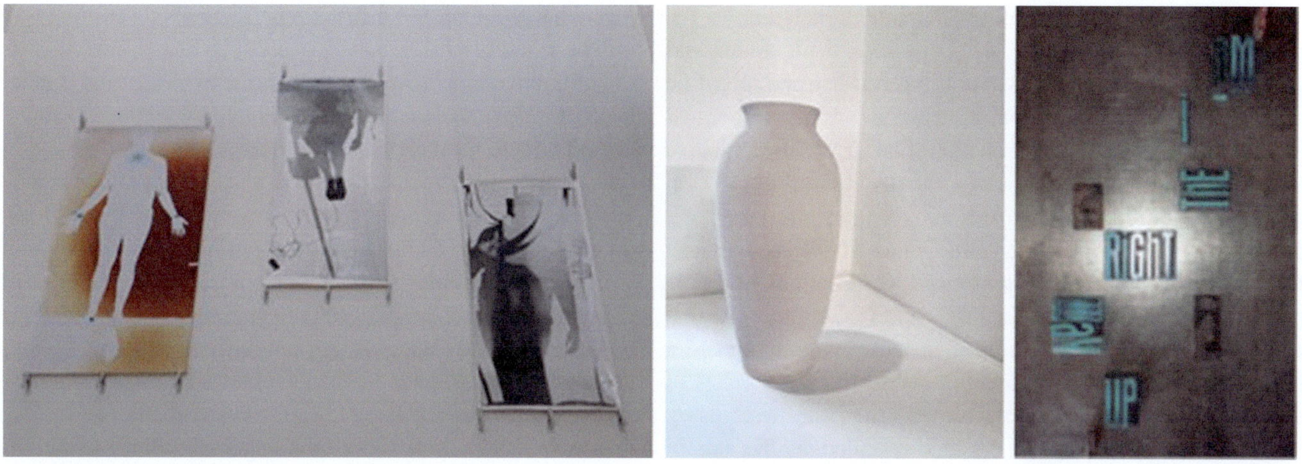

Fig. 2 Left panel: Charlie Murphy's 'balance' photograms. Middle panel: Rachel Wilcock's Illusion Vase. Right panel: Murphy's typographic experiments with Gaynor Hulme's statement provided the stimulus for participatory letter press workshops with people with dementia and care partners during 2017 Dementia Awareness Week

5 Transdisciplinary Extension of the Am I the Right Way Up Project: Single Yellow Lines

As described above, one comment from a person living with a dementia can be enough to stimulate a project that improves our understanding, detection and management of balance problems in PCA, and provides a rounded account of its impact on people's lives. However, the number and variety of responses to the intriguing uncertainty also gave rise to secondary, transdisciplinary pieces of research, as scientists and artists found common ground in their areas of enquiry.

Single Yellow Lines was a separate Created Out of Mind project, led by the artist Charles Harrison and borne out of a desire to investigate forms of creativity used in painting in a way that would be open to everyone and not reliant on previous skill or interest [6, 9]. Anyone can paint a single yellow line on a piece of grey card, and in many cases, people are happy to do this without giving it a second thought.

At the time of the projects' conception the Created Out of Mind team were discussing 'in the moment' experience and how a 'moment' might be captured. Among the various possible technological means available, paint seemed like an interesting approach: the painted mark would describe a specific creative moment for each individual, and everyone would run out of paint eventually. Added to this, there is a theoretical and expressive weight placed on a single brush-stroke or action in art history, commonly referred to as 'the mark of the artist'.

Initial interest centered on whether there might be a difference between expressive and controlled painted gestures so each participant was invited to paint both straight lines and expressive lines. There was only one rule—each line had to be a single movement, as if a move in chess. The activity was trialed at Rare Dementia Support groups for those whose conditions primarily affect visual perception (PCA), language (primary progressive aphasia; PPA) and behavior (front temporal dementia; FTD) and from around 300 lines painted at these groups we conducted conversations and further painting activities with scientists, artists, people with dementia, carers and the general public to begin to understand how the lines might be interpreted. These conversations allowed us to see possible differences between diagnoses: people at the PCA group perhaps creating the most 'de-centred' lines, people from the PPA group often using icons to communicate something, people from the FTD group showing the most repetition or rigidity, and people painting lines at public events frequently breaking rules or being particularly creative (especially when alcohol was being served!).

One of Harrison's conversations was with Professor Brian Day, the neuroscientist leading the 'Am I the right way up?' study of balance and movement difficulties. Together, Harrison and Day became particularly interested in the whole-body movements that precede, accompany and follow on from the execution of a single painted line. The researchers wanted to address questions such as how does the context of research and the materials used affect participation & behavior? What can the data reveal about creative movements, intentionality and unintentionality? What does the qualitative data reveal about subjective experience of the activity? What does the physiological data suggest about bodily 'moments'? What is the relationship between these different datasets? Furthermore, Harrison and Day were intrigued by what happens when an artist works in a research lab?

Although the painted line is the intentional aim of the painter, it occurred that the less-conscious manoeuvres and gestures that surround and support the brushstroke might be revealing. Therefore a line painting activity was incorporated into the research protocol which had been designed to understand better how the disease can affect one's perception of self-orientation in the world, as well how the automatic balance system might be simultaneously affected (see Fig. 3a–b). To do this a paintbrush was adapted to incorporate motion sensors and a larger canvas frame built that could be rotated so that when tilted the canvas mimicked a tilted square that was employed in one of the formal tests of self-orientation (see Fig. 3c). We recorded the trials using the CODA motion-capture system installed at the lab in conjunction with wearable Empatica devices, which measured physiological responses, and anecdotal feedback from the participants (see Fig. 3d). 45 participants were each invited to walk up to the canvas and paint 5 separate lines with the canvas either vertical or tilted by 18° (see Fig. 3e). Four of the lines were instructed to be straight and the fifth was an unconstrained continuous expressive line.

The research trials have now been completed and the process of analysing the data is underway. Initially Harrison and Day have been looking at the trials from a qualitative point of view focusing on where something curious appears to have happened. For example, one participant dances back to the starting point after painting their line, another rushes up to the canvas and seems to paint in one continuous body movement and another seems to walk like a cowboy. There are also more troubling examples of people waiting for a long time before painting, getting lost on the way, brushing the air, or having problems retaining the instructions. As analysis continues, it will be of interest to see how, if at all, such behaviours relate to quantitative performance not only on the experimental balance tasks but also on standard neuropsychological measures of episodic memory, language and other cognitive skills.

Through this process there has also been an opportunity to think about the relationship between our perception of the activity as observers and the perception of the participants. To give an example: one of the control participants after painting their line returned to the starting point by walking backwards (rather than turning around), presumably so they could continue to gaze at their work. This is the only time this happened in all the recorded trials and this action was interpreted to be a rather deep engagement with the activity—a desire to appraise/consider the line they had just painted. The observer notes commented that the participant appeared to be 'prowling' towards the canvas, again giving this sense of deep engagement. However, when asked "how did it feel to paint these lines" the participant responded with a disinterested "fine", and when asked which their favourite line was, rather than identifying the line under our scrutiny they said "the last one because it was slightly less boring than the others". There appeared to be nothing particularly special about the experience if the participant's comments are taken at face value even though the observers may have perceived a deep engagement. Conversely, there were examples where the data showed very little engagement, but the participant expressed how much they enjoyed themselves. Does the truth of these creative and/or expressive experiences lie with what is seen in the data, with what the participant said or with the painted line itself?

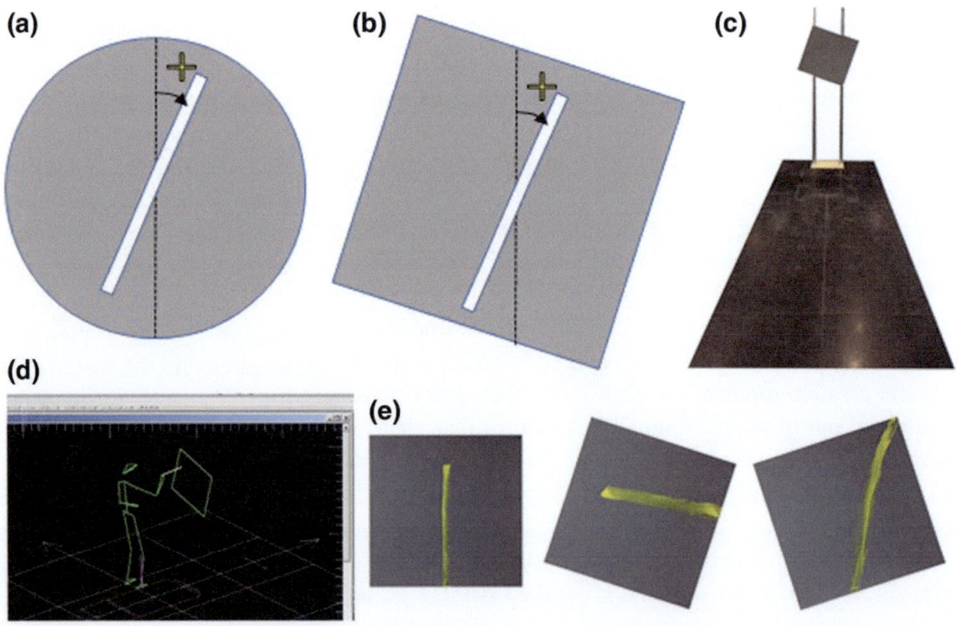

Fig. 3 **a–b** Example rod and frame test stimuli from the original Am I the Right Way Up neuroscience study, to assess the impact of the absence (**a**) or presence (**b**) of visual cues on perceived verticality. **c** Example Single Yellow Lines task canvas with matching 18° tilt. **d** CODA motion-capture data from a participant painting a single line on a tilted canvas. **e** Example lines produced during the experiment

Painting a line takes no time at all, in most cases just a few seconds. Even so, during the trials answers have been recorded of participants who relay that they feel ownership over the lines—saying things like "that line is mine" or there is "something of myself in it". If this is the case, that each line is indeed 'the mark of the artist', then it also raises questions about ownership in other aspects of the research. For example, who 'owns' the bodily movements made in the lab, or the various other forms of record made in research working with people with a dementia? There is an understanding that if a researcher is taking something from or asking something of the participant, then that research will lead to some future benefit either for that participant or for a broader group. Do research activities that involve arts-based methods and processes function with the same sort of promise, and where should the line be drawn if participants (or accompanying family members or friends) feel increased ownership over the products and processes of their participation?

6 Opportunities with Challenges

Although one of the most valuable outcomes from the Single Yellow Line/Am I the Right Way Up work so far has been in exposing the many creative intersections between artists, scientists and research participants, the project does face challenges when it comes to interpreting and communicating the results. It may be difficult to resolve whether the essential value of the project lies in the advocation for creativity in person-centred approaches to testing; in presenting new group data that expands our understanding of these rarer aspects of dementia; or in the possibility that although the artistic outputs are rich with information, attempting any definitive interpretation of the single yellow lines is absurd. It may be that these positions can sit happily alongside one another, or that the frictions undermine efforts to communicate and develop transdisciplinary methods going forward.

More generally, whilst we advocate for the value of transdisciplinary approaches to the generation of new knowledge inspired by subjective experience of those living with health conditions, a number of issues and challenges are inherent in such work. First, it is important to appreciate different ontological and epistemological perspectives across various disciplines. Second, it is vital to develop an appreciation of disciplinary vocabularies [e.g. activities (artists) vs. interventions (researchers); audience participants (cultural sector) vs. patients (health care practitioners); positive engagement (arts organisations) vs. statistically significant results (researchers)]. Third, 'see where it goes' research' requires methodological flexibility to be able to address a range of questions about potentially complex activities/interventions, and benefits from a structured approach to such challenges (e.g. the Medical Research Council [MRC] complex interventions guidance (https://mrc.ukri.org/documents/pdf/complex-interventions-guidance/).

7 Concluding Remarks

The opportunities provided through the Hub residency and the approach adopted by the Created out of Mind team highlight the huge potential for subjective, lived experiences to inspire and shape transdisciplinary research into health conditions. There is a genuine opportunity to create spaces that will demonstrate and develop best practice within the realms of transdisciplinarity. To that end, one can envisage spaces that are not just a laboratory, a clinic, a public venue nor a gallery but something quite different—greater than the sum of its parts. A place where conversations and creative collisions can enrich the lives of all who share the space and where experiences shared may inspire new strands of enquiry, research, better teaching and artistic expression. A place where the arts can connect the isolated and communicate both the experience and biology of the human condition.

Although it is clear that the Wellcome Hub methodology will continue to evolve and there is much to continue to learn about how to conduct transdisciplinary research, enough is known to take existing learning and practice into new arenas and explore a variety of research topics and questions in a multitude of research settings. It is hoped that research funders will take on this challenge to realise the potential of funding transdisciplinary research in the healthcare setting. Another vital part of Created Out of Mind's success was the academic freedom they were afforded in terms of not needing to be prescriptive about their methodology, detailed objectives, potential outputs and outcomes at the application stage. The Hub grant process required high level aims only —the team was then free to develop their programme of activity once they had taken up residence and crucially, had the opportunity to change their minds and follow new, richer veins of research. We believe this model is highly beneficial and is an approach that should be used more widely across the sector—the key message to funders being have confidence (and trust) in those you fund to deliver. The unexpected is, more often than not, more exciting than the expected.

Our work through the Created Out of Mind project has reinforced our belief that the exploration of the experiential nature of health through a transdisciplinary approach encompassing medical sciences, the arts, and humanities, is invaluable to our pursuit of health and wellbeing. Cultures of medicine possess a blend of 'scientific' and 'non-scientific' meanings and it is therefore necessary to examine health and illness from a multitude of perspectives. For example, the overwhelming majority of funded research studies in the field of dementia are quite rightly located within objective science and clinical medicine. However, improvements in health and wellbeing are not solely achieved through research into neurons, genes, microbes and large data sets. Everyone has different journeys when living with chronic disease and we believe that creative activity has more flexibility to address that complexity than generic therapies or drugs. The subjective experience ultimately adds immense value to the objective learning. Hence, we believe it is timely to capitalise on our knowledge and the recognition of others who regard this transdisciplinary approach to be of great value and ever-increasing importance. This vision of a new approach is outlined in Box 1.

In our own work in relation to those living with rarer dementias, we believe we are now at a tipping point that will result in the realisation of this vision. The proof of concept for scalability at local level has been made and national impact is the new aim.

Box 1. Vision for Transdisciplinary, Person-centred Research

- To promote the creation of a sustainable model for 'people-centred' research hubs that address different chronic conditions by combining research, artistic practice, engaging education and therapeutic support.
- To see a shift in research culture where outcomes and measures are more aligned to a people-centred approach and shift the 'centre of gravity' of research away from the laboratory to the communities that need the outcomes.
- To increase collaboration between the university, healthcare and cultural sectors through a shared commitment to people-centred practices.
- To establish hubs in a variety of 'healthy' spaces that will afford artistic and cultural input into subjective, people-centred research.
- To provide education and training to equip leaders with the skills and experience to run successful transdisciplinary Hubs.
- To ultimately improve health by involving and applying the human, subjective perspective. Therefore, promoting the re-personalization of health services and countering the modern preoccupation with biological dysfunction in isolation from a wider concern with the effects of disease on patients' lives and social functioning.

Acknowledgements We are indebted to members of Rare Dementia Support for their inspiration and generous contributions to this work. We would particularly like to thank Gaynor Hulme (1948–2018), Pam Hulme and Isabel Phillips for sharing their 'Am I the right way up?' conversation, Trina and Graeme Armstrong, John Williamson and Lori Hayden, Pam and Richard, Raena and Nick Frankel-Pollen, Geoff and Joan Moffatt and Valerie Blumenthal for their contributions to Simon Ball's film *Do I See What You See?*, and Charlie Murphy and Rachel Wilcock for images of their artworks. This work is presented on behalf of the Created Out of Mind team. This work was supported by the Wellcome Trust [200783] and the Alzheimer's Society [AS-PG-14-022]. The Dementia Research Centre receives support from the NIHR Queen Square Dementia BRU and the NIHR UCL/H Biomedical Research Centre.

References

1. Arends, B., Thackara, D.: Experiment: conversations in art and science. Wellcome Trust, London (2003)
2. Benson, D.F., Davis, R.J., Snyder, B.D.: Posterior cortical atrophy. Arch. Neurol. **45**(7), 789–793 (1988)
3. Brotherhood, E., Ball, P., Camic, P.M., Evans, C., Fox, N., Murphy, C., Walsh, F., West, J., Windle, G., Billiald, S., Firth, N., Harding, E., Harrison, C., Holloway, C., Howard, S., McKee-Jackson, R., Jones, E., Junghaus, J., Martin, H., Nolan, K., Rollins, B., Shapiro, L., Shapiro, L., Twigg, J., van Leeuwen, J., Walton, J., Warren, J., Wray, S., Yong, K., Zeilig, H., Crutch, S.: Preparatory planning framework for created out of mind: shaping perceptions of dementia through art and science. Wellcome Open Res. **6**(2), 108 (2017). https://doi.org/10.12688/wellcomeopenres.12773.1
4. Callard, F., Fitzgerald, D.: Rethinking interdisciplinarity across the social sciences and neurosciences. Palgrave Macmillan, UK (2015). https://doi.org/10.1057/9781137407962
5. Callard, F., Staines, K., Wilkes, J. (eds.): The restless compendium: interdisciplinary investigations of rest and its opposites. Palgrave Macmillan, Cham (2016). https://doi.org/10.1007/978-3-319-45264-7
6. Camic, P.M., Crutch, S.J., Zeilig, H., Harrison, C., Strohmaier, S., Harding, E., Firth, N., West, J., Wray, S., Murphy, C., Windle, G.: Conceptualising artistic creativity in the dementias: interdisciplinary approaches to research and practice. Front. Psychology **9**, 1–12 (2018). https://doi.org/10.3389/fpsyg2018.0182
7. Crutch, S.J., Schott, J.M., Rabinovici, G.D., Murray, M., Snowden, J.S., van der Flier, W.M., Dickerson, B.C., Vandenberghe, R., Ahmed, S., Bak, T.H., Boeve, B.F., Butler, C., Cappa, S.F., Ceccaldi, M., de Souza, L.C., Dubois, B., Felician, O., Galasko, D., Graff-Radford, J., Graff-Radford, N.R., Hof, P.R., Krolak-Salmon, P., Lehmann, M., Magnin, E., Mendez, M.F., Nestor, P.J., Onyike, C.U., Pelak, V.S., Pijnenburg, Y., Primativo, S., Rossor, M.N., Ryan, N.S., Scheltens, P., Shakespeare, T.J., Suárez González, A., Tang-Wai, D.F., Yong, K.X.X., Carrillo, M., Fox, N.C.: Alzheimer's association ISTAART atypical alzheimer's disease and associated syndromes professional interest area. Consensus classification of posterior cortical atrophy. Alzheimer's & Dementia. **13**(8),870–884 (2017). https://doi.org/10.1016/j.jalz.2017.01.014
8. Crutch, S.J., Yong, K.X., Peters, A., Ocal, D., Kaski, D., Gonzalez, A.S., Ryan, N., Ball, S., Harrison, C.R., Murphy, C., Hulme, P., Phillips, I., Hulme, G., Brown, A., Brown, L., Riley, P., Ramsey, L., Woods, A., Day, B.: Contributions of patient and citizen researchers to 'Am I the right way up?' study of balance in posterior cortical atrophy and typical Alzheimer's disease.

Dementia (London). **17**(8), 1011–1022 (2018). https://doi.org/10.1177/1471301218789296
9. Harrison, C.R., Carton, A.M., Brotherhood, E.V., Hardy, C.J.D., Cohen, M.H., Warren, J.D., Crutch, S.J.: Profiles in paint: contrasting responses to a common artistic exercise by people with different dementias. Arts Health (2017). Epub ahead of print. https://doi.org/10.1080/17533015.2017.1402795
10. Thomas, G.E.C., Crutch, S.J., Camic, P.M.: Measuring physiological responses to the arts in people with a dementia. Int. J. Psychophysiol. **123**, 64–73 (2018). https://doi.org/10.1016/j.ijpsycho.2017.11.008

Sebastian J. Crutch is a professor of Neuropsychology at UCL. His research focuses on rare and young onset dementias, exploring topics including dementia-related visual impairment. He directed the Created Out of Mind 2016–2018 dementia and arts residency at The Hub, Wellcome Collection, a collaboration between artists, scientists, and people living with dementia.

Charles R. Harrison has been collaborating with researchers and building projects that foreground the creative strengths of people with different forms of dementia for the past four years. Charlie's current research is focused on characterizing creativity through singular gestures while studying the curious intersections between neuropsychological testing and conceptual art.

Emilie V. Brotherhood is the Project Manager for Created Out of Mind, and a Research Associate at the Dementia Research Centre, UCL. Emilie is working toward a part-time Ph.D. investigating physiological responses to music in people living with dementia.

Paul M. Camic is a Consultant Clinical Health Psychologist, Professor of Psychology and Public Health at Canterbury Christ Church University and Research Director at the Salomons Centre for Applied Psychology in Tunbridge Wells. His research interests include investigating the social and psychological impact of community-based programmes involving the arts, galleries, and museums with older people and those with dementia.

Brian Day is a professor of Motor Neuroscience at the University College London where he directs the Whole-body Sensorimotor Laboratory. He has studied many different aspects of motor control in a wide range of neurological diseases and currently works on multisensory processes that control and integrate whole-body actions.

Anthony J. Woods manages The Hub at Wellcome Collection—a transdisciplinary research space embedded within Wellcome. Tony is a neurophysiologist by training, having been a neuroscience lecturer at SGUL before his career at Wellcome. His interests now lie in promoting and fostering transdisciplinary, co-creative research in the healthcare sector.

Part V

How Disruptive Neurotechnologies Are Changing Science, Arts, and Innovation

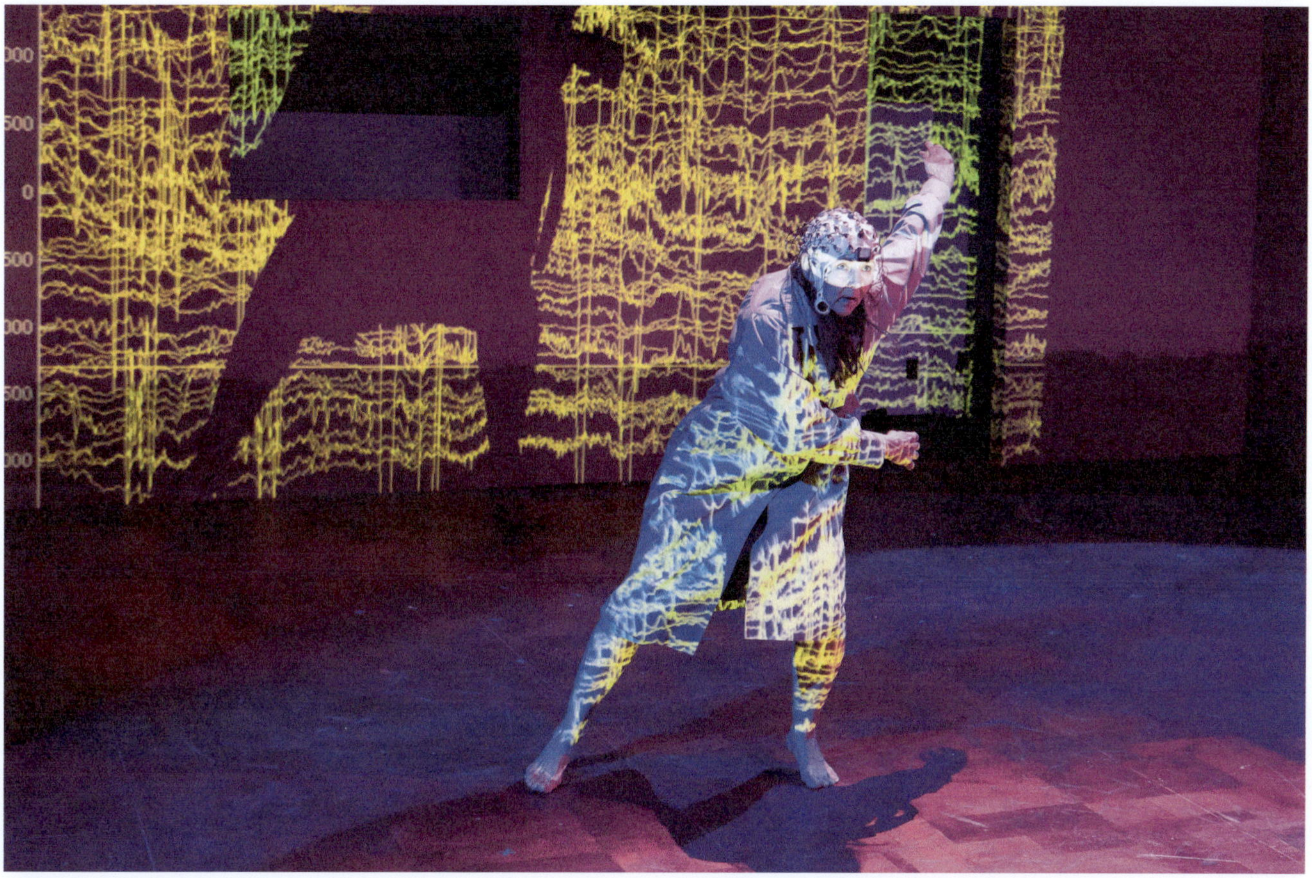

Red Square by Rebecca Valls. Blaffer Art Museum Innovation Series
Performer: Rebecca Valls
Technical Director: Jose L Contreras-Vidal
Technical staff: Zachery Hernandez, Jesus G. Cruz-Garza, Andrew Paek
Photo credit: Lynn Lane
2015
Still photography documenting "Red Square", a solo performative collaboration between Rebecca Valls and the Laboratory for Noninvasive Brain–Machine Interface Systems at the University of Houston. Valls's brain waves were projected in real time as a backdrop to the perfomrance. The information from Valls' brain was mapped to affective states based on Laban Movement Analsysis effort qualities, and used to control the stage lighting, an experiment in creating an empathetic room

The Art, Science, and Engineering of BCI Hackathons

Mario Ortiz, Eduardo Iáñez, Christoph Guger, and José M. Azorín

1 Introduction

When people think about hackathons, the first image that comes to their minds is an exhausted young computer fan sleeping over his laptop after a long day of work and fast food eating. However, a hackathon is far away from a pizza LAN party. The hackathon concept had its origin on computer programming. It came up as a collaborative experience with the aim to develop a computer application or software in a short period of time. First hackathons were held in the 1999, related to open-source software developers of the OpenBSD computer operating system and Sun Microsystems [1]. 'Hackathon' is a closed compound word that includes the words hack and marathon. The term 'hack' is related to the exploratory aspect of programming, instead of its traditional use in computer security. The second term, 'marathon', refers to the intense periods of exhausting work in which the participants need to give a 200% of themselves to carry out a project in a competitive environment.

However, why hackathons are becoming more and more popular and are even promoted by different companies? Reasons are multiple. From the point of view of the participants, the incentives are numerous: the excellent learning experience, the social interaction possibilities, the prizes awarded, and the excitement associated with competition are some of them. Companies have also discovered hackathons as a way to find new talents and launch innovative projects and ideas. As it can be seen, hackathons excel in innovation and competition excitement as well as a learning experience. This is part of the reason for their success.

Due to its collaborative nature, hackathons are extending as an excellent learning tool. The participants are distributed by the different groups depending on their qualifications. The learning process is related to the do-it-yourself approach [2]. This way, it is the group of participants who have to carry out a project guided only by some basic restraints using the materials and equipment which they have at their disposal. This favors the transfer of knowledge between the participants and encourages the creativity of the proposed solutions.

In the case of the *2017 International Conference of Mobile Brain–Body Imaging and the Neuroscience of Art, Innovation and Creativity* ("Brain on Art (BOA) Conference"; http://yourbrainonart2017.egr.uh.edu/), the multidisciplinary nature of the students made this aspect even more relevant. As it will be explained later, the teams were formed by one engineer, one scientist, and one artist, and they were able to come up with working prototypes of an artistic Brain–Computer Interface in only three days. The interaction of these three participants enriched their creative thinking due to the contact between areas of knowledge traditionally non-related. This helps to make up new creative ideas thanks to study the process from a different perspective.

The first day the teams were settled and have a first approach to the hardware, establishing the initial ideas for their projects. They also started recording some data to analyze it and make the initial adjustments. The second day, the teams went on working on improving their development thanks to the experimental feedback. In the last day, projects were defended against the jury and the attendant public of the conference, with real-time performances in some of the cases.

2 Brain to Art Interfaces (B2AI)

The goal of the BOA conference hackathon was to develop artistic Brain–Computer Interfaces. A Brain–Computer Interface (BCI) is a device that captures the electro-encephalographic (EEG) biosignals of the brain and

M. Ortiz (✉) · E. Iáñez · J. M. Azorín
Brain–Machine Interface Systems Lab, Miguel Hernández University of Elche, 03202 Elche, Alicante, Spain
e-mail: mortiz@umh.es

C. Guger
g.tec Medical Engineering GmbH, Herbersteinstrasse 60, 8020 Graz, Austria

J. L. Contreras-Vidal et al. (eds.), *Mobile Brain–Body Imaging and the Neuroscience of Art, Innovation and Creativity*, Springer Series on Bio- and Neurosystems 10, https://doi.org/10.1007/978-3-030-24326-5_15

translates them in order to be able to communicate with an external device [3, 4]. The BCI collects the EEG signals of a subject in the range of microvolts and processes them in order to detect different patterns associated with a certain mental task. The acquisition can be done by invasive or non-invasive electrodes [5]. Although invasive electrodes allow obtaining higher quality signals, they require surgery which makes them not suitable for its use due to the risks associated and ethical issues. The non-invasive EEG equipment uses scalp electrodes that can acquire the EEG signals from the skin of the subject. The contact with the skin of the user can be dry or wet using conductor gel. Dry electrodes are easier and faster to use, but usually provide a higher noise to signal ratio than wet ones.

However, a BCI is not only a cap with multiple electrodes. In order to have a working BCI, EEG signals must be amplified, treated, and processed by hardware and software. As the brain processes are complex, sometimes it is harder to obtain the patterns of a certain mental task without further analysis. Each electrode captures the signals of a certain part of the brain. Some zones are more related to movement, vision, or other actions, so position must be considered. However, in the case of creativity tasks, due to the multiple processes involved, it is not easy to enclose which brain part, and therefore electrodes, should be considered. Additionally, the appearance of artifacts is possible. An artifact is an increment of the signal noise due to a non-related task [6, 7]. For instance, frontal electrodes are susceptible to suffer artifacts due to blinking, because of the little movement of the skin. These undesirable perturbations must be filtered or eliminated from our analysis in order to avoid errors in the modeling of the mental task.

BCIs are traditionally focused on motor neuroprosthetics [8]. They are used to help individuals in their rehabilitation process or to provide mobile assistant to people that suffer from disabilities. In order to build a BCI, the next steps are typically followed: (1) a certain number of trials must be accomplished in order to have enough data of the mental task that has to be decoded; (2) the EEG signals are processed and classified by its features creating a model; and (3) new trials are tested, identifying the mental tasks depending on the classification provided by the model. This classification allows to execute a command to control an external device. For instance, the decision to move or stop an exoskeleton just with the mental intention of execution of a movement or its motor imagery.

The BOA conference hackathon project changed this traditional acceptation, introducing the artistic creation as the external device of a BCI, allowing a new way of communication and expression from the brain biosignals of an artist to its artistic creation, that it could be called Brain to Art Interface (B2AI). Thus, the goal of the BOA conference hackathon was to design and develop new B2AIs.

3 The Design of B2AI Hackathons

3.1 Teams and Projects

As it has been previously indicated, a B2AI hackathon can be defined as a design event in which a group of novel researches in different fields collaborate intensively in the development of an artistic project, BCI related, in a short period of time. Therefore, the design of a hackathon begins with the composition of the teams. Among all participants, homogenous working groups must be composed regarding their fields of experience. The teams must have a great commitment to be able to develop and present a project within the scheduled time. The design must be innovative, creative, and functional.

Although hackathons are generally performed in software fields, where teams are mainly composed by programmers and engineers, in the case of a conference related to B2AI, like BOA conference, other types of researches must be considered. Thus, the groups in the BOA hackathon were constituted by artists, scientists, and engineers. This multidisciplinary nature made possible to have different points of view and expertise, which contributed to generate new and innovative ideas that would not have been possible otherwise. For example, the engineer could provide a more analytical analysis and help with the implementation of the software and hardware; the artist could give an artistic vision to the project looking for a more creative approach in the design of the application; and the scientist would analyze the social, emotional, and cognitive impacts that the application could have on the subjects that use the developed tools.

In the BOA conference hackathon, 9 teams were composed to develop different projects based on same basic ideas. Table 1 shows the list of the projects and a description.

3.2 Materials Available

In order to develop the projects of a B2AI hackathon, a certain set of materials within the scope of the hackathon has to be provided to the teams. One of the main materials is the BCI equipment, which allows to obtain information from the person's thoughts and transform them into commands, artistic expressions, or interactions. This equipment should be as plug and play as possible, reducing the time needed to set it up and to start working quickly with the recorded information. In addition, to be able to express such information, either with movement, painting, music, or multimedia, other materials will be necessary, such as mobile robots, artistic kits, or even software.

One important thing in order to assure the successful flow of the hackathon is the technical support and supervision given by the sponsor companies that provide the

Table 1 List of projects and teams that participated in BOA conference hackathon

Project	Description
Brain and painting	Is there a relationship between EEG and painting? Can we modify some EEG bands during the painting process? Show it!
Do, Re, Mi and EEG	This team did an analysis of imagined and performed musical settings with BCI
EEG del Sol[a]	The team designed a minimalistic, non-threatening EEG headcap that offers comfort and make children feel "cool" while wearing it
Dream painting[a]	For the dream painting app, team members slept with a BCI headset on their heads. When they woke up, they got an image created according to their EEG signals
intendiX painting[a]	Create images by using your thoughts only!
Waves of creation [a]	Whale figures changed colors according to the BCI users brain activity
The art of war[a]	A BCI mixed-reality collaborative strategy game
Sphero SPRK control[a]	Social interactions using the Shpero robot, P300, and music
Artistic BCI (dance)	Partner with a professional dancer attending the conference and make your brain-based choreography controlling lights and music using your EEG signals

[a]Based on BR41 N.IO (see Sect. 4)

hardware/software. This way, the teams can put their focus on the analysis of the projects and their artistic expressions, being released of the implementation problems related to the registration of data and basic procedures. The hardware approach should be eased for the researchers from unrelated fields allowing them to center their capabilities in the innovative use of these tools. Researchers should also have at their disposal a set of application program interfaces (APIs) that allow to work at a lower level if necessary.

For the analysis of brain signals, basic treatment procedures should be available, from basic filters to avoid noise, such as temporal or frequency analysis, to artificial intelligence tools to help with the classification of the different extracted patterns used to generate the output of the BCI. In addition, it is necessary an easy interconnection between the BCI and the external systems to interact with. A basic software architecture should be provided that can be adapted and improved by the teams in order to achieve the specific objectives of their projects.

In the case of BOA conference hackathon, teams used materials provided by Brain Products (GmbH, Germany) and g.Tec (GmbH, Austria) companies, which also provided technical support. On one hand, Brain Products provided 5 LiveAmp 32 channel mobile wireless amplifiers, three of them with active dry electrode systems and caps (actiCAP Xpress Twist), and the other two with slim active gel based-electrode systems and caps (actiCAP snap). On the other hand, g.Tec provided four wireless devices with 16 channels and six wireless devices with 8 channels. Moreover, g.Tec also provided a 3D printer (to allow printing prototypes for placing EEG electrodes), and a Sphero SPRK, which is a ball with a motor that can be controlled from BCI software. Several artistic kits were also provided to the teams (in order to be combined with BCI interfaces). Software to register EEG signals from the different equipment as well as to control the 3D printer and the Sphero ball was also provided.

3.3 Development

The development of hackathons usually takes place in very short periods, where teams work tirelessly to develop, to implement, to test and to present an idea. Sometimes, as it was the case of the BOA conference, the duration of the hackathon extends to the time of the conference. This way, the team participants can have more time to develop their projects and also to attend the different lectures held in the conference and that are related to their research and interests. Figures 1 and 2 show images of some teams of the BOA conference hackathon working in their projects.

During the development of the projects, as it has been indicated previously, it is very important the technical support of the companies providing the hardware and software in order to overcome any technical difficulty.

Once the projects are finished, it is time to present them. It is not only about presenting a result, but also the background, the ideas, and the concepts that have led to its development. It is also important to remark how the interactions among these homogeneous participants have helped to join knowledge and create something new.

The projects were defended against a jury and exposed to the whole conference attendees. The jury must be also constituted in the same way as the teams, that is, by a group of professionals from different fields, so that the evaluation can take into account the different aspects of the project developed by the teams (artistic, scientific, social and technical impact). Figures 3, 4 and 5 show different teams presenting their projects and the jury.

In the case of the BOA conference, the jury was formed by all the members of the organizing committee. Each one voted privately the best candidate for each category. The votes were accounted and the teams with most votes were awarded with the following prizes:

- **Best artistic prototypes**: Three awards sponsored by the IEEE Brain Initiative were given to the best artistic BCI prototypes.

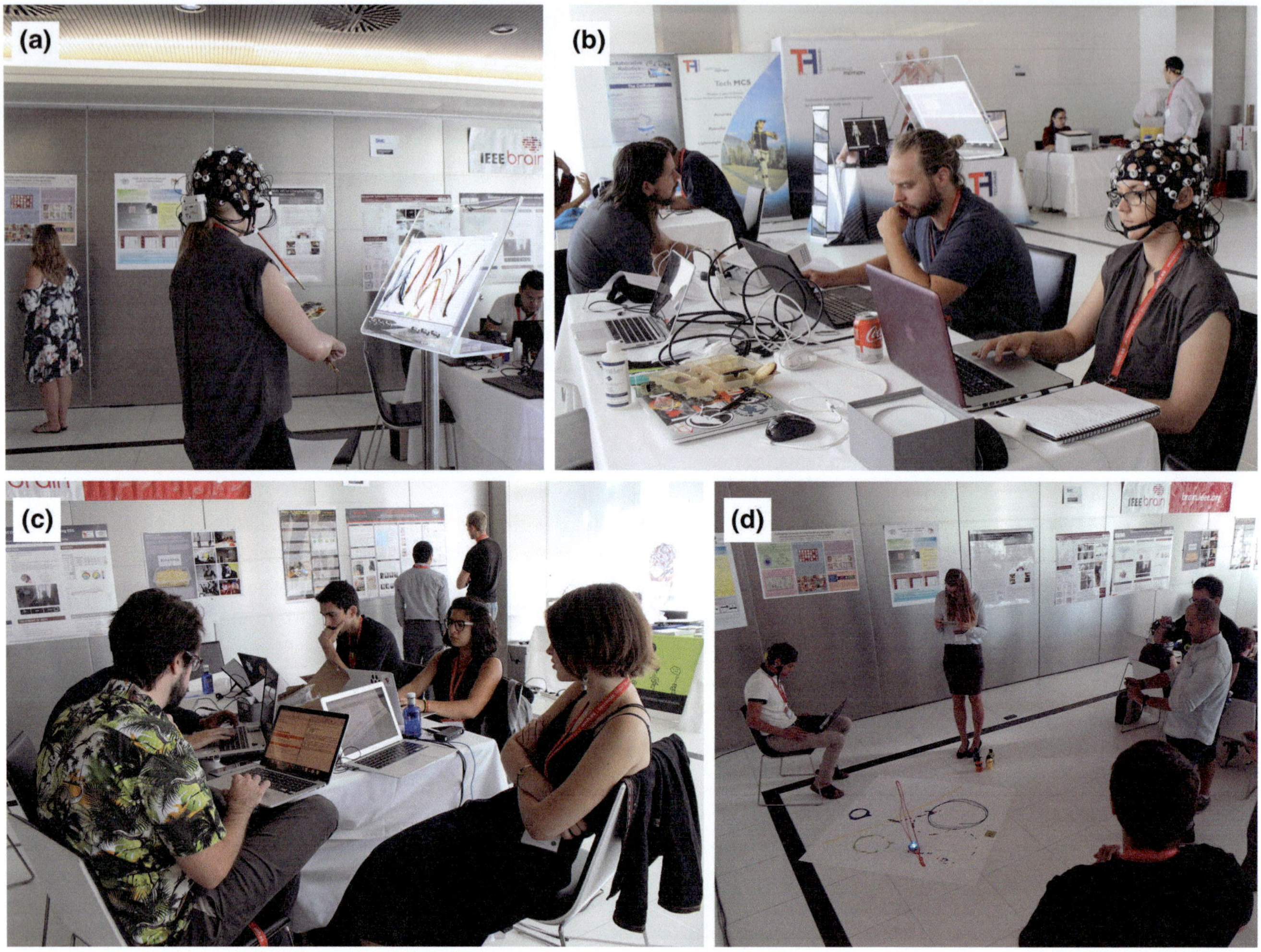

Fig. 1 Some teams developing and trying their projects. **a-b** The team "Brain and Painting" is developing and testing their project. **c** Shows some of the multidisciplinary participants discussing possible approaches to their projects. **d** The team "The Art of War", awarded with first prize to the Best Artistic Prototype is testing their project in real time

- **Most innovative prototype**: An Emotiv Epoc + donated by Emotiv was given to the most innovative prototype.
- **Most disruptive prototype**: A Muse device donated by InteraXon was given to the most disruptive prototype.

In addition, three awards were given by g.Tec to the projects based on BR41 N.IO (see Sect. 4). Figure 6 shows a picture of all the winners.

4 The BR4IN.IO Hackathons

As it was indicated in Table 1, some of the projects of the BOA conference hackathon were based on BR41 N.IO. BR41 N.IO (www.br41n.io) is a series of hackathons launched in 2017 that brings together engineers, programmers, designers, artists or enthusiasts, to collaborate intensively as an interdisciplinary team to program or build their own fully functional EEG-based BCI. Some of these hackathons have been stand-alone events, while others occurred in collaboration with major conferences, festivals or other activities. In 2017 the hackathons took places in Dublin (Ireland), Linz and Graz (Austria), Valencia (Spain) and Banff (Canada), while in 2018 the hackathons took place in Asilomar (USA), Honolulu (USA), Berlin (Germany), Linz (Austria), Poznan (Poland), Prague (Czech Republic) and Miyazaki (Japan). Participants from all over the world have worked together and achieved innovative and playful BCI headsets and applications. The BR4IN.IO hackathons relied on a jury of experts to score the projects at the end of each hackathon and select the winners.

BR4IN.IO hackathons include four different types of projects:

- PROGRAMMING PROJECTS: These projects challenge programmers to code an interface to control

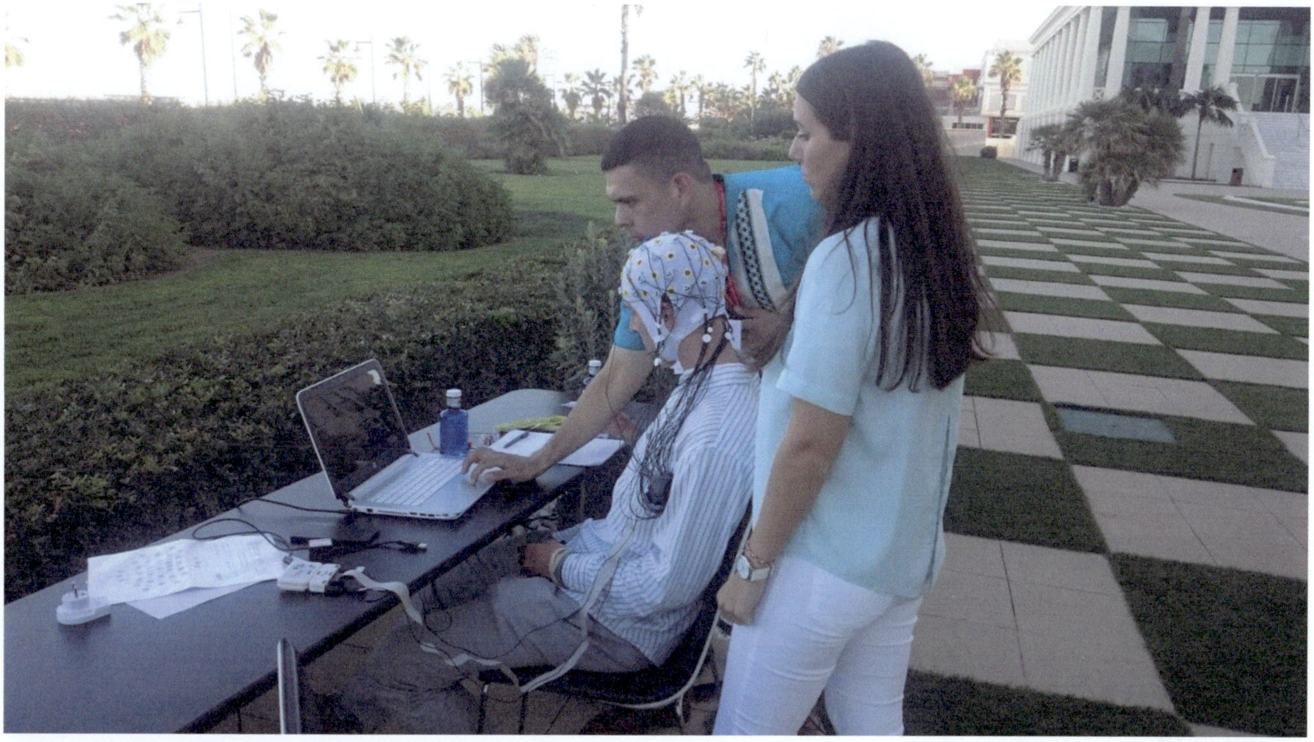

Fig. 2 The team "Do, Re, Mi and EEG" is working on their project in a great environment

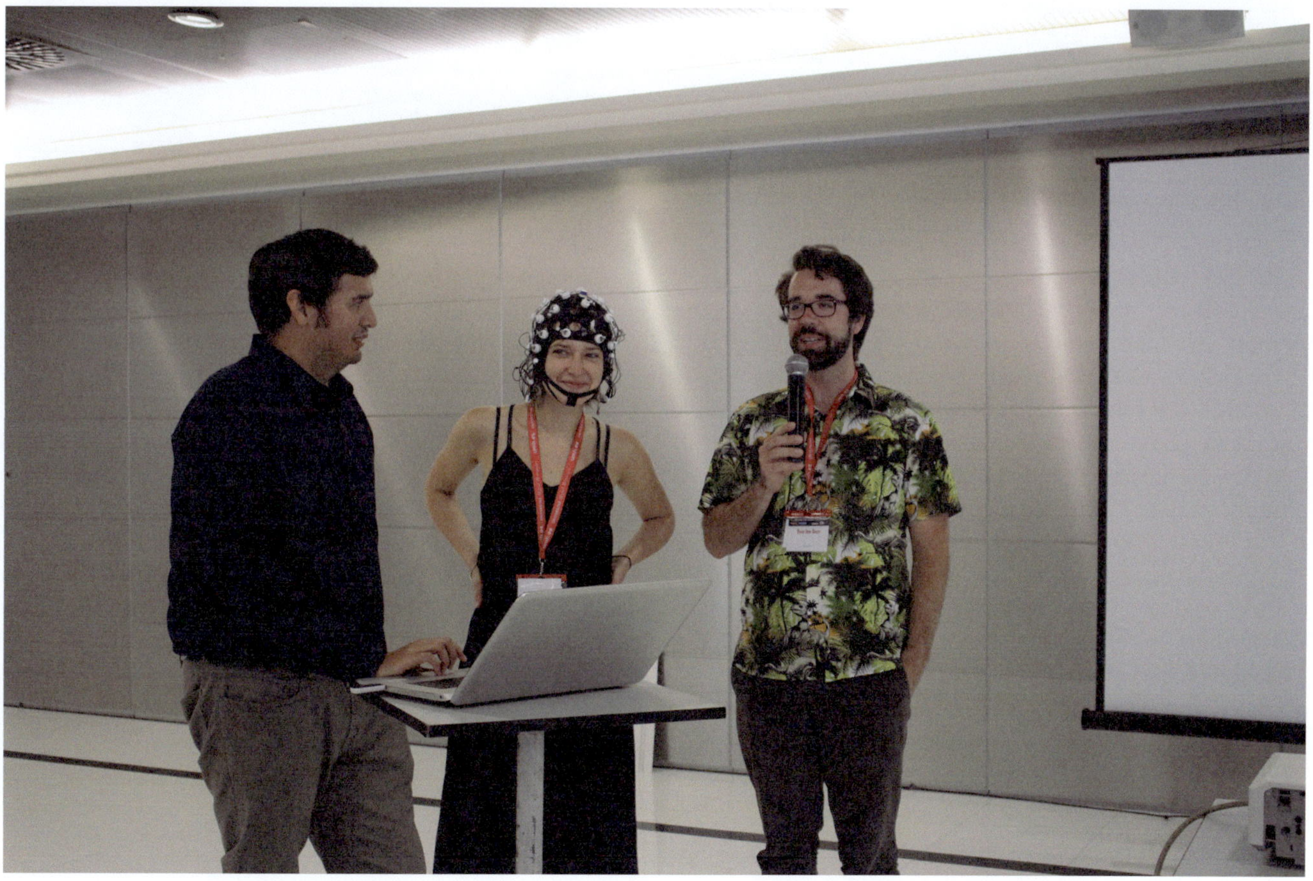

Fig. 3 The team "Artistic BCI" is presenting their project against the jury

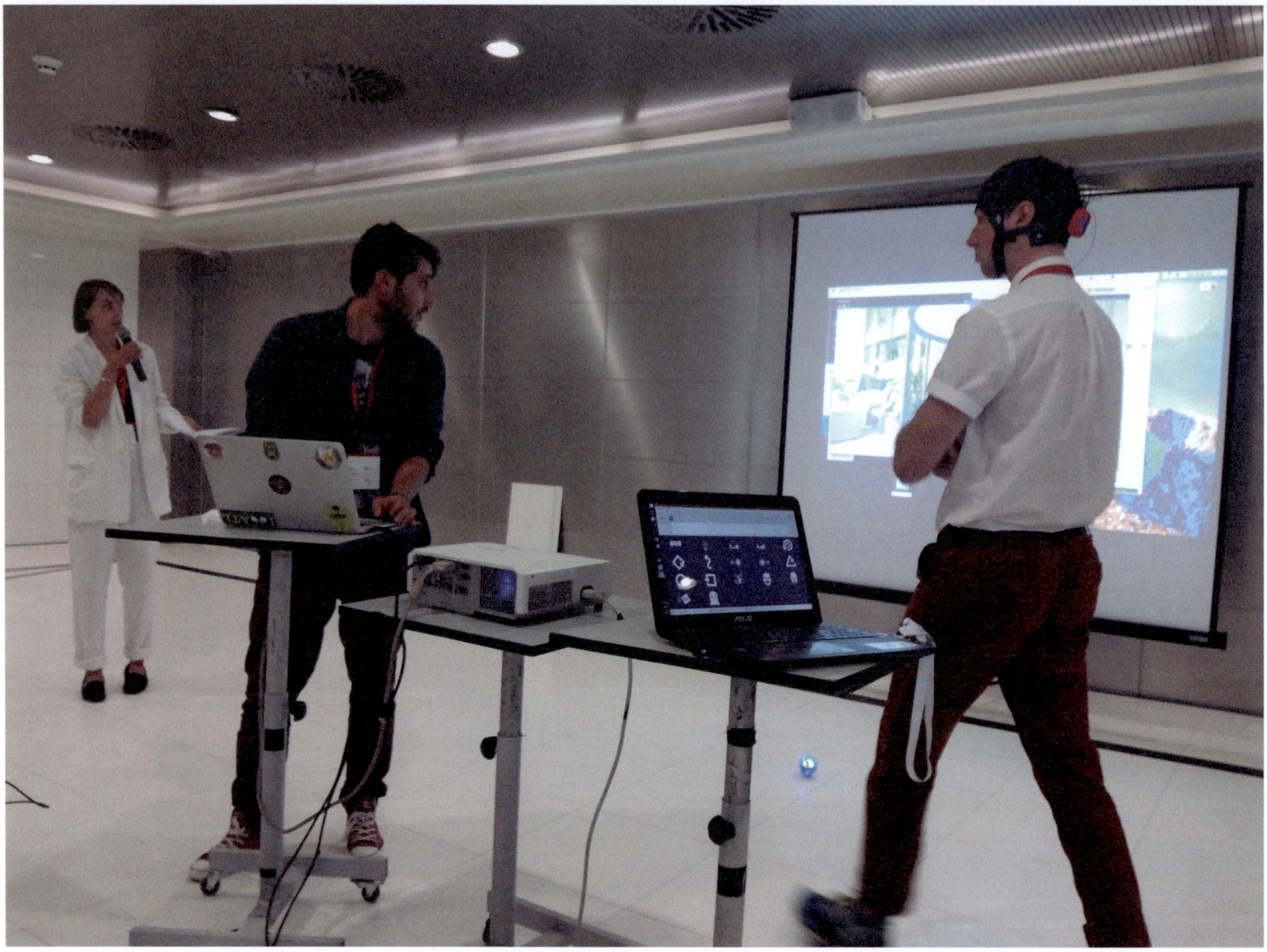

Fig. 4 The team "Sphero SPRK Control" is performing a demonstration during their presentation against the jury

devices, robots or applications, write messages or draw paintings by using their thoughts alone. These are the most common projects in BR4IN.IO and other BCI hackathons like P300 smart home control, Sphero SPRK control, dream painting, orthosis control, flight control, camera control, e-puck control, social media control, functional near infrared spectroscopy and EEG control, unity games.

- ARTISTIC PROJECTS: BR41 N.IO challenges creative minds to design and build a unique, playful, and wearable headpiece that can measure useful EEG signals in real-time to create any sort of interaction. For the development of these projects, 3D printers, handcraft materials, and sewing machines are provided at BR4IN.IO hackathons, so the teams can design and prototype their own BCI headpieces. Teams have also used BCIs to produce artful paintings or post a status update in their Social Media accounts. FLAGSHIP PROJECTS: A few BR4IN.IO hackathons have featured "Flagship projects" with special devices that are not available in most hackathons. For example, hackers have used BCIs to control heavy equipment for excavation and massive robot arms, see Fig. 7.
- KIDS' PROJECTS: Some kids also have participated in BR4IN.IO hackathons. In these projects, kids have created their own head accessories that are inspired by animals or mythical creatures. These projects are most similar to artistic projects.

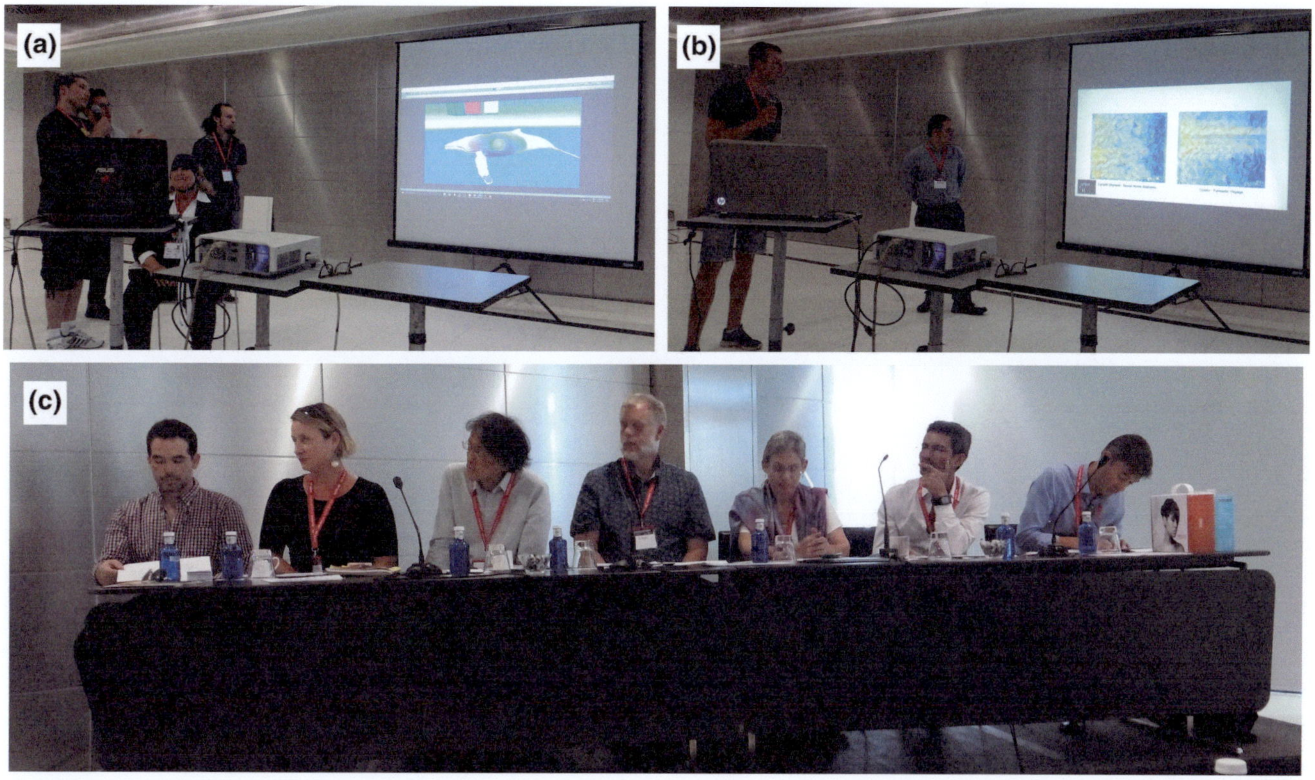

Fig. 5 **a** The team "Waves of Creation" shows a real-time performance of the project painting a whale with the EEG signals. **b** The team "Dream Painting" explains before the jury how they create images with brain waves obtained while sleeping. **c** The seven member of the jury attending the presentation of the hackathon teams

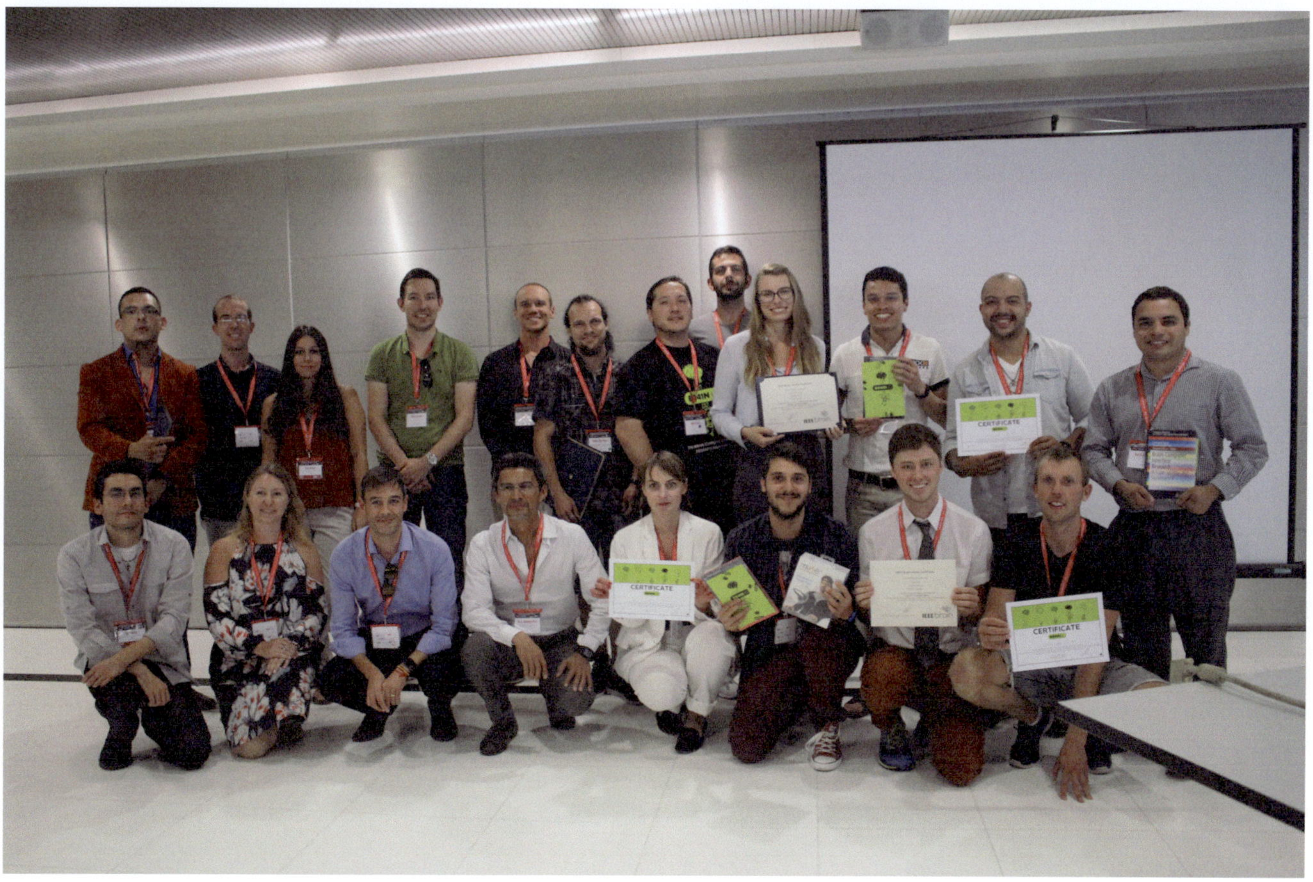

Fig. 6 Picture of all the teams awarded in the hackathon with the chairs of BOA conference (3rd and 4th position from the bottom left)

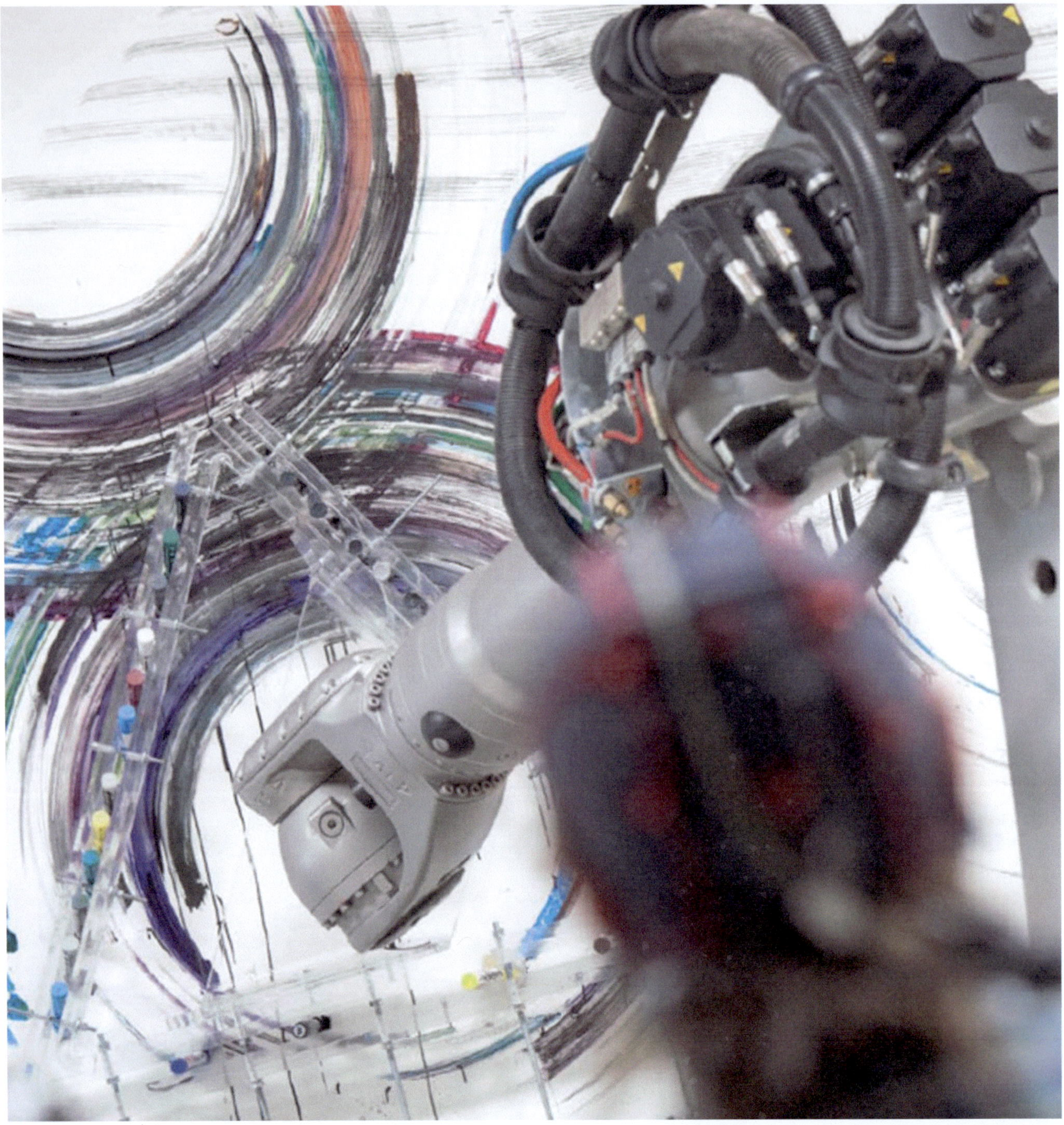

Fig. 7 An artist uses a robot arm to draw

5 Conclusion

Hackathon success is based on the interesting approach to the scientific method it provides. In the case of B2AI hackathons, the interaction between the different backgrounds of the participants allows an enriched observation of reality. Because, is not art perceived by our brain as an alternative vision of reality indeed? The starting hypothesis is that our EEG brain signals can be used to express our creativity. From this hypothesis, different ideas and concepts are developed and tested through the hackathon by the teams. The present chapter has introduced the elements that make this possible. Following chapters will show some successful examples developed in the BOA conference hackathon held in Valencia in September 2017.

Acknowledgements The authors would like to thank all the sponsors that made possible the 2017 BOA conference hackathon: IEEE SMC, IEEE brain, Brain Products, g.Tec, Emotiv, OpenBCI, and BR41 N.IO.

References

1. Briscoe, G., Mulligan, C.: Digital Innovation: The Hackathon Phenomenon (2014)
2. Grey, F.: Creativity unleashed. Nat. Nanotechnol. **10**, 480 (2015)
3. Guger, C., Allison, B.Z., Müller-Putz, G.R.: Brain-computer interface research: a state-of-the-art summary. Brain-Computer Interface Research **4**, 1–8 (2015)
4. Pfurtscheller, G., Neuper, C., Guger, C., et al.: Current trends in Graz d-computer interface (BCI) research. IEEE Trans. Rehabil. Eng. **8**, 216–219 (2000). https://doi.org/10.1109/86.847821
5. del Millán, J.R., Carmena, J.M.: Invasive or noninvasive: understanding brain-machine interface technology. IEEE Eng. Med. Biol. Mag. **29**, 16–22 (2010). https://doi.org/10.1109/MEMB.2009.935475
6. Costa, Á., Salazar-Varas, R., Úbeda, A., Azorín, J.M.: Characterization of artifacts produced by gel displacement on non-invasive Brain-Machine Interfaces during ambulation. Front Neurosci **10**, 60 (2016). https://doi.org/10.3389/fnins.2016.00060
7. Urigüen, J.A., Garcia-Zapirain, B.: EEG artifact removal - State-of-the-art and guidelines. J. Neural Eng. **12**, 031001 (2015). https://doi.org/10.1088/1741-2560/12/3/031001
8. Müller-Putz, G.R., Scherer, R., Pfurtscheller, G., Rupp, R.: EEG-based neuroprosthesis control: A step towards clinical practice. Neurosci. Lett. **382**, 169–174 (2005). https://doi.org/10.1016/j.neulet.2005.03.021

Mario Ortiz is an Assistant Professor and Researcher in the Brain–Machine Interface Systems Lab at Miguel Hernández University of Elche (Spain). He holds a Ph.D. in Industrial Technologies. His research activities focus on mathematical transforms applied to Brain-Machine Interfaces. He is also a passionate oil painter.

Eduardo Iáñez is an Assistant Professor and Researcher in the Brain–Machine Interface Systems Lab at Miguel Hernández University of Elche (Spain). He holds a Ph.D. in Communication and Industrial Technologies. His research interests are Brain–Computer Interfaces (BCI) to improve rehabilitation therapies.

Christoph Guger holds a Ph.D. about brain–computer interfaces. He works for Guger Technologies OEG and g.tec medical engineering Gmbh, both based in Austria. The companies develop hardware and software for biosignal acquisition and analysis. Christoph Guger is partner in several EU-funded projects.

Jose M. Azorin is the Director of the Brain–Machine Interface Systems Lab and Full Professor at Miguel Hernández University of Elche (UMH), Spain. He holds a Ph.D. in Robotics. He has been a visiting professor at the University of Houston (USA) and at Imperial College London (United Kingdom).

True Integration; the MoBI Hackathon for STEM Informing Arts and Arts Informing STEM

Tess Torregrosa

1 Active Learning in Science Education Pedagogy and Its Need in Higher Education

As a chemical engineering Ph.D. candidate, one of the important takeaways that I have learned from my doctoral experience is that scientists do not expect others outside of our respective fields or broader audiences to fully understand our work. In lectures, conferences, and posters we communicate our science through passive learning, a teaching strategy that presents the audience with information without taking into account audience feedback. This kind of communication is akin to throwing balls or ideas at an audience without realizing if they are catching them. A better way to approach sharing our knowledge is to create a dialogue of teaching and learning with our audience, realizing if they are catching the balls and allowing them to throw them back. The back and forth sharing of ideas or learning by doing is called active learning or immersive learning.

Active learning is highly supported by the American Association for the Advancement Science [6], which publishes teaching materials that anyone can access at www.scienceintheclassroom.org. In the classroom, it is been shown that students and teachers (namely graduate students) who participate in active learning gain skills that drive analytical thinking [4, 7]. In museums, passive learning with guided tours results in lower levels of learning for children on the tour [3], whereas a well-designed activity supported by active learning pedagogy allows children to come to learning on their own [5]. Active learning fosters creativity which is not inherently part of our thinking and is mastered through diverse thought experiments and reiterations [8, 10]. Creativity is a broad term that takes into account the following attributes: problem identification and observation, detecting new patterns or combinations, originating new patterns through analogies, body thinking and empathizing, being comfortable with no one correct answer, and articulating new ideas [1, 8–10]. Creativity is highly sought in the workforce and both employers and educators agree that educational degrees demanding abstract or critical thought is the best indicator [2].

Teaching awareness in visual cues, tone of voice, and language to connect to a broader audience through active learning takes practice for scientists that we may not have in our everyday lives, myself included. Working in a lab surrounded by my peers in the field of neuroscience does not lend many opportunities to practice bouncing ideas around with other people with different perspectives on life through culture, careers, or ways of thinking. It is imperative that scientists have supplementary experiences to practice what engaging in active learning means for them and learning the foundational communication skills such as listening, empathizing, breaking down technical language to continue practicing active learning and active teaching. I have found that hackathons are one of the best ways to do this. Hackathons challenge all participants to teach and learn from each other using active learning all within a day or two in order to reach their goals.

2 Active Learning in the Hackathon Environment

Hackathons are a deep dive into active learning for a condensed time period. Hackathons are known in the computer science community and have been extended to other disciplines. They are featured at college campuses and tech companies that attract people who are ready and willing to take an idea, iterate it into a prototype, and then sell their idea to an audience. A hackathon brings people together either from different disciplines or those who normally do not work together to build a prototype of an idea. There are

T. Torregrosa (✉)
Chemical Engineering Department, Northeastern University, Boston, MA, USA
e-mail: torregrosa.t@husky.neu.edu

J. L. Contreras-Vidal et al. (eds.), *Mobile Brain–Body Imaging and the Neuroscience of Art, Innovation and Creativity*, Springer Series on Bio- and Neurosystems 10, https://doi.org/10.1007/978-3-030-24326-5_16

three different phases that rely on clear communication in hackathons. First, there is the ideas phase. The hackathon starts with pitches or ideas that people already have for a problem in the world that they want to solve. In the ideas phase, the solutions to that problem are more fleshed out, experiencing pivots or changes along with the way in addition to building on layers of logistical details. The teams then enter the prototyping phase where individuals work on parts of the project; coding, gathering data, and testing the prototype. Both of these phases can move back and forth between each other through the bulk of the hackathon. When the individuals come together in the end because of the time constraints, the team enters the selling phase. This phase is especially important to clearly convey to the other teams and judges about their particular solution to their problem. Teams who have the most persuasive arguments backed by their prototype win the hackathon. At the MoBi conference, the hackathon component went further than a traditional hackathon both in the participants, bringing together international scientists who were interested in the arts and neuroscience, and the call for solutions to use technology that were at the intersection of art and science.

3 A Personal Experience from the MoBI Hackathon

I study and work with the sympathetic and parasympathetic parts of the autonomic nervous system, a subcategory of the peripheral nervous system, outside of the brain of which this conference was based on. I did not know what to expect coming to this conference or hackathon except that I would be surrounded by people like me who were interested in the intersection of art and science. I was paired with Fabio, an engineering Ph.D. student doing research on brain–computer interfaces at the Universtitat de Girona in Catalona, Spain. We were given the challenge to create images using our thoughts and we had a little under 36 hours to go from an idea to a working prototype (Fig. 1).

Right away one of the hurdles Fabio and I had to navigate was a language barrier between the two of us with my limited knowledge of Spanish and his limited knowledge of English. We really needed to rely on the words we knew in neuroscience and working closely together on our different components. We found that the technology that was provided to us was limited for the scope of the prototype we wanted to create. The technology allowed us to draw circles and squares with our thoughts by recognizing patterns in our brain waves when we thought of a square or circle but was limited in freeform drawing. We wondered if we could create a prototype that was able to draw a smooth line with our thoughts for people who may be paralyzed or for ourselves who couldn't communicate with words. Freeform drawing gives a user much more space to create rather than squares and circles. And so Fabio got to work building the code to recognize patterns in brain waves to control the direction of a point based off of the background code of in the technology that we were given. I learned how to put on an electroencephalogram (EEG) headset that measured my brain activity and had to train the program that connected to

Fig. 1 Hackathon teams working together with Brain Products staff to experiment with electroencephalography (EEG) caps to control movements of a ball or a virtual paintbrush with thoughts. From left to right; Julia, Guillermo, Martin, Tess, Fabio, and Javier

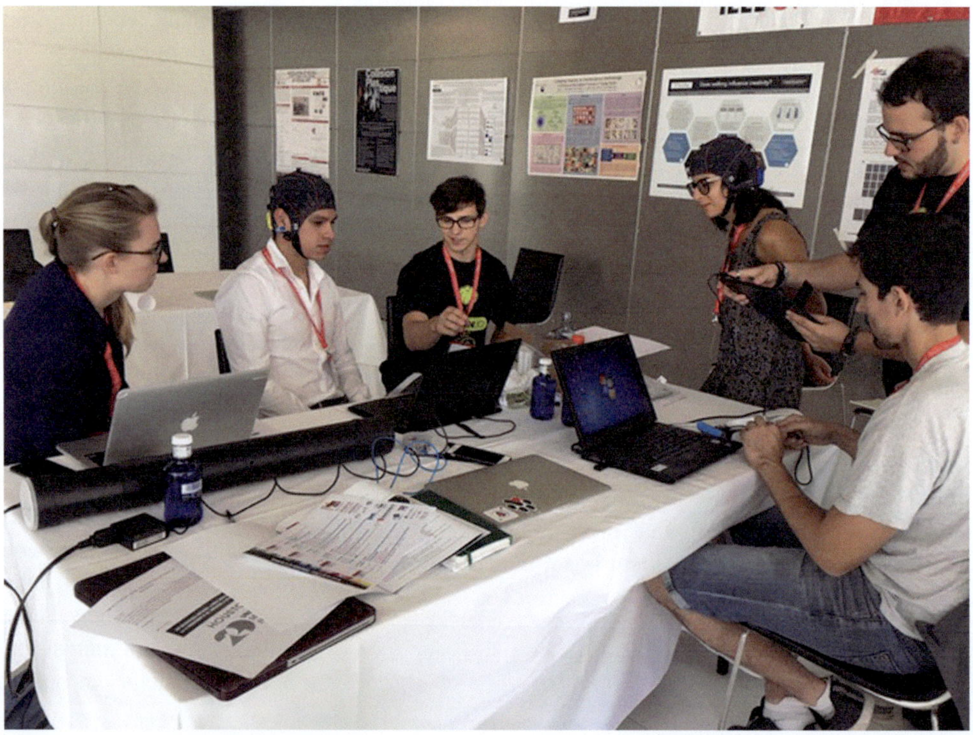

the EEG to recognize patterns in my thinking. Even though Fabio was a stronger coder, we had to work out theoretical problems of how the code could function within our time constraint of the hackathon and how the prototype would run. This is when we could "speak" the same language of science. Fabio could show me his code and go through the logic behind it and I could make suggestions. Even though I am not a coder, Fabio and I both engaged in active learning from each other because we could both relate to our shared idea of what a solution would look like. My other experiences with coding were in college classes that had us follow a guide or recipe of what we were supposed to produce. Helping Fabio code, I was in a completely different mindset because the end product was unknown. We did not know if certain strategies in the code would work for our solution (and more often they did not) so we had to keep approaching the code from different perspectives. This approach was more like research in the lab where the process is often more important rather than the deliverable.

Another unique characteristic of this particular hackathon was that art was always at the forefront of our minds from pitching ideas, to prototyping, and lastly communicating our solution. Art was woven into the pitches of solutions because the problem was inherently art related. More interestingly was the iterative approach as a feedback loop between art and science thinking, a true integration. Iterations of our theory were also influenced by the capability of the software to maximize the potential of art to be created and how the artist could interface with the technology. Both of these features informed how we wanted the experience to feel like which was translated into code. The silo between who was an artist and scientist was blurred in this hackathon. Fabio and I both happened to be trained as scientists but this context we were both creators in the sense of purity of artistic ideas and technical support.

At the end both the ideas and prototyping phases, we had a rough prototype that was able to draw a line in a box with code in Matlab. The last part of any hackathon is selling your idea to judges. This is where Fabio and I came up short. We spent so much time communicating in our little bubble to ourselves that we did not spend time thinking about communicating to a broader audience. Upon reflection, it makes sense to me why this was our weakest part of the hackathon because communicating to a broader audience takes practice and we do not normally have that in our lives as Ph.D. students. I took change away from the MoBI hackathon the awareness that I should seek experiences, like hackathons, saturated in opportunities to practice communication and actively learn from others. Engaging in active learning and teaching can help any STEM student clearly share their ideas in a team, translate those ideas into results, and be able to promote those results to other people.

4 Creating a Successful Integrative and Collaborative Hackathon

Hackathons are usually marketed towards computer science, design, and business savvy students. I believe the key to a successful cross-discipline hackathon is to have a foundation that anyone who wants to participate thinks that they can contribute valuable information no matter their discipline. One of the limitations of the MoBi hackathon is that to be accepted into the entire conference, including the hackathon, graduate students needed to apply self selecting those who participated. Students accepted were already in the mindset that both the STEM (Science Technology Math Engineering) and art and design students had equally valid and important ideas. Conferences that do not have this self-selecting process in the beginning should be aware of the messaging to potential participants to be inclusive of all majors. Additionally, staff working the hackathon including the volunteers, judges, and mentors should be supportive of art and design integration into an event that is normally STEM heavy. Once the hackathon starts, students will find out on their own that communicating to each other may be difficult at first but working toward a common goal, a successful prototype, and pitch, they may ultimately realize that success comes with true collaboration rather than staying siloed.

5 Summary: Lessons Learned

Traditional hackathons have the capacity to become more inclusive to other disciplines such as art and design. Hackathons promote active learning, a type of structure that enhances learning through a feedback loop with the participants and the constant information during the ideas, prototype, and selling phases. Individuals get feedback through iterative design of a prototype, working within a team, and proposing the design to a broad audience. This hackathon through the MoBi conference is an example of successful integration of art and design into the traditional hackathon structure by advertising a this event to both artists, designers, and STEM majors and encouraging equal importance on the contributions made by all disciplines in all levels of the support staff and prototype judges. Inclusive hackathons like this one demonstrate that collaboration between disciplines and integration of theories and ideas from design and art into STEM and vice versa can create novel and creative solutions.

The largest take away that I can share from this hackathon is the importance of flexible communication. Artists and designers have similar ways of thinking like STEM majors, we all go through an iterative process. Realizing this basis of thinking that is more similar than what we may perceive may help individuals from multiple disciplines acknowledge that different ideas in the group have the same validity. The key for sharing ideas for true integration and collaboration is the use of broad language so that all members of the team may learn from each other rather than having a bottle-neck for understanding jargon.

References

1. Ainsworth, S., Prain, V., Tytler, R.: Drawing to learn in science. Science **333**(6046), 1096–1097 (2011). https://doi.org/10.1126/science.1204153
2. Conference Board, I.: Ready to Innovate: Conference Board, Incorporated (2008)
3. Cox-Petersen, A.M., Marsh, D.D., Kisiel, J., Melber, L.M.: Investigation of guided school tours, student learning, and science reform recommendations at a museum of natural history. J. Res. Sci. Teach. **40**(2), 200–218 (2003). https://doi.org/10.1002/tea.10072
4. Freeman, S., Eddy, S.L., McDonough, M., Smith, M.K., Okoroafor, N., Jordt, H., Wenderoth, M.: Active learning increases student performance in science, engineering, and mathematics. Proc. Natl. Acad. Sci. **111**(23), 8410–8415 (2014). https://doi.org/10.1073/pnas.1319030111
5. Gentner, D., Levine, S.C., Ping, R., Isaia, A., Dhillon, S., Bradley, C., Honke, G.: Rapid learning in a children's museum via analogical comparison. Cogn. Sci. **40**(1), 224–240 (2016). https://doi.org/10.1111/cogs.12248
6. Luken, J.: Universities and the teaching of science. Science **306** (5694), 229–230 (2004). https://doi.org/10.1126/science.306.5694.229
7. Roden, J.A., Jakob, S., Roehrig, C., Brenner, T.J.: Preparing graduate student teaching assistants in the sciences: an intensive workshop focused on active learning. Biochem. Mol. Biol. Educ. (2018). https://doi.org/10.1002/bmb.21120
8. Root-Bernstein, R.: Arts and crafts as adjuncts to STEM education to foster creativity in gifted and talented students. Asia Pac. Educ. Rev. **16**(2), 203–212 (2015). https://doi.org/10.1007/s12564-015-9362-0
9. Root-Bernstein, R.S.: The arts' unsung role in supporting science. The Educ. Dig. (1997)
10. Root-Bernstein, R.S., Root-Bernstein, M.: Sparks of Genius: The Thirteen Thinking Tools of the World's Most Creative People. Houghton Mifflin Co., Boston, MA (1999)

Tess Torregrosa is a chemical engineering Ph.D. candidate at Northeastern University. Her research focuses on the mechanisms in the autonomic nervous system behind heart failure. Additionally, she is passionate about education policy and art theory and design principle integration into the STEM curriculums that engage students in the STEM pipeline.

Neurofeedback During Creative Expression as a Therapeutic Tool

Stephanie M. Scott and Lukas Gehrke

1 Introduction

The relationship between art and medicine has a long-documented history. Dating back as far as the sixteenth century, artists such as Leonardo da Vinci used knowledge of human anatomy to depict the human form more accurately. Although the practice of integrating artistic elements with health and science concepts translates across cultures, it was not until the early twentieth century that, for example, the United States began to incorporate art within hospitals, mainly through the Works Progress Administration (WPA) efforts. Initial creative arts therapy practices began in the 1940s as a means for recovery of soldiers returning home from the second world war. Since then, various interdisciplinary and collaborative efforts have been put forth that focus on enhancing and humanizing the health care experience through integrating art within these endeavors [1]. Projects that incorporate the arts in ways that promote public health and enhance community engagement continue to be explored, but many arts and health scholars argue that there is a need for more research and exploration to be conducted within this field [2].

2 Embracing Social Technologies

Technological development presents opportunities for researchers and developers alike to explore innovative ways for increasing the likelihood of adoption and application of new methods of communication. Efforts directed towards improving the utility, efficiency, function, and design of new tools often support these efforts; however, these new technologies often undergo modifications and alterations which impact the ways in which a user participates and interacts with a given tool [3]. This fluctuation that occurs within the user/interface relationship depends not just on the properties of the technology itself, but also on the user's own needs and abilities. The individual variance emphasizes the importance of evaluating society's use of newer and more advanced methods of communication, as well as the ways in which this continual development can directly influence the ways we interact with one another. As such, it is especially important to assess how these tools are both designed and applied, because, as the technical properties of the tools themselves change, so too do our responses to them [4]. This presents the collective social with the possibility to create a new language through how we design the new tools that guide new types of interactions accompanied by new sets of meanings [5].

Although an initial objective of BCI research and technological development was to enable basic forms of social interaction for patients, the potential exists to not only restore and enhance communication for the motor-impaired, but to extend them to include opportunities for creative expression [6] and therapeutic care. This epistemological approach into the intersection of how these sophisticated technologies mediate communication, enable cognitively embodied interactions, and afford users the ability to share subjective and collective experiences through artistic interventions, can encourage new conceptual understandings as to how new boundaries of digital and physical user-system interactions can explored and further applied. Scholars argue that science and art can enrich and interact in ways that are meaningful and contribute towards positive social and cultural progress. They suggest that art provides a tool that enables us to enhance our knowledge about how aspects of how our minds work, and that this knowledge becomes realized through our experiences and the ways in which we interact with the world [7]. Similarly, they argue that this

S. M. Scott
Department of Media Communications,
Colorado State University, Fort Collins, CO, USA

L. Gehrke (✉)
Biological Psychology and Neuroergonomics, Technische
Universität Berlin, Berlin, Germany
e-mail: Lukas.gehrke@tu-berlin.de

J. L. Contreras-Vidal et al. (eds.), *Mobile Brain–Body Imaging and the Neuroscience of Art, Innovation and Creativity*,
Springer Series on Bio- and Neurosystems 10, https://doi.org/10.1007/978-3-030-24326-5_17

process is the key to observing and responding to art, in that it is not necessarily in our awareness, but that our reaction to art engages us [8, p. 13].

3 Towards Neurofeedback Drawing Canvases

Rehabilitative trainings and interventions potentially involve a high level of repetitions and may be accompanied by frustrations on both, the patients as well as the caregivers side. In this project, we explored the challenge to use neurofeedback, specifically the visualization of EEG band power, in a setting of artistic self-expression. Art therapy makes use of artistic expression as a therapeutic means and we aimed at addressing two points, (1) to promote (artistic) self-awareness, rehabilitation and recovery as well as (2) extend the storytelling lens for users in a way that impacts understanding and perspective of family and friends towards complex health technologies and complicated health issues. Recently, first evidence of the real-life applicability of neurofeedback therapeutic interventions has been provided [9]. Using a neurofeedback setup relying on motor imagery signals, the authors report that "patients enjoyed the training and were highly motivated throughout". These behavioral effects were observed alongside long term effects in both functional (EEG) as well as structural (MRI) measures comparing pre-and post-measurements. Self-directed plasticity [10, 11], the idea that intentionally perturbing distributed brain system dynamics in a desired direction or way of functioning, is the core concept of the benefits of neurofeedback training, see [11] for a comprehensive review. Framing neurofeedback training efficacy in this way provides a foundation to assess the effectiveness of training interventions by investigating pre- and post-effects (a) behaviorally (b) using measures of EEG, such as functional connectivity, as well as functional MRI and (c) structural MRI effects [10]. Moving towards establishing these metrics of training effectiveness is an important step for neurofeedback training towards widespread acceptance across the expert as well as the general population.

Our primary objective was to promote (artistic) self-awareness during the process of painting. The concept originates in art therapy and we challenged ourselves to think of potential use-cases: (1) we conceived of patient populations, e.g., after stroke, potentially benefiting from having their own electrical brain activity visualized as a response to, or cause of, their own actions during therapeutic interventions; (2) to promote awareness of brain damage conditions in the medical field, patients social surroundings as well as in therapeutic care, and lastly (3) to situate this emerging technology in the realm of (art) therapy thereby hoping to alter potential misconceptions and fears by showing an alternative use as an interaction modality.

4 Proof-of-Concept

We recorded EEG data from 32 active dry electrodes (actiCAP Xpress Twist, Brain Products, Gilching, Germany) with the LiveAmp compact wireless amplifier (Brain Products, Gilching, Germany) sampled at 250 Hz. The data was streamed to the network using LabStreamingLayer's LiveAmp Plugin[1] from the recording computer. A LabStreamingLayer inlet on the presentation computer received the raw data in python,[2] data of 2s was buffered, then a bandpass filter (1–125 Hz) was applied on the 2s data window with a subsequent time-frequency decomposition using fast fourier transform to estimate power spectral density. Subsequently, we extracted power values for five typically selected EEG bands (delta, theta, alpha, beta, gamma)[3] and fed them to a visualization scheme, see Fig. 1.

Here, we used the power values to set the line height of consecutive lines to get an effect similar to Joy Division's Unknown Pleasures[4] album cover. The visualization output was projected onto a transparent podium paper holder using a projector connected through HDMI completing the closed-loop neurofeedback setup, see Fig. 2.

For the drawing, we gathered watercolors and brushes as well as a white drawing sheet put up on the paper holder. With this setup, we could maintain the desired see through effect. In this prototype, the participant did not receive any instruction concerning the drawing.

In future revisions, the following two points may be considered depending on patient condition and therapy goals. Firstly, we propose using source level instead of sensor level EEG dynamics using appropriate spatial filtering techniques [12]. To best target EEG features and/or source locations primarily affected by patient's condition, a good understanding of the affected EEG signatures is of high importance in maximizing intervention outcomes [13, 14]. Secondly, designing interfaces, here drawing surfaces, taking into account specific challenges various patient populations may face will have a significant effect on user's acceptance. We point out the possibility to use modern tablets with drawing pens as individualized drawing canvases.

[1] https://github.com/sccn/labstreaminglayer.
[2] Python Software Foundation. Python Language Reference, version 2.7. Available at http://www.python.org.
[3] https://en.wikipedia.org/wiki/Electroencephalography.
[4] https://en.wikipedia.org/wiki/Unknown_Pleasures.

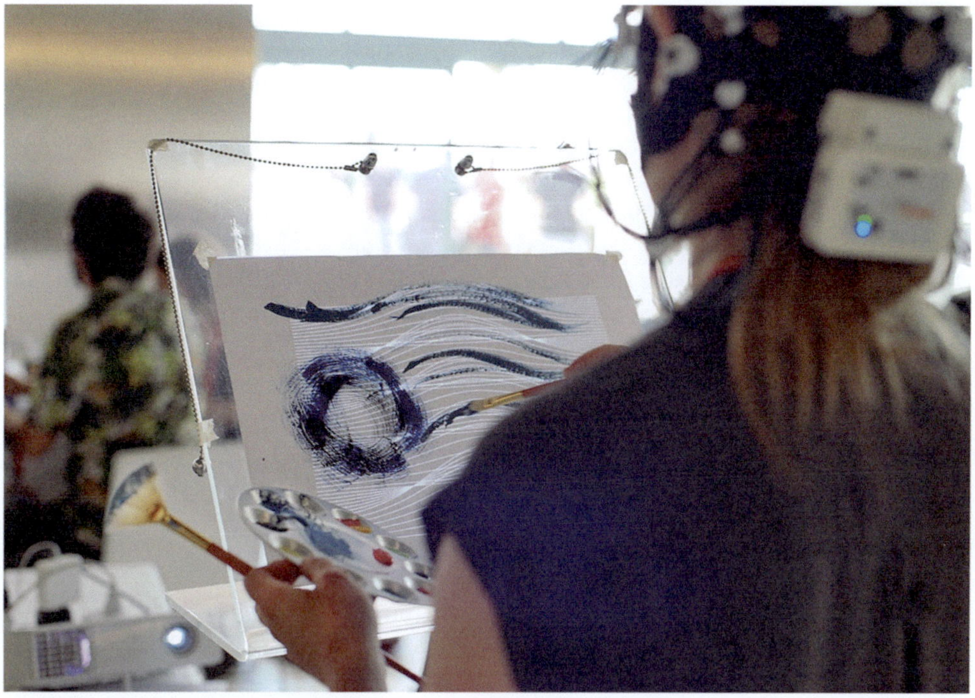

Fig. 1 Hackathon participant, Stephanie Scott, wears a mobile EEG cap equipped with 32 dry electrodes. EEG activity is processed online and extracted band power values are used in a closed-loop neurofeedback application. The background of a drawing canvas is updated live in response to, as well as a cause of her drawings

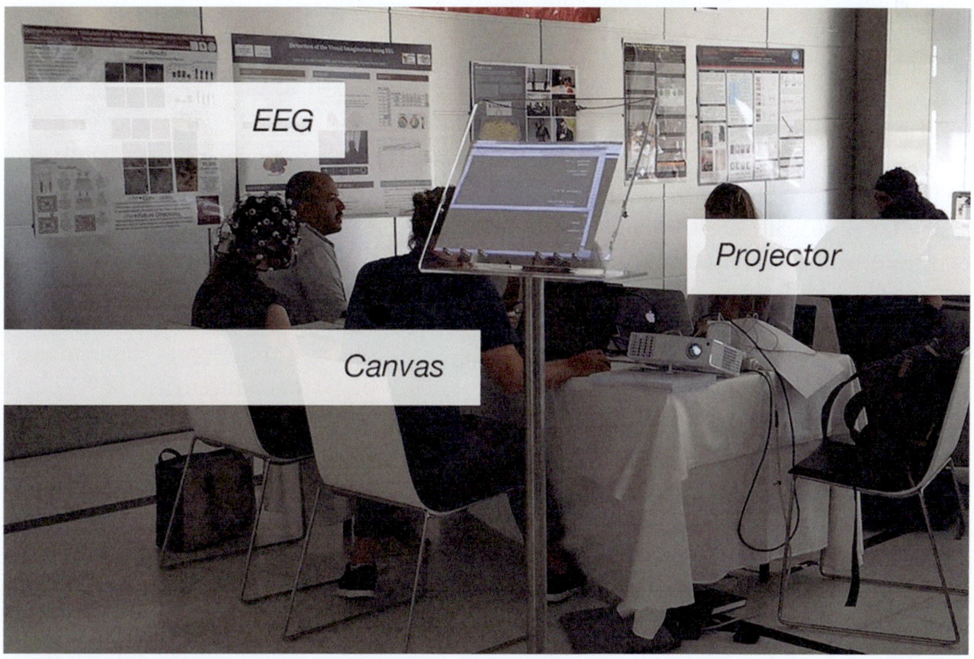

Fig. 2 Participant's EEG activity (32 channels) is wirelessly transmitted, preprocessed to extract band power and subsequently projected onto a transparent drawing canvas using the back projection setting of a projector. Band power values were displayed using a simple visualization adapted from the "Unknown Pleasures" album of British rock band Joy Division (not pictured here)

5 Outlook: The Future of "Neurodata" in Therapeutic and Rehabilitative Settings

Moving forward, development and implementation of new systems should consider a shift towards ideas that support technological mediation as a framework for design, and include analytical methods that support inclusive interactions. This type of shift would acknowledge the roles of technological, semiotic, and economic processes, as characteristics that constitute these mediums, and through a reflexive approach towards human and technological interactions, progress could focus on considerations that evaluate users' experience and agency through action. Reconfigurations and subsequent applications of the technology should explore new tools and boundaries of interface structures that emerge from conscious spatial design, allowing for participatory engagement with technology. This approach has the potential to great benefit current and future users of brain–computer interface (BCI) systems mobile brain-body imaging (MoBI) systems.

Scholars argue that the arts are effective at communicating across language and other cultural divides, resulting in improved social learning. Studies have demonstrated that the professionals interviewed feel that the arts empower health communication by engaging with people's emotions, attracting attention, focusing and clarifying messages, facilitating dialog, and cultivating solidarity [15]. Science and health communication scholars posit the idea that science, to some degree, is shaped by social forces, thus suggesting, that a "multivalent" approach would be the most successful for being able to interpret community concerns and "understand their effects on the practices and policies of science" [16]. Parrott and Kreuter [17] propose a similarly constructed transdisciplinary approach to communicating health issues that allow for the "intellectual integration" of medical, epistemological and behavioral approaches that can "transcend disciplinary boundaries." Integrating this type of approach to BCI and MoBI technologies and subsequent digital interaction spaces would allow researchers to identify the various micro-level indicators that are encompassed within the more macro-level concerns.

Through combining new technologies with visual educational strategies, along with the integration of knowledge from other disciplines, more innovative strategies towards communicating complex information about new health communication technologies can be developed and implemented. The visual feedback loop that neurofeedback offers, extends a user's vision from discovering what is present in the world and where it is, i.e., this "is why appreciation of the music or painting or other works of the creative person is also a creative act on our part" [8, p. 22]. Biological information presented through a feedback loop can be thus considered as representative of a different identity; it offers a view into an individual biological identity, thus encouraging identity construction through unique representations. Feedback enables creative engagement through an interpretive and reciprocal learning process between user and system.

This type of system could encourage and empower health communication by engaging with people's emotions, attracting attention, focusing and clarifying messages, facilitating dialog, and cultivating solidarity. Some research has found that art can bridge understanding of specific health conditions [8]. Integrating the process of making art with biological feedback allows the space for different types of information to be expressed, exposed and combined, leading to the possibility of new interpretations. This type of application could offer users a unique type of self-reflexive lens as well as an alternate perspective towards the user and system relationship. Embodiment is not just a state of being, but an emerging quality of interactions, and conscious technical design can provide innovative, inclusive and engaging spaces for users. Forward thinking, this type of application could be set to other forms of artistic engagement, such as music, and could also be extended and implemented in Brain–Computer Interface systems, thus offering users a full feedback loop through participation [18]. Additionally, these types of applications could eventually be paired with virtual environments that allow users to interact with one another in a gallery setting, allowing individuals to display and share their creative expressions. Facilitating the development of these types of digital spaces could allow new narratives and dialogues to emerge, and thus, mirroring the overarching goal of what transdisciplinary collaboration strives to create.

Personal statement and reflection of impact of engaging in art while having direct feedback from my individual neural signals-

"It was representative of information I had not been granted access to before. It served as an extended lens into information my body was creating to and responding to, but that I had never been privilege to. It was a tool that helped me identify with my health situation, but it also enabled me with a sense of self-efficacy through changing my perception of my own brain's mobility. The crude brain signals meant more to me than just being representative of raw data and signal acquisition."

-Traumatic brain injury (TBI) and Post-traumatic stress disorder (PTSD) patient/user-

6 Summary: Experiencing Multidisciplinary

This report summarizes the hackathon project entitled "neuroCanvas" at "Your Brain on Art Conference 2017" held in Valencia Spain. Throughout the four-day event, our project conceptualization and participation was focused on answering questions about how to effectively generate a multidisciplinary collaboration between our different concentrations. Including both scientific and artistic approaches towards creative problem-solving, helped us to identify the importance of combining artistic expression within educational and scientific research endeavors. This process helped us realize that the ways in which we perceive, respond and react to information and experiences are different, and although more difficult, it resulted in designing a project that we feel can have a definitive positive impact on users emotional and biological responses to their surrounding environments. The exercise of coding and decoding our different methodologies with one another through dialogue and experiment helped us to design an application that embodies a holistic approach to exploring the intricate intersections of our interpersonal experiences with one another, as well as with the world around us. Our participation has led us towards a better understanding of how important it is for scientists, artists, educators and therapists alike to recognize that a shared space for trust and exploration should be established in all collaborative endeavors. These communities must work together to identify research goals and objectives, and clearly identify the intent behind their efforts in order to create meaningful research and generate positive impact. Both scientists and artists need to reimagine and modernize their boundaries to create shared meanings and to have supportive spaces created and designed to foster these important types of multi-modal discourses.

7 Summary: Lessons Learned

This chapter summarizes the efforts and methods used to design an innovative neurofeedback application that integrates EEG and neurofeedback technologies with art therapy techniques. This proof of concept aims to provide users with a tool for exploring their individual biological data through creative means. It also extends the lens of self-discovery by pairing neurofeedback technologies with art therapy interventions. Additionally, this tool can be applied to existing training techniques and serve as a point of entry learning approach for users of Brain–Computer Interface systems.

This project was conceptualized and tested by the authors while participating in a sponsored Hackathon. Engagement with this event provided the authors with valuable insights; primarily the need for development of a roadmap necessary for promotion of future collaborative and interdisciplinary efforts. It also provided the space to explore how this could best be created between the authors. Likewise, it illustrated the unique value that each participant had to offer in a multidisciplinary setting. By allowing participants to share ideas and varied approaches to the assigned task, it initiated creative thinking as well as a willingness to expand existing frameworks towards the design process. Most importantly, this event highlighted the need for understanding the level of commitment, patience, and respect that is needed to take part in interdisciplinary endeavors. Working with individuals within different academic and professional fields can be challenging; however, truly innovative and inspiring work can result from participants taking the time to listen to and engage one another.

References

1. Sonke, J.: Professionalizing the arts in healthcare field. In: Lambert, P.D. (ed.) Managing Arts Programs in Healthcare, pp. 32–44. Routledge, New York (2016)
2. National Organization for Arts in Health (2017) Arts, Health, and Well-Being in America. https://thenoah.net
3. Cox, R., Depoe, S.: Emergence and growth of the "field" of environmental communications. In: Hansen, A., Cox, R. (eds.) The Routledge Handbook of Environment and Communication, pp. 13–25. Routledge, London (2015)
4. Webster, A.: Innovative health technologies and the social: redefining health, medicine and the body. Curr. Sociol. **50**(3), 443–445 (2002)
5. Winograd, T., Flores, F.: Understanding computers and cognition: a new foundation for design. Ablex Pub, Norwood, New Jersey (1986)
6. Zickler, C., Halder, S., Kleih, S.C., et al.: Brain painting: usability testing according to the user-centered design in end users with severe motor paralysis. Artif. Intell. Med. **59**, 99–110. https://doi.org/10.1016/j.artmed.2013.08.003
7. Kandel, E.: Reductionism in Art and Brain Science: Bridging the Two Cultures. Columbia University Press, New York (2016)
8. Potash, J.S.: Guided Relational Viewing: Art Therapy for Empathy and Social Change to Increase Understanding of People Living with Mental Illness. Dissertation, University of Hong Kong (2010)
9. Zich, C., Debener, S., Schweinitz, C., et al.: High-intensity chronic stroke motor imagery neurofeedback training at home: three case reports. Clin. EEG Neurosci. 48:403–412 (2017). https://doi.org/10.1177/1550059417717398
10. Orndorff-Plunkett, F., Singh, F., Aragón, O.R. et al.: Assessing the effectiveness of neurofeedback training in the context of clinical and social neuroscience. Brain Sci. (2017). https://doi.org/10.3390/brainsci7080095
11. Ros, T., Munneke, M.A., Ruge, D., et al.: Endogenous control of waking brain rhythms induces neuroplasticity in humans. Eur. J. Neurosci. (2010). https://doi.org/10.1111/j.1460-9568.2010.07100.x
12. Cohen, M.X.: Comparison of linear spatial filters for identifying oscillatory activity in multichannel data. J Neurosci. Meth. **278**, 1–12 (2017). https://doi.org/10.1016/j.jneumeth
13. Monge-Pereira, E., Molina-Rueda, F., Rivas-Montero, F.M., et al.: Electroencephalography as a post-stroke assessment method: an updated review. Neurol. English Edn. **32**(1), 40–49 (2017). https://doi.org/10.1016/j.nrleng.2014.07.004
14. Lizio, R., Vecchio, F., Giovanni, B.F., et al.: Electroencephalographic rhythms in alzheimer's disease. Int. J. Alzheimer's Dis. (2010). https://doi.org/10.4061/2011/927573
15. Sonke, J., Pesata, V., Nakazibwe, V., Ssenyonjo, J., et al.: The arts and health communication in Uganda: a light under the table. Health Commun. **33**(4), 401–08 (2018). https://doi.org/10.1080/10410236.2016.1266743
16. Gilbert, J.K., Stocklmayer, S.: Preface: the changing face of science communication. In: Gilbert, J.K., Stocklmayer, S. (eds.) Communication and Engagement with Science and Technology: Issues and Dilemmas, a Reader in Science Communication. Routledge, New York, pp viii–xi (2013)
17. Parrott, R., Kreuter, M.W.: Multidisciplinary, interdisciplinary, and transdisciplinary approaches to health communication: where do we draw the lines? In: Thompson, T.L., Parrott, R., Nussbaum, J.F. (eds.) The Routledge Handbook of Health Communication, 2nd edn, pp. 3–17. Lawrence Erbaum Associates, New York (2011)
18. Scott, S.M., Raftery, C., Anderson, C.: Advancing the rehabilitative and therapeutic potential of BCI and noninvasive sensing systems. In: Nijholt, A. (ed.) Brain Art. Springer, Cham (2019)

Stephanie M. Scott serves a communications analyst for The Brain–Computer Interface Laboratory at Colorado State University, and is a graduate student researcher within the Journalism and Media Communications Department at CSU. Her work involves developing more effective communication strategies related to neuroadaptive technologies and neurosensing systems.

Lukas Gehrke is a cognitive scientist at TU Berlin. He researches human spatial navigation using rich virtual environments and mobile EEG. Tinkering around with state-of-the-art VR technology, he strives to advance neuroscientific research methods. Like all other millennials, he sometimes updates his work on a personal webpage at lukasgehrke.com.

Do, Re, Mi, and EEG: An Analysis of Imagined and Performed Musical Settings

Jorge A. Gaxiola-Tirado, Aaron Colverson, and Silvia Moreno

1 Introduction

The study of brain connectivity in the neuroscience of music has been increasing in recent years. This approach aims to understand the emergence of functional networks in the brain. Diverse studies have reported that music listening is traceable in terms of network connectivity and activations of target regions in the brain [10]. Listening, however, is complementary to the act of performing. Recently, researchers have explored the positive effects of music on our health and well-being, specifically when music leads to an aesthetic experience [2]. Under this concept, the cerebral mechanisms of musical improvisation have been studied, considering the improvisation as an instantaneous creative behavior [12]. With this in mind, we have created and carried out a pilot study to investigate brain connectivity based on electroencephalography (EEG) signals acquired during imagined and performed musical settings. The method utilized is based on Partial Directed Coherence (PDC), with a simple implementation and an easy interpretation of the revealed interconnections. Thereby, knowing more about the mental states in music we could generate therapies, educational methods, as well as using imagined music as a possible paradigm for a brain–computer interface (BCI).

2 Intersection of Art, Science and Engineering/Technology: Challenges and Solutions

Creating music according to EEG signals is a difficult task to accomplish, particularly when the collaborating individuals tasked to do so come from very different disciplinary backgrounds. Initially, communication barriers were a significant challenge, with team members including a biomedical engineer, musician, and computer scientist. Our respective languages seemingly precluded our collective success, but as the project progressed, our individual agency, creative intuition, and timeline to deliver drove us to produce a product greater than the sum of its parts.

We jumped in the deep end at the outset of the Hackathon and within an hour had created a topic, methodology, individual assignments, plan B, C, D, and perhaps most importantly, established a fervent belief in our collective efforts. The environment at the Hackathon bolstered this belief, with eight other teams participating, roaming advisers, technical support, and a general air of excitement looming in the space. Coffee was also much appreciated, but overall, we were truly excited to collaborate at these intersections of art and science.

Our topic—measuring and analyzing functional networks in the brain during varying modes of music performance—required us to move quickly to achieve our deadline. Just two days from meeting one another for the first time, we had to present a final product. Therefore, designing and running multiple trials of our experimental paradigm, recording data, and analyzing and reporting those data needed to be expedited. Nevertheless, our efficiency perhaps proved more resilient through our collaboration than individual disciplinary pursuits alone, as we moved past communication barriers and produced in record time.

Interdisciplinary team work at the intersections of art and science requires each member to perform their best, as their contribution to the whole is the only representative

J. A. Gaxiola-Tirado (✉)
CINVESTAV, Monterrey's Unit, Apodaca, NL, Mexico
e-mail: jgaxiola@cinvestav.mx

A. Colverson
Department of Ethnomusicology, University of Florida, Gainesville, FL, USA

S. Moreno
Brain–Machine Interface Systems Lab, Miguel Hernandez University of Elche, Elche, Alicante, Spain

J. L. Contreras-Vidal et al. (eds.), *Mobile Brain–Body Imaging and the Neuroscience of Art, Innovation and Creativity*, Springer Series on Bio- and Neurosystems 10, https://doi.org/10.1007/978-3-030-24326-5_18

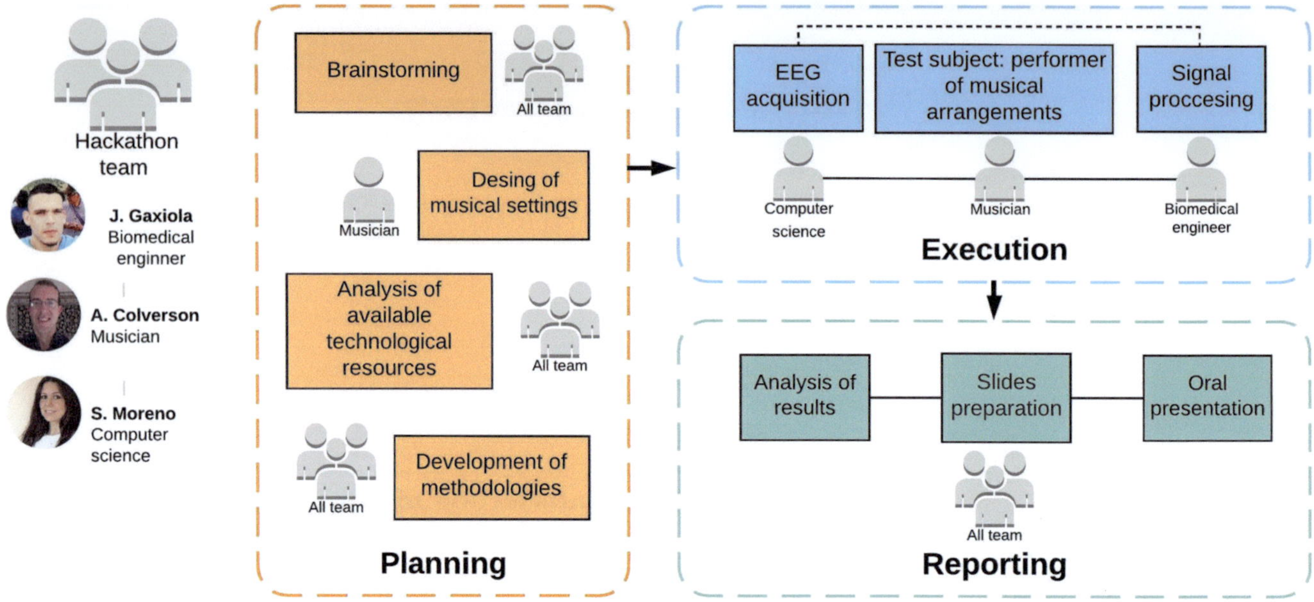

Fig. 1 Collaborative methodology used in our Hackathon project. The three phases involved in this process (planning, execution, and reporting) are shown. In each phase, the tasks performed by each team member are presented

component of that aspect of the whole. In our case, however, the alignment of each member's performance produced another level of required collaborative effort in order for the final product to be presentable. This alignment defines the intent behind our collaborative methodology to a certain extent (see Fig. 1), such that the placement of the EEG cap on the musician's head, individual channel readings from the EEG cap properly coordinated with computer software, and execution by the musician to perform the required musical tasks fed into one another as a circular feedback loop. Thankfully, we did not face any execution errors on the part of any of our members contributing to this whole and we were able to successfully reach our required timeline.

The above paragraph spurs consideration of a multitude of contributing perspectives specific to the agency, confidence, self-esteem, and belief members of interdisciplinary teams establish for themselves and their teammates through their collaborative efforts. In their systematic review on interdisciplinary teamwork, Nancarrow et al. [9] include a quote from Xyrichis and Ream [13] to highlight the establishment of these qualities, stating that interdisciplinary teamwork is "a dynamic process… accomplished through interdependent collaboration, open communication and shared decision-making" (p. 238). Henneman et al. [6] focus on the value of collaboration supports these claims, stating that collaboration "requires competence, confidence and commitment on the part of all parties. Respect and trust, both for oneself and others, is key to collaboration".

Our group members' collaborative experience of producing, executing, and reporting on the intersections of art and science was accomplished through the concepts discussed by Xyrichis and Ream [13] and Henneman et al. [6]. Individual competence, confidence, commitment, respect, and trust was supplemented by the open-communication and shared decision-making required of us to produce within the confines of our 48-h deliverable. Perhaps it was divine intervention to a certain extent for all participants of the Hackathon, but deeper still, our collective synchronization on a human level to appreciate the value of creativity towards a previously unknown end yielded a beautiful collaboration.

3 Objective

The main objective of our group's hackathon project was to analyze the EEG-based brain connectivity presented during imagined and performed musical settings, in order to compare the brain dynamics between three distinct modes created for this project: (1) Imagination; (2) Strict mode; and (3) Improvisation.

4 Experimental Procedure

On the first day of the hackathon, we brainstormed and planned our project's direction. Experimentation of our project model was completed on the second day of the hackathon (Fig. 2), wherein our test subject sat in front of a laptop screen and performed specific instructions while his EEG signals were recorded (Fig. 2c). Our test subject was a member of our team, Aaron Colverson, who is a professional musician.

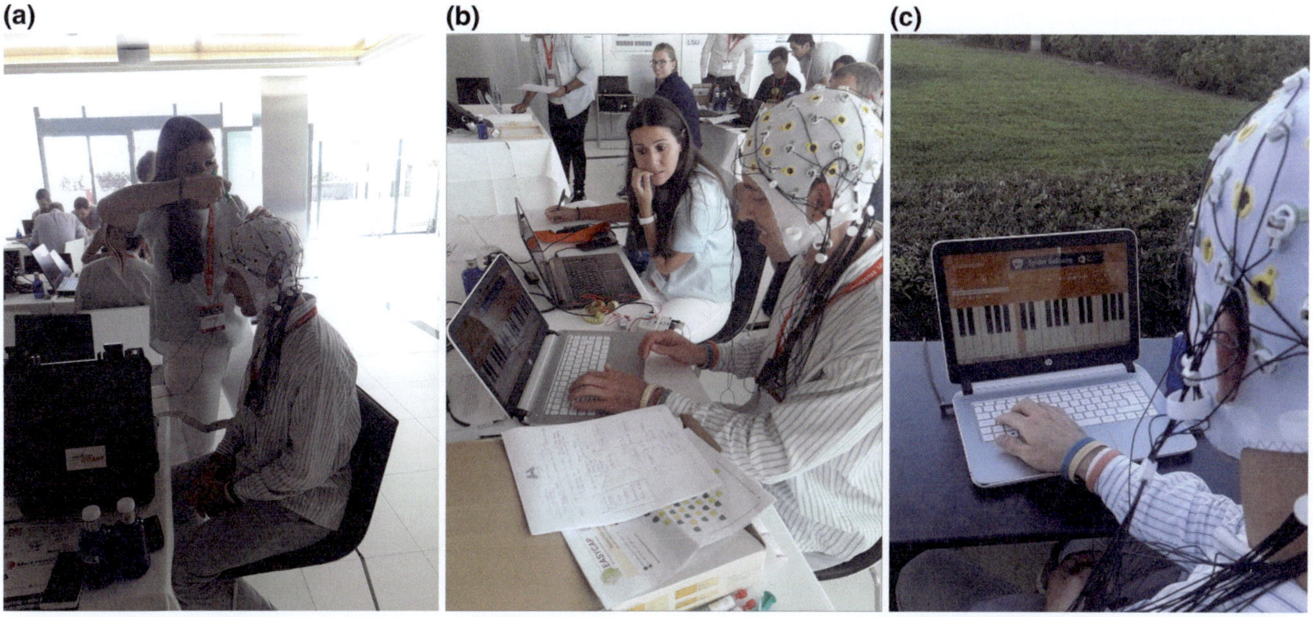

Fig. 2 General experimental design. **a** *EEG cap placement:* a member of our team, Silvia Moreno, positioned the cap on Aaron's head. Once the cap was in place, the electrodes were prepared with gel and conductive paste to lower impedance. **b** Platform familiarization: Aaron practiced use of the digital musical platform, as well as familiarized himself with the "QWERTY" computer keyboard. **c** Experimental environment: Aaron sat upright in front of the laptop screen and performed the required verbal instructions, indicated by our third group member, Jorge Gaxiola, while his EEG signals were recorded

Three types of instructions in order to assess imagined and performed musical settings were indicated:

- *Imagination* (*IM*): Aaron imagined the melody of "Frère Jacques" from his own memory of the melody without listening to it in advance. The key center (D major), tempo, dynamics, articulations, nor timbre were given as references in advance to his imagining the melody while being recorded by the EEG;
- *Strict mode* (*SM*): Aaron read letters from the Standard English alphabet (i.e., A, S, D, F, G, H, and B) that were encoding musical pitches defining the melody of "Frère Jacques" (Fig. 3). While reading the letters, Aaron simultaneously used only his right hand to perform the melody on the laptop keyboard, with real-time audible feedback provided through the computer's speakers;
- *Improvisation* (*CD*): Aaron played an improvised melody (instantaneous creative performance) using the same encoded pitches, tempo, dynamics, articulations, and timbre as described in the SM description.

Each activity (*Imagination*, *Strict mode* and *Improvisation*) was carried out separately in 24-min sessions. Each session was composed of three trials. Two sets of instructions were indicated in each session: action and relax. During action periods, Aaron performed the indicated instruction. The sessions were carried out separately for each musical setting.

80 bpm

ASDA | ASDA
DFG- | DFG-
GHGFDA | GHGFDA
ABA- | ABA-

Fig. 3 Encoded letters from a standard "QWERTY" computer keyboard defining the melody for the song "Frère Jacques" used to perform the *Strict mode* setting of the experimental protocol. The horizontal dashes next to the "G" on the second row and "A" on the fourth row of the figure, respectively, indicate pitches needing to be held for twice the amount of time required to play pitches present within specifically the first row of the figure (i.e., A, S, D, A). The sequence "G, H, G, F" present on the third row of figure indicates pitches needing to be played twice as fast as the pitches present within the first row of the figure. The vertical dashes present within each row indicate measure line markers, delineating individual groupings of four-beat sets defining the rhythmic sequencing of "Frère Jacques".

5 EEG Processing

The actiCHamp 32 system (Brainproducts) was used to acquire EEG signals from 32 channels placed on the scalp according to the extended 10–20 system at a sampling frequency of 1000 Hz. Signals were processed in 3 s epochs (180 epochs for each task). A digital band-pass filter between 5 and 50 Hz and a notch filter with 50 Hz cut-off frequency were applied to the data. Since the analysis of all possible directed interconnections of the full array of 32 electrodes is exhaustive, we decided to analyze the directed interconnections in a set of nine electrodes: F3, Fz, F4, C3, Cz, C4, P3, Pz and P4. We chose this set of electrodes considering that these zones are involved in musical structure processing and information integration processes [11].

The Partial Directed Coherence (PDC) analysis was used to assess the exchange of information flow between brain regions (brain connectivity) during the different musical settings. The PDC ($\pi_{i\leftarrow j}$) is a frequency-domain metric that provides information about directionality in the interaction between signals recorded at different channels [1]. This metric measures the outflow of information from the signal acquired on the channel j to i in relation to the total outflow of information from j to all channels.

The PDC analysis used in this project was based on a method similar to the one proposed by [4]. We analyzed the frequency range of 1–30 Hz. For the given set of frequencies, the PDC values from electrode j to electrode were obtained for each 3 s epoch. Statistical significance for the PDC threshold in all cases was set at $\alpha = 0.05$ (for details see [4]. Any epochs with a higher PDC value than the significance threshold were retained in our calculations.

6 Results

The results are shown in Fig. 4 for the cases of *Imagination* (Fig. 4a), *Strict mode* (Fig. 4b) and *Improvisation* (Fig. 4c). For each case, we show the directional connectivities that resulted with at least 70% of significant epochs. From these results, it is clear that the brain connectivity seems to favor the right hemisphere. We can observe that for the case of the *Imagination* the patterns of connectivity are widely distributed in centro-parietal (C3 $\rightarrow$ P4, C3 $\rightarrow$ Pz and Pz $\rightarrow$ C3), parietal (P3 $\rightarrow$ P4, Pz $\rightarrow$ P4) and fronto-central (Fz $\rightarrow$ C4) regions.

During *Improvisation* we found connections widely distributed in fronto-central (Fz $\rightarrow$ C4, F4 $\rightarrow$ C4 and C3 $\rightarrow$ Fz), central (Cz $\rightarrow$ C4) and centro-parietal (Cz $\rightarrow$ P4, C4 $\rightarrow$ Pz) regions especially towards the right hemisphere. While, in *Strict mode* we note centro-parietal (C3 $\rightarrow$ P4, Cz $\rightarrow$ P4 and P4 $\rightarrow$ C4) and parietal (P3 $\rightarrow$ Pz) connections, without frontal connections.

In all cases, centro-parietal connections were presented. It has been reported that the parietal lobe is a zone of information integration. Therefore, the centro-parietal coherence is related to the integration of exteroceptive and proprioceptive information. This zone has also been related to emotional arousal [7].

Previous studies have reported the implication of inferior frontal cortex in musical structure processing [11]. Furthermore, it has been reported that frontal function integrates reasoning, learning and creative abilities at the service of decision-making and adaptive behavior [3]. In this context, as we mentioned above, in *Strict mode* no frontal connections were obtained. Furthermore, this may be an indication that effectively, the frontal zone is activated during creative processes such as *Improvisation*.

Regarding the clear influence of the connections obtained on the right hemisphere, it has been reported that neural processes of music are lateralized to the right hemisphere.

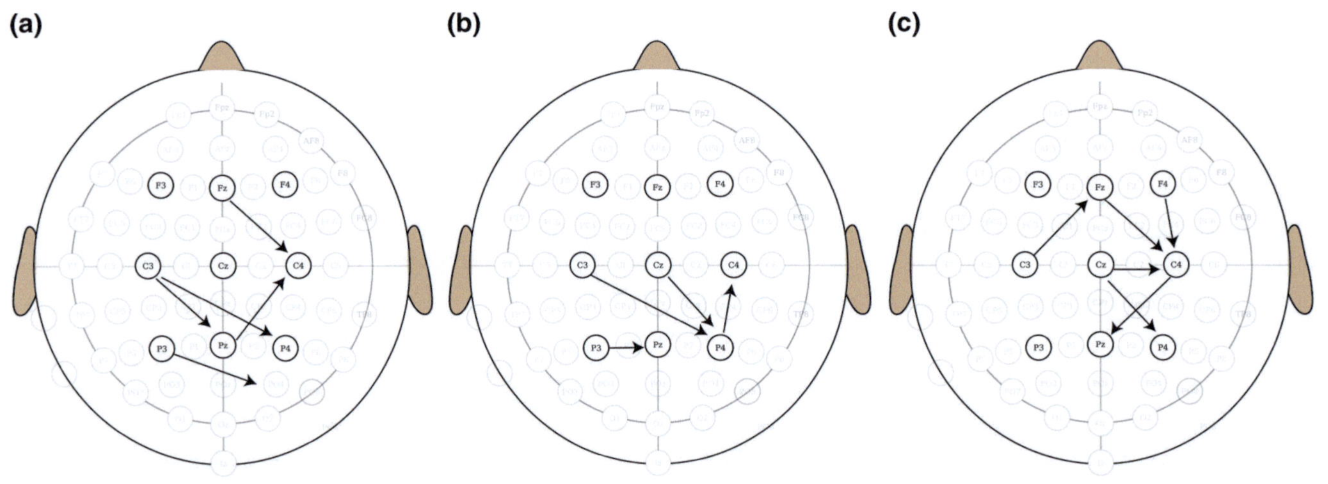

Fig. 4 Brain connectivity maps for **a** *imagination*, **b** *strict mode* and **c** *improvisation*. The brain regions are labeled by the position of nine EEG electrodes: F3, Fz, F4, C3, Cz, C4, P3, Pz, and P4. The arrows indicate the flow of information between signals acquired in the respective regions. Results show right-hemispheric dominance in all musical settings, with centro-parietal networks concurrently active in all settings. Medial-frontal and right medial-frontal activation are reported in the imagined and improvised musical settings, respectively, but not the strict mode

In the last two decades the right hemisphere has also been implicated with neural processes of creativity. However, more recent research indicates the opposite [5, 8], suggesting that the brain mechanisms of music are not entirely elucidated and should be further investigated.

7 Summary: Lessons Learned

As described in our group's objective and experimental procedure, we used EEG-based brain connectivity to compare the brain's activity between three distinct musical modes. Results show that centro-parietal connections were presented in all modes, suggesting integration of exteroceptive and proprioceptive information in these regions. However, brain mechanisms involved with the production and processing of music are not well-understood and need further investigation.

Regarding an overarching goal of the conference to facilitate interdisciplinary collaboration, our group's efforts produced a novel experimental design, execution of that design, analysis, and presentation of the results in a very short period of time. The environment surrounding our group's collaboration instilled a level of urgency deeply tied to our productivity. However, our collective creativity, shared confidence, and appreciation for one another's individual skills powerfully complimented our ability to perform within this environment.

This suggests that interdisciplinary collaboration between artists and scientists stimulates ecological exchange of ideas. Ecological exchange of ideas promotes the creation of holistic approaches to research questions, perhaps even grand challenge questions such as solutions to climate change, poverty, or cancer. Therefore, research projects crafted together by artists and scientists are encouraged, to promote further humanistic inquiry into what makes us human.

8 Conclusion

In this preliminary study, the PDC-based EEG analysis allowed our group to detect brain connectivity patterns during imagined and performed musical settings. The experimentation was carried out in one day and only one subject was involved. Replication of this experiment in a more controlled environment including more test subjects and experimental sessions as well as standardization of the experimental process is advised. It is premature and ill-advised to compare our results with specific brain functions, as well as connectivities reported in similar research published in the literature.

Acknowledgements Gaxiola-Tirado thanks Dania Gutiérrez (https://gutierrezruiz.com/) and CINVESTAV for the opportunity to participate in the conference Doctoral/Postdoctoral consortium. A. Colverson also thanks the Doctoral/Postdoctoral consortium for the opportunity to participate, the staff of the University of Florida's School of Music, and Dr. John Williamson, for his continued advice and guidance.

References

1. Baccalá, L.A., Sameshima, K.: Partial directed coherence: a new concept in neural structure determination. Biol. Cybern. **84**(6), 463–474 (2001)
2. Brattico, E., Bogert, B., Jacobsen, T.: Toward a neural chronometry for the aesthetic experience of music. Front. Psychol. **4**, 206 (2013)
3. Collins, A., Koechlin, E.: Reasoning, learning, and creativity: frontal lobe function and human decision-making. PLoS Biol. **10** (3), e1001293 (2012)
4. Gaxiola-Tirado, J.A., Salazar-Varas, R., Gutiérrez, D.: Using the partial directed coherence to assess functional connectivity in electroencephalography data for brain-computer interfaces. IEEE Trans. Cogn. Develop Syst. **10**(3), 776–783 (2018)
5. Habib, M., Lardy, C., Desiles, T., et al.: Music and dyslexia: a new musical training method to improve reading and related disorders. Front. Psychol. **7**, 26 (2016)
6. Henneman, E.A., Lee, J.L., Cohen, J.I.: Collaboration: a concept analysis. J. Adv. Nurs. **21**(1), 103–109 (1995)
7. Lang, P.J., Bradley, M.M.: Emotion and the motivational brain. Biol. Psychol. **84**(3), 437–450 (2010)
8. Mihov, K.M., Denzler, M., Förster, J.: Hemispheric specialization and creative thinking: a meta-analytic review of lateralization of creativity. Brain Cogn. **72**(3), 442–448 (2010)
9. Nancarrow, S.A., Booth, A., Ariss, S., et al.: Ten principles of good interdisciplinary team work. Hum. Resour. Health **11**(1), 19 (2013)
10. Reybrouck, M., Vuust, P., Brattico, E.: Brain connectivity networks and the aesthetic experience of music. Brain Sci. **8**(6), 107 (2018)
11. Tillmann, B., Koelsch, S., Escoffier, N., et al.: Cognitive priming in sung and instrumental music: activation of inferior frontal cortex. Neuroimage **31**(4), 1771–17872 (2006)
12. Wan, X., Crüts, B., Jensen, H.J.: The causal inference of cortical neural networks during music improvisations. PLoS ONE **9**(12), e112776 (2014)
13. Xyrichis, A., Ream, E.: Teamwork: a concept analysis. J. Adv. Nurs. **61**(2), 232–241 (2008)

Jorge A. Gaxiola-Tirado received the B.Sc. degree in Biomedical Engineering (University of Guadalajara, Mexico, 2013), and the M.Sc. degree in Biomedical Engineering and Physics (Cinvestav Monterrey, Mexico, 2016). Currently, he is a Ph.D. student in Biomedical Engineering and Physics. He is interested in Statistical Signal Processing for Biomedical applications and BCI.

Aaron Colverson is a Ph.D.-degree seeking student at the University of Florida, studying ethnomusicology and neuroscience. His MM thesis focused on the intersections of music, social engagement, and empathic decision-making. His dissertation work intends to investigate phylogenetic and cross-cultural analysis of music-making as an explanation for human consciousness.

Improving EEG Form Factor in Order to Alleviate Pediatric Anxiety in Diagnostic Settings

Justin Tanner, Shane Esola, and Kenneth Veldman

1 Problem Description

At the Brain–Computer Interface (BCI) designer's hackathon carried out during the *2017 International Conference on Mobile Brain–Body Imaging (MoBI) and the Neuroscience of Art, Innovation, and Creativity*, the authors were tasked to design a "fancy and futuristic EEG headpiece". The specific project was sponsored by BR41N.IO and the assignment required the use of 3D printed parts with encouragement to let the device move, blink, or just look unique and interesting. While there was a strong temptation to design a purely "fun" EEG headpiece, the group pondered what purpose an aesthetically pleasing device could serve, and if artistic form combined with engineering fit and function could yield something useful to the MoBI field (Fig. 1). Based on decades of prior research, EEG is a robust tool for measuring brain activity, but consumer EEG products that are designed primarily for aesthetics, such as the Emotive EPOC (San Francisco, CA) and the Neurosky MindWave (San Jose, CA), are not particularly adequate quantitative research tools. The listed devices "exhibit high variability and non-normality of attention and meditation data" and were at best 75% accurate when used to determine if a subject blinked—a low bar for EEG/EMG systems [1]. The authors wanted their device to maintain clinical quality by integrating with existing 10–20 EEG systems and the respective electrodes and electrode placement protocols. The team sought to identify and alleviate a present issue with clinical EEG systems and capitalize on the free-form design benefits of 3D printing. In the end, the team created a simple but sleek headset design that can mount onto a subject's favorite pair of sunglasses, focuses on a specific EEG purpose to limit cumbersomeness, and allows pediatric subjects to feel in control and comfortable during the EEG procedure.

2 Challenges

To scope the problem relevant to a present issue with clinical EEG systems, the group identified pediatric anxiety as a major hurdle in clinical EEG compliance [2]. The authors opted to design a headpiece dedicated to reducing this anxiety where EEG may be used for pediatric diagnostic or treatment purposes, such as ADHD, epilepsy, or neuro-feedback [3–6]. It is fairly easy to imagine that medical settings, in general, produce long-lasting anxiety in children, often more anxiety than they or their parents comprehend [7–9]. Foundational methods to help reduce anxiety exist and are commonly employed when working with a pediatric population. To improve the overall setting, Child-Centered Play Therapy (CCPT), a process that allows the child to lead through play, can decrease pre-neurosurgical patients' level of anxiety; CCPT allows the child to feel comfortable and in control [10]. Additionally, this can be supplemented by the CARE strategy: Choices, Agenda, Resources, and Emotions. Respectively, these refer to letting the patient have choices, informing them of what to expect and what the medical agenda is, identifying the subjects' resiliencies or strengths to help limit the negatives of the situation, and providing emotional support and normalizing common fears [11]. All of these considerations are necessary when collecting data from children, but even something as simple as providing toys or other playful accessories is a common consideration for acquiring pediatric EEG data [12].

J. Tanner (✉)
School of Biological and Health Systems Engineering,
Arizona State University, Tempe, AZ, USA
e-mail: jctanner@asu.edu

S. Esola
Department of Mechanical Engineering and Mechanics,
Drexel University, Philadelphia, PA, USA

K. Veldman
Vlamm Inc., Groningen, The Netherlands

J. L. Contreras-Vidal et al. (eds.), *Mobile Brain-Body Imaging and the Neuroscience of Art, Innovation and Creativity*,
Springer Series on Bio- and Neurosystems 10, https://doi.org/10.1007/978-3-030-24326-5_19

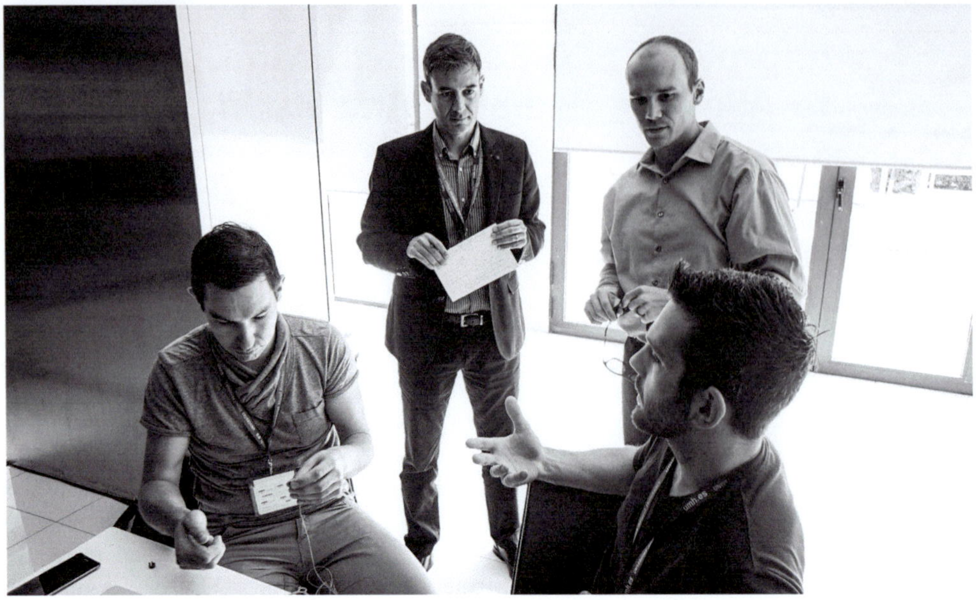

Fig. 1 The authors discussing hackathon plans with Dr. José M. Azorín, the director of the Brain–Machine Interface Systems Lab and Associate Professor of the Systems Engineering and Automation Department at Miguel Hernández University of Elche, Spain. From left to right: Kenneth Veldman, José M. Azorín, Shane Esola, and Justin Tanner. *Photo Credit* Gintare Minelgaite

Pediatric anxiety can be induced by EEG equipment and invoke low compliance with the study, considering a standard 10–20 EEG headpiece has 21+ electrodes across the entire scalp with wires protruding [2]. Wearing a device with all the wires attached can be unfamiliar and uncomfortable and can require a significant amount of time to set up appropriately, increasing anxiety in children. A smaller, simpler headpiece is paramount to reducing anxiety. Minimizing electrode count for specific tasks in an easily applied accessory can help and is used in other settings, such as drowsiness detection [13, 14]. With respect to pediatric populations, epilepsy and ADHD are common foci in pediatric EEG, specifically observing the P300 response [4, 6]. These studies typically follow a common electrode placement consisting of 6–7 electrodes on the back of the head (Fig. 2) where sensory event responses are readily recorded. While the design described in this research focuses on these specific purposes, many EEG applications do not utilize the entire scalp, and a similar approach can be applied to these other situation-specific electrode arrangements.

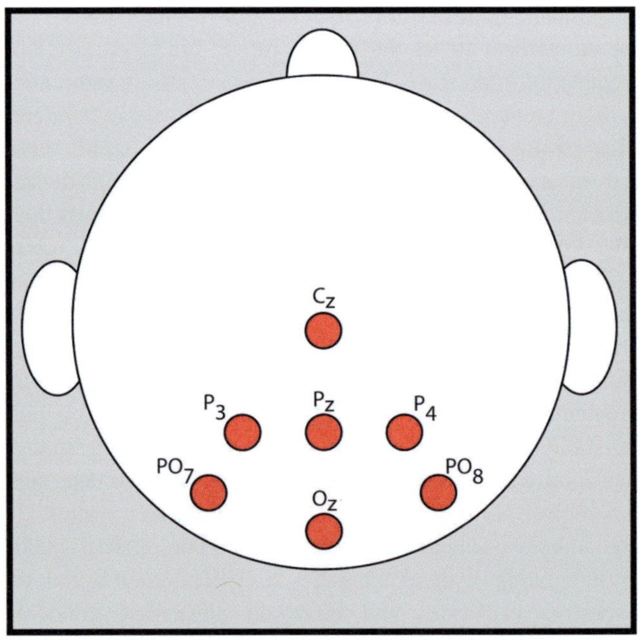

Fig. 2 Electrode placement for ADHD and epilepsy diagnostic tests. These placements are based on 10–20 system that base electrode location on relative skull measures

The final part of the design challenge required the use of a supplied 3D printer. This production technique affords two benefits, which are often opposed: complexity of design that cannot be achieved in standard manufacturing and on-site rapid construction of a prototype. More complex designs often require longer print times and support material, so balance is necessary while making design tradeoffs. Immediately, the design freedom benefit of 3D printing for headsets is obvious as EEG electrode placement depends on an individual subject's head and pediatric head-size is highly variable across covered age groups (e.g., infants to pre-teens). Headset customization to account for idiosyncrasies is a priority to ensure data reliability. Regardless of size customization, the printed design must be able to position electrodes in the right location, require minimal design time, and require minimal printing time.

3 Solution

The hackathon team's printed headpiece design attempted to solve the outlined challenges by introducing a more playful aspect, appealing to the CARE strategy of reducing anxiety, and striking a balance between part complexity and 3D printer speed. By designing the headpiece around everyday sunglasses, the authors introduce playful atmosphere to the traditionally intimidating EEG wire harness while gaining a design advantage by utilizing existing framework to mount the electrodes (Fig. 3). The ACTION therapy program from Stark and Kendall [15] empowers children to bring themselves out of a depressed mood by engaging in pleasant, light-hearted activities. The ACTION strategy teaches children to recognize emotional challenges and places them in control of overcoming the issue themselves. Hannesdottir and Ollendick later noted with respect to the ACTION program that "[A] simple game of putting on different colored sunglasses demonstrate for children how easily they can change their mood and see the world as 'dark' or 'bright'" [16, pg. 285]. A child may feel considerably more comfortable with the opportunity to wear familiar or "cool" sunglasses, providing a positive reference in the unfamiliar situation and even a way to feel secure by "hiding". This leads directly into the CARE and ACTION strategies, as the child is immediately given choice and possesses apparent control of the situation. The sunglasses can be the child's own or from a provided selection, but the opportunity for choice is essential. Next, the authors limited the number of electrodes by designing around a targeted data collection protocol (e.g., P300), rather than a whole cap of unnecessary electrode sites. A child may not understand exactly what the P300 potential represents, but they will feel less burdened by a less invasive electrode set, especially one that is contained behind the head (i.e., mostly out of sight). Third, the solution approach identifies the child's resiliency and strength by reframing the negative as a positive, that is, reframing the clinical procedure as a potential to look cool, feel comfortable, and wear something playful and fun. This could help the child feel like less of a patient and more like a person actively participating in the data collection process. While the design does not directly address the emotional support part of the CARE strategy, the better mood may help the child feel more comfortable asking questions or being open, potentially indirectly improving the data quality.

Using 3D computer-aided design (CAD) software, the team designed a device that localized the appropriate electrode sites according to the physical measurements of one hackathon team member's head. The greatest design challenge for the team was conformal flexibility to provide sufficient contact between the electrodes and the scalp. By necessity to limit print time and accommodate the small print space, the design was overall flat and was split into multiple components that were printed separately. Early designs included chain mail inspired links and long, slender connection elements (Fig. 4). The slender-connection-element

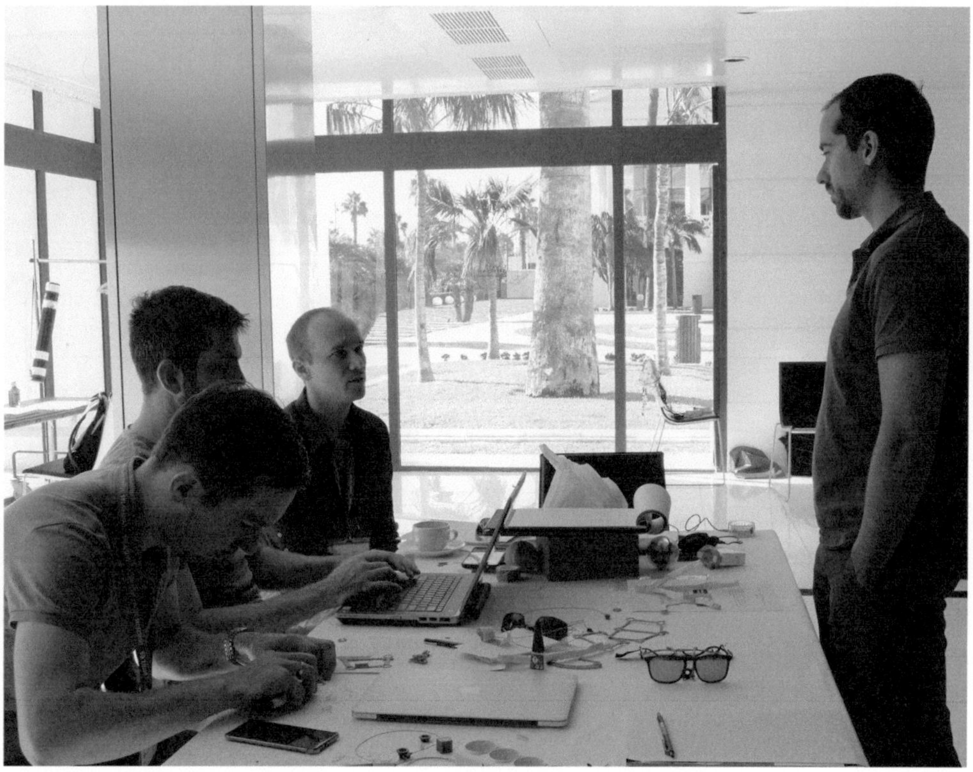

Fig. 3 The authors in the midst of collaborating, designing, and prototyping. On table are initial designs alongside various sunglasses, sketches, and electrodes. *Photo Credit* Miguel Hernandez

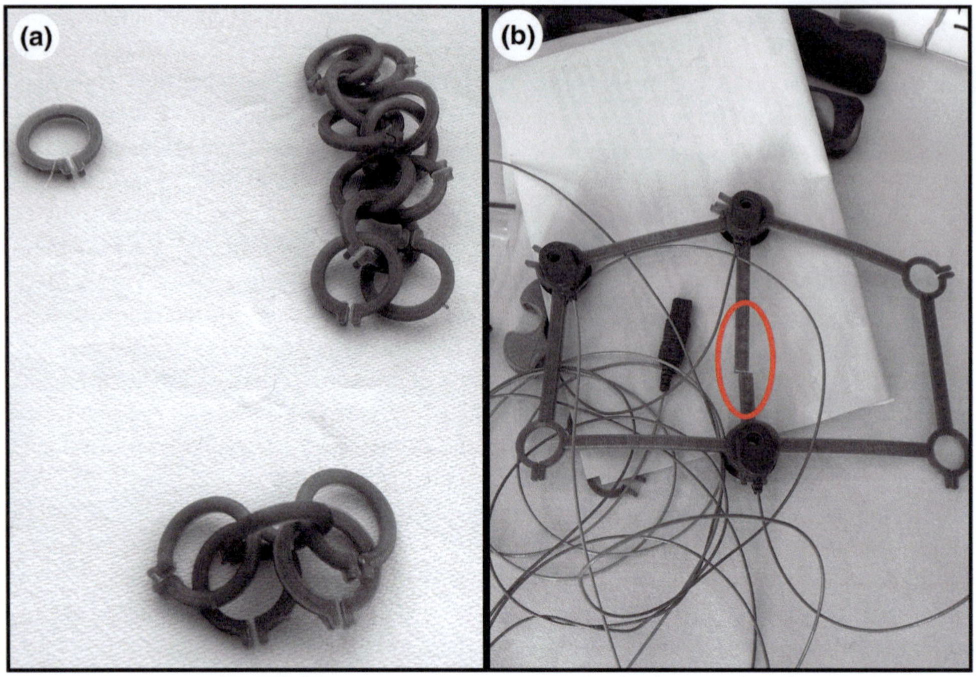

Fig. 4 Early designs. **a** Chain mail inspired design that would allow for maximum fit flexibility, but required increased construction time. **b** Basic connection elements that did not allow for the flexibility necessary, which caused a break indicated in the red circle

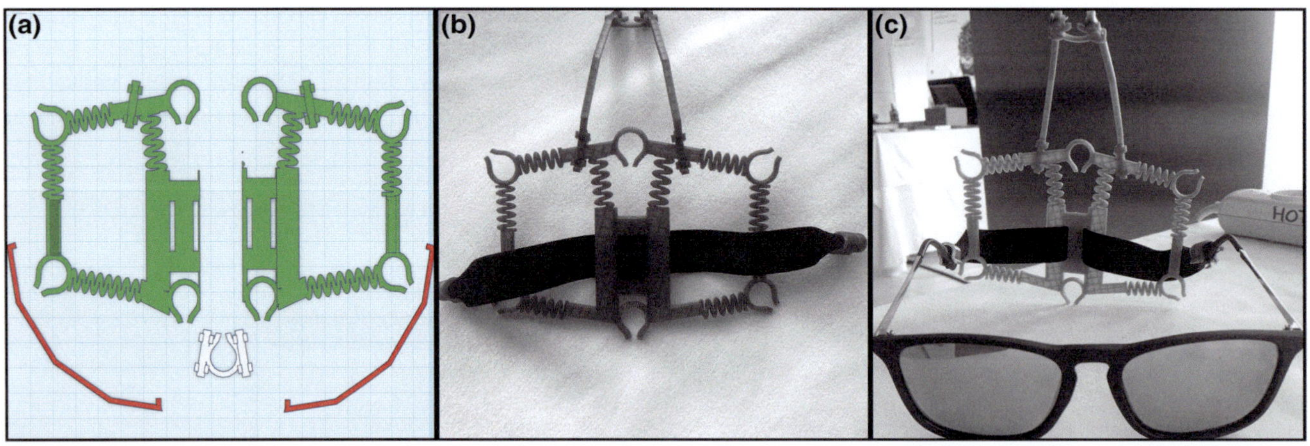

Fig. 5 Final design and construction. **a** CAD model of flexible design and necessary components. **b** Assembled design and elastic strap. **c** Assembled design mounted onto an example pair of sunglasses

design ultimately fractured during the large deformation required to fit the head and the chain mail design took too long to print, too long to assemble, and did not accommodate the electrodes easily. Eventually, in order to fit the curve of the head, "kerf bends" were introduced to the design where flexibility was necessary [17]. A "kerf bend" is a 3D printing and laser cutting strategy that consists of a pattern of parallel lines, often in a sinusoidal pattern, that allow a rigid material to become more flexible—like a spring. The device also required some flexible clips to attach to sunglasses and arms to reach a distant electrode site, but the complete CAD design can be seen rendered, printed, and assembled in Fig. 5.

4 Summary: Bridging Art and Science

While this hackathon project did not produce a flashing and moving headpiece, the result incorporates vital scientific and artistic considerations. There are many opportunities to pursue when designing a "fancy and futuristic" EEG headpiece, and many of those can bridge the intersecting artistic, clinical, and engineering ideologies. Initial considerations were made into whimsical designs that lit up. The team's ideas quickly iterated into neuro-feedback tools that could indicate when certain cognitive regions become active or certain EEG metrics were met; a potential area of future work to aid MoBI researchers. Despite limitations in time and resources, the hackathon team successfully blended important aspects of EEG equipment placement and pediatric anxiety-reducing strategies with common "hacker" and "maker" aspects of 3D printing such as kerf bends. All in all, a useful and practical MoBI device was achieved that provides a minimalistic and decisively unobtrusive design. Art should affect emotion, and the authors hope this design would help improve the emotions of children in stressful circumstances while improving the reliability of data collection that may lead to brain research advancement.

Acknowledgements Thank you to Dr. José L. Contreras-Vidal and Dr. José M. Azorín for the opportunity to participate in the conference Doctoral/Postdoctoral consortium. Thank you to BR4IN.IO for sponsoring this project and providing the 3D printing capabilities.

References

1. Maskeliunas, R., Damasevicius, R., Martisius, I., Vasiljevas, M.: Consumer grade EEG devices: are they usable for control tasks? PeerJ **4**, e1746 (PeerJ, Inc.) (2016). https://doi.org/10.7717/peerj.1746
2. Benore, E., Enlow, T.: Improving Pediatric Compliance with EEG: Decreasing Procedural Anxiety and Behavioral Distress (2013). https://doi.org/10.1016/j.yebeh.2013.01.011
3. Kubik, P., Stanios, M., Iwan, E., Kubik, A.: Neurofeedback therapy influence on clinical status and some EEG parameters in children with localized epilepsy. Przeglad Lekarski **73**(3), 157–160 (2016). http://www.ncbi.nlm.nih.gov/pubmed/27349045
4. Kuperman, S., Johnson, B., Arndt, S., Lindgren, S., Wolraich, M.: Quantitative EEG differences in a nonclinical sample of children with ADHD and undifferentiated ADD. J. Am. Acad. Child Adolesc. Psychiatry **35**(8), 1009–1017 (Elsevier) (1996). https://doi.org/10.1097/00004583-199608000-00011
5. Moreno-García, I., Delgado-Pardo, G., Camacho-Vara de Rey, C., Meneres-Sancho, S., Servera-Barceló, M.: Neurofeedback, pharmacological treatment and behavioral therapy in hyperactivity: multilevel analysis of treatment effects on electroencephalography. Int. J. Clin. Health Psychol. **15**(3), 217–225 (Elsevier Doyma) (2015). https://doi.org/10.1016/j.ijchp.2015.04.003
6. Sunaga, Y., Hikima, A., Otsuka, T., Nagashima, K., Kuroume, T.: P300 event-related potentials in epileptic children. Clin. EEG Neurosci. **25**(1), 13–17 (SAGE Publications Sage CA: Los Angeles, CA) (1994). https://doi.org/10.1177/155005949402500106
7. Kain, Z.N., Mayes, L.C., O'Connor, T.Z., Cicchetti, D.V.: Preoperative anxiety in children. Arch. Pediatr. Adolesc. Med. **150**(12), 1238 (American Medical Association) (1996). https://doi.org/10.1001/archpedi.1996.02170370016002
8. Tiedeman, M.E., Clatworthy, S.: Anxiety responses of 5- to 11-year-old children during and after hospitalization. J. Pediatr. Nurs. **5**(5), 334–343 (1990). http://www.ncbi.nlm.nih.gov/pubmed/2213477
9. Wachtel, J., Rodrigue, J.R., Geffken, G.R., Graham-Pole, J., Turner, C.: Children awaiting invasive medical procedures: do children and their mothers agree on child's level of anxiety? J. Pediatr. Psychol. **19**(6), 723–35 (Oxford University Press) (1994). https://doi.org/10.1093/jpepsy/19.6.723
10. Lerwick, J.L.: The Impact of Child-Centered Play Therapy on Anxiety Levels in Pre-neurosurgical Pediatric Patients, Oct 2011. https://ir.library.oregonstate.edu/concern/graduate_thesis_or_dissertations/q237hv93w
11. Lerwick, J.L.: Minimizing pediatric healthcare-induced anxiety and trauma. World J. Clin. Pediatr. **5**(2), 143 (2016). https://doi.org/10.5409/wjcp.v5.i2.143
12. Puce, A., Hämäläinen, M.: A review of issues related to data acquisition and analysis in EEG/MEG studies. Brain Sci. **7**(12), 58 (2017). https://doi.org/10.3390/brainsci7060058
13. Chiou, J.-C., Ko, L.-W., Lin, C.-T., Hong, C.-T., Jung T.-P., Liang S.-F., Jeng, J.-L.: Using novel MEMS EEG sensors in detecting drowsiness application. In: 2006 IEEE Biomedical Circuits and Systems Conference, pp. 33–36. IEEE (2006). https://doi.org/10.1109/biocas.2006.4600301
14. Tyagi, A., Semwal, S., Shah, G.: A review of EEG sensors used for data acquisition. In: National Conference on Future Aspects of Artificial Intelligence in Industrial Automation. Proceedings Published by International Journal of Computer Applications. IJCA, pp. 13–18 (2012). https://doi.org/10.1007/s10228-015-0496-1
15. Stark, K.P., Kendall, P.C.: Taking Action : A Workbook for Overcoming Depression. Workbook Pub (1996). https://books.google.com/books/about/Taking_Action.html?id=55I7AAAACAAJ
16. Hannesdottir, D.K., Ollendick, T.H.: The role of emotion regulation in the treatment of child anxiety disorders. Clin. Child Fam. Psychol. Rev. **10**(3), 275–293 (2007). https://doi.org/10.1007/s10567-007-0024-6
17. Gomes, A.C.: Kerf Bending: Técnica de Flexão de Madeira Aplicada Ao Mobiliário, Nov 2017. Florianópolis, SC (2017). https://repositorio.ufsc.br/handle/123456789/181957

Justin Tanner received his BS in Bioengineering and Neuroscience from Washington State University in 2010 and his Ph.D. in Biomedical Engineering from Arizona State University in 2017. His work focuses on sensory feedback and modulating its perception through context, stimulation paradigms, and neuromodulation. Currently, he is a Post-Doctoral Researcher with the Sensory Motor Research Group at Arizona State University.

Shane Esola received his BSME from the University of Miami in 2007, MSME from Carnegie Mellon University in 2009, and his Ph.D. in Mechanical Engineering & Mechanics from Drexel University (Philadelphia, PA) in 2019. His research interests include waves in solids, signal processing, and novel sensing networks. He continues to pursue multidisciplinary research opportunities where applied mechanics may yield insight.

Kenneth Veldman is currently working as a Technical Lead and Software developer at Dairy Data Warehouse, a company that collects data from farmers all over the world to provide detailed reports and predictions to consultants of those farmers. He resides in the Netherlands (Groningen) where he grew up, he has a bachelors in IT with specializations in Agile Software development and IT Service Management. Vlamm Inc, Groningen, Netherlands.

Analyzing EEG During the Painting Process

Shane Esola, Justin Tanner, and Kenneth Veldman

1 Introduction

During the *2017 International Conference on Mobile Brain-Body Imaging and the Neuroscience of Art, Innovation, and Creativity*, Ortiz et al. [9], Ubeda, Iáñez, and Azorin conducted a live performance and presented an overview of their ongoing work, "Analyzing EEG During the Painting Process". The research presentation centered on a real-time electroencephalogram (EEG) demonstration that served as a bridge between art and science. Blending the art and science communities is a unique attribute of MoBI conferences. The demonstration was carried out as shown in Fig. 1a, an instrumented artist painted while live-streaming EEG data was displayed for conference participants to view and comment on during an interactive question and answer session. The authors of this chapter recorded, synthesized, and analyzed participant comments in order to assist MoBI conference organizers in identifying momentum-gaining methodologies, remaining technical challenges, lessons learned, and agreed-upon ways to stimulate future innovation in the field.

Prior to the live demonstration, the presenters gave an overview of their ongoing research, which seeks to answer two hypotheses: Can we evaluate artistic expression from EEG signals (H1)? What is the difference in brain activity between artists and non-artists (H2)? The overview briefing provided background and context to the diverse audience comprised of artists, engineers, industrial practitioners, educators, medical professionals, and scientists—some of whom may have been viewing EEG data for the first time.

For the demo, researchers fitted a 16-channel EEG cap onto an artist who was actively engaged in the painting process (reference Fig. 1). The subject painting, shown in Fig. 2, was a recreation of Edouard Manet's *Portrait of Irma Brunner*, a masterpiece circa 1880. For the purposes of the demonstration, the painting was partially complete prior to starting the demo. During the performance, the artist focused on adding details to the painting, specifically articulating the subject's feather headdress. EEG signals were live-streamed, Fast Fourier Transform (FFT) performed, filtered for α-band (8–12 Hz), normalized and averaged over a temporal window. Power spectral density (PSD) was projected onto a map of corresponding sensor locations, forming a spatial topography for the viewing audience. No active artifact filtering was employed.

2 Discussion

The performance presenters were challenged with communicating complex brain activity data and the potential link to artistic cognitive processes to an audience whose experience ranged from science and engineering researchers to art therapists to educators to various artists including musicians, dancers, and painters. Live performance methodologies are gaining momentum at interdisciplinary conferences like MoBI and are an important communication tool that demonstrates the state of the art. The performances serve as a physical blending of art and science, which uncomfortably mashes two very different schools of thought that often struggle to communicate—like two subject matter experts that speak different languages but are searching for common ground to share their exciting ideas. Both the art and science communities seek answers to some of the same fundamentally human questions that define the essence of art, the brain, perception, expression, and emotion.

S. Esola (✉)
Department of Mechanical Engineering and Mechanics,
Drexel University, Philadelphia, PA, USA
e-mail: sae47@drexel.edu

J. Tanner
School of Biological and Health Systems Engineering,
Arizona State University, Tempe, AZ, USA
e-mail: jctanner@asu.edu

K. Veldman
Vlamm Inc, Groningen, The Netherlands
e-mail: veldman@vlamm.nl

J. L. Contreras-Vidal et al. (eds.), *Mobile Brain–Body Imaging and the Neuroscience of Art, Innovation and Creativity*,
Springer Series on Bio- and Neurosystems 10, https://doi.org/10.1007/978-3-030-24326-5_20

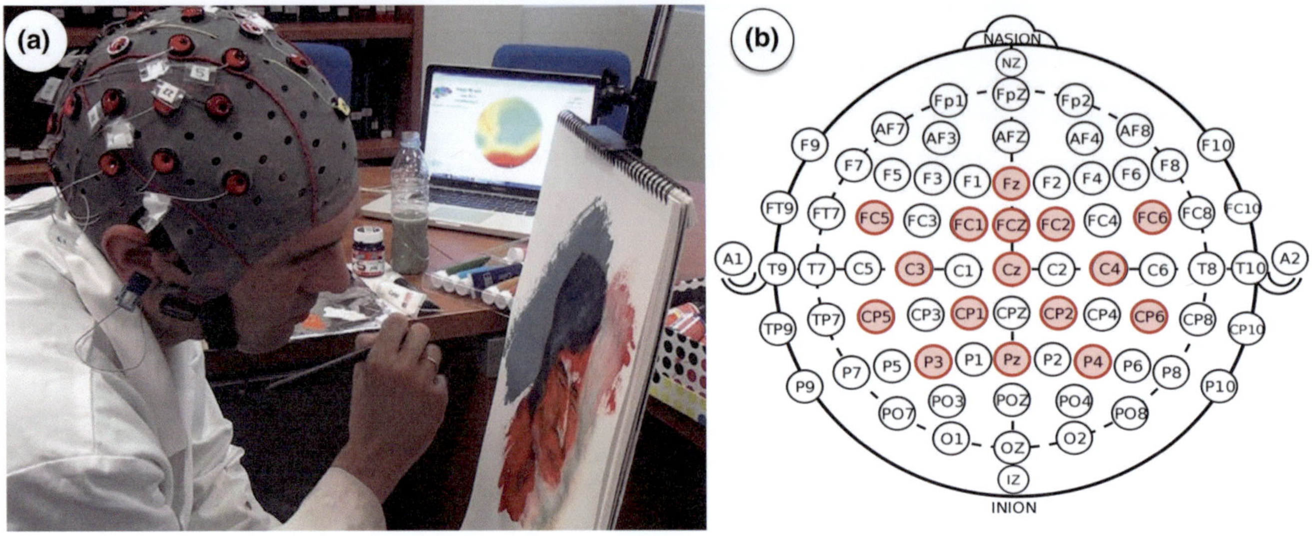

Fig. 1 An artist instrumented with EEG electrodes displaying live results while painting

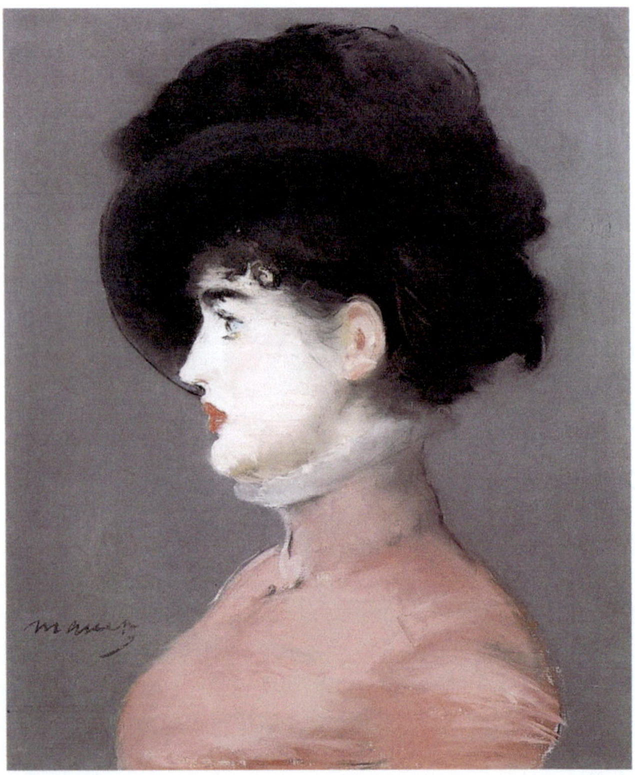

Fig. 2 Edouard Manet, 'La Viennoise, Portrait d'Irma Brunner', 1880. Pastel on canvas. In the collection of Musée d'Orsay

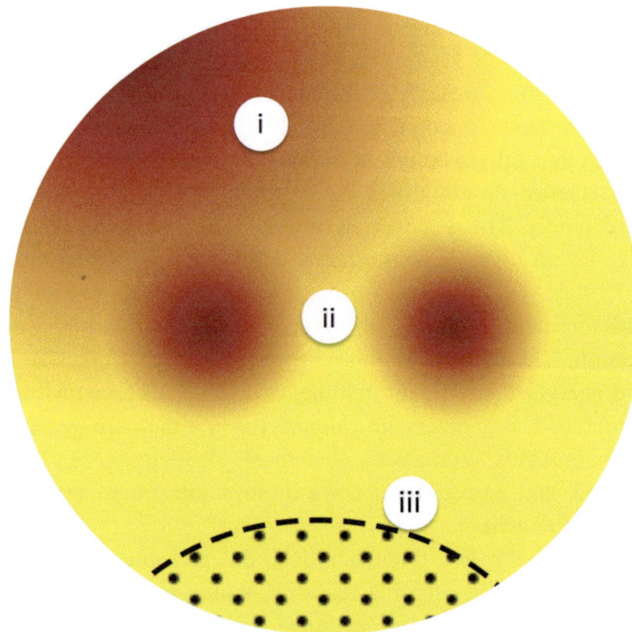

Fig. 3 Rendering of EEG signal observations during live painting demonstration

The live performance provided data for the audience to collaboratively evaluate H1. Prior research from Fink and Benedeck [5] showed that α-power is positively correlated with creative ideation, which suggests that differences in α-power may give insight into artistic expression. Figure 3 is a rendering of observations from the EEG α-band output during the demo. As can be seen from Fig. 3 area [i], frontal lobe activity was elevated and may be indicative of the creative process. Upon further investigation, the persistent signal intensity in this area was attributed to a faulty electrode, a risk during live demonstrations. Bilateral motor activity was observed as well (Fig. 3 area [ii]), manifested by the two symmetric, concentrated areas of activity near the lateral mid-plane. This is characteristic of muscle movement and motor activity; it was intensely present during painting motions. Some observers reported perceiving a stronger signal on the left-hand side, presumably due to the right-handedness of the painter.

A study from Belkofer et al. [1] supports that artists display an α-power increase in the visual/spatial centers of the brain after a drawing activity. During the demo, researchers expected to see visual cortex activity (noted by the pattern area [iii] in Fig. 3), but it was not present during the demonstration. This is likely due to the signal saturation from the bad sensor impacting the normalization; subtle signals may be washed out in the presence of such a persistent and intense signal.

At this early stage in the study, evidence for H2 was supported through literature review and discussion with conference attendees. Shourle et al. [11] observed differences in EEG signal data between artists and non-artists. Specifically, approximate signal entropy revealed population differences during visual perception and may indicate the effect of prior knowledge and training in visual arts. Artists were reported to demonstrate more complex EEG activity during visual perception and mental imagery than non-artists. Bhattacharya and Petsche [2, 3] reported differences between artist and non-artist groups when composing original drawings. Differences in alpha, beta, gamma, and delta band synchronization were reported as well as differences in regional activation. In contrast, Belkofer et al. [1] reported no statistical difference in average α-power between artists and non-artists; however, it was suggested that differences in regional activation patterns were qualitatively observed and the authors attributed the explanation to learned behavior differences between the groups.

Ortiz et al. are currently collecting a broader data set to compare artist and non-artist brain activity. Quantifiable differences are anticipated based on prior research and the general opinion of conference attendees. The work from Ortiz et al. will advance current understanding, yet technical and scientific challenges remain when utilizing EEG to evaluate artistic expression or assess artist's brain activity in order to gain insight into art-creativity-brain connections. Table 1 outlines the challenges based on feedback from conference participants.

Table 1 Challenges when using EEG during artistic activities

Artist perspective	Engineering/science perspective
EEG may serve as a distraction from the artistic process. It is viewed as confusing, cumbersome, and restrictive	Brain feedback during artistic process may be a confound that can influence the artists behavior and thoughts during the experiment [7]
EEG measurement may inhibit artist participation in research. Some artists expressed the belief that creativity cannot be quantified	There is a need for a common analytical approach to noise filtering, digital signal processing, and data characterization
The artist may not understand EEG feedback or why they are being told to perform certain actions during a test. The test design may inhibit the creative process	The inverse problem of inferring cognitive function (bio-physical processes) via EEG observation needs further exploration [4, 6]
Does EEG reveal anything the artists did not already know about themselves?	What is the appropriate metric to characterize brain activity? One participant argued, contrary to Frink and Benedeck, reduced alpha power may be correlated with increased brain activity [8]. Perhaps entropy [11] or synchrony [2, 3] is a better measure?
Creating original work vs. adding details to a partially completed work versus recreating a masterpiece may influence results since the tasks may require different mental approaches	How do we objectively evaluate EEG measure effectiveness [10]? One dimensional (1-D, single feature) representations of n-D (complex) signal spaces may insufficiently capture nonlinear and significant interaction effects that make up cognitive processes
A representative sample population is needed to generalize results—all artists are not necessarily represented by painting, which is only one of many art mediums	Artifacts (blinking, muscle movements) may create high amplitude, irregular noise during a motor-intense activity like painting; there is a need for an intelligent filtering algorithm [12]

3 Summary: Bridging Art and Science

In summary, Ortiz et al. developed an experimental program during MoBI 2017 to advance the state of the art in understanding artistic expression and differences in artist brain activity compared to non-artists. Researchers demonstrated the proposed experimental method for conference participants, which enabled constructive dialogue among the attending subject matter experts. Feedback from the conference attendees, artists, and scientific personnel alike, may lead to the identification of agreed-upon ways to stimulate future innovation in the field necessary to overcome identified technical challenges. Close collaboration between scientists and artist is vital to achieve desired outcomes. Specifically, there is a need for a common (1) language to communicate findings and related concepts; (2) understanding and interpretation of EEG feedback; (3) approach to experimental design with collaborative participation from the artists and the scientific researchers. While the live performance was not expected to solve the questions posed by the researchers, it begins and facilitates the necessary conversation between artists and scientists that may lead to the eventual solution.

Acknowledgements Thank you to Dr. José L. Contreras-Vidal and Dr. José M. Azorín for the opportunity to participate in the conference Doctoral/Postdoctoral consortium. Additional thank you to Mario Ortiz, Andrés Úbeda, Eduardo Iáñez, and the Brain–Machine Interface Systems Lab at Miguel Hernández University of Elche, Spain for putting on the live performance.

References

1. Belkofer, C.M., Van Hecke, A.V., Konopka, L.M.: Effects of drawing on alpha activity: a quantitative EEG study with implications for art therapy. Art Therapy **31**(2), 61–68 (2014)
2. Bhattacharya, J., Petsche, H.: Shadows of artistry: cortical synchrony during perception and imagery of visual art. Cogn. Brain. Res. **13**(2), 179–186 (2002)
3. Bhattacharya, J., Petsche, H.: Drawing on mind's canvas: differences in cortical integration patterns between artists and non-artists. Hum. Brain Mapp. **26**(1), 1–14 (2005)
4. Dietrich, A., Kanso, R.: A review of EEG, ERP, and neuroimaging studies of creativity and insight. Psychol. Bull. **136**(5), 822 (2010)
5. Fink, A., Benedek, M.: EEG alpha power and creative ideation. Neurosci. Biobehav. Rev. **44**(Supplement C), 111–123 (2014). https://doi.org/10.1016/j.neubiorev.2012.12.002
6. Grech, R., Cassar, T., Muscat, J., Camilleri, K.P., Fabri, S.G., Zervakis, M., Xanthopoulos, P., Sakkalis, V., Vanrumste, B.: Review on solving the inverse problem in EEG source analysis. J. Neuroeng. Rehabil. **5**(1), 25 (2008)
7. Gruzelier, J.H.: EEG-neurofeedback for optimising performance. I: A review of cognitive and affective outcome in healthy participants. Neurosci. Biobehav. Rev. **44**, 124–141 (2014)
8. Klimesch, W.: EEG alpha and theta oscillations reflect cognitive and memory performance: a review and analysis. Brain Res. Rev. **29**(2–3), 169–195 (1999)
9. Ortiz, M., Contreras-Vidal, J.L., Azorín, J.M.: Can Creativity be Assessed? Your Brain on Art, When Art and Science Meet. BrainInsight. IEEE Brain Initiative (2018)
10. Pesaran, B., Vinck, M., Einevoll, G.T., Sirota, A., Fries, P., Siegel, M., Truccolo, W., Schroeder, C.E., Srinivasan, R.: Investigating large-scale brain dynamics using field potential recordings: analysis and interpretation. Nat. Neurosci. **21**(7), 903–919 (2018)
11. Shourle, N., Firoozabadi, M., Badie, K.: Analysis of EEG signals related to artists and nonartists during visual perception, mental imagery, and rest using approximate entropy. Biomed. Res. Int. **2014**, 10 (2014). https://doi.org/10.1155/2014/764382
12. Urigüen, J.A., Garcia-Zapirain, B.: EEG artifact removal—state-of-the-art and guidelines. J. Neural Eng. **12**(3) (2015). https://doi.org/10.1088/1741-2560/12/3/031001

Shane Esola received his BSME from the University of Miami in 2007, MSME from Carnegie Mellon University in 2009, and his Ph.D. in Mechanical Engineering & Mechanics from Drexel University (Philadelphia, PA) in 2019. His research interests include waves in solids, signal processing, and novel sensing networks. He continues to pursue multidisciplinary research opportunities where applied mechanics may yield insight.

Justin Tanner received his BS in Bioengineering and Neuroscience from Washington State University in 2010 and his Ph.D. in Biomedical Engineering from Arizona State University in 2017. His work focuses on sensory feedback and modulating its perception through context, stimulation paradigms, and neuromodulation. Currently, he is a Post-Doctoral Researcher with the Sensory Motor Research Group at Arizona State University.

Kenneth Veldman Kenneth Veldman is currently working as a Technical Lead and Software developer at Dairy Data Warehouse, a company that collects data from farmers all over the world to provide detailed reports and predictions to consultants of those farmers. He resides in the Netherlands (Groningen) where he grew up, he has a bachelors in IT with specializations in Agile Software development and IT Service Management.

Paint with Brainwaves—A Step Towards a Low Brain Effort Active BCI Painting Prototype

Andreas Wulff-Abramsson, Adam Lopez, and Luis Antonio Mercado Cerda

1 Introduction

Active Brain–Computer Interface (BCI) systems are classified as systems which relate the processed EEG signal directly with what is in the head of the user, while the produced EEG is independent from the external events [2]. In contrast to a reactive BCI system which relies on learned patterns such as the P300 to guess what the subject wants to do based on the attention of the subject; e.g. the P300 spelling machine [3, 10]. Since the reactive BCI systems are based on machine learning and patterns, the controllability of said systems are rather high. This is, however, at the cost of not being directly related to the objectives at hand and high latency between wanting to do a command, and the execution [8]. Again, with the spelling machine one does not think of the wanted letter to be written but rather look at it. Then the subjects waits for the system to discover which letter is attended, by searching for the P300 expectancy signal when the said letter is flashing [3, 10]. On the contrary active BCI systems are less controllable, but the feedback is faster, as the system relates the signal from the subject directly to the interpret command. An example can be seen in a ball moving application, in which the amplitude of alpha brainwaves (relaxation) provokes a ball to move away from the subject towards their opponent [4]. For our prototype we will exploit the possibility of utilizing an active BCI system for painting purposes, as painting is a continuous creative act where the brush floats free as the creator imagines how the painting shall unfold, both with respect to how the brush strokes are applied and color selection. Before reaching the prototype though, it is important to understand if human brains universally elicit the same EEG patterns across gender, age, and culture when viewing and thinking about colors, thus enabling this prototype to rely on an EEG language rooted in the signals from the human brain. Just like Colocalization analysis (see Fig. 1), which is found within immunocytochemistry, it is expected that not the whole EEG signal can be related to each color, but components such as certain frequency bands or amplitude peak values found in the time domain.

2 Related Works

Yoto et al. [7] presented a work where they investigated the physiological effects of color using electroencephalogram (EEG) as subjects looked at sheets of paper of color red, green, and blue. These colors showed different effects on the mean power of the alpha band, theta band, and the total power in the theta–beta EEG bandwidth (i.e., theta + alpha + beta) and the alpha attenuation coefficient. Results of the power densities of the alpha band for red and green were greater than blue at Fp1. Also from their results, the EEG Alpha band power at Fp1, F7, T5, and Fz; and the theta band at Pz showed higher power during the red color presentation than blue. This suggests that looking at the red paper had a less arousing effect than looking at the blue paper. On the other hand, no significant differences appeared in the beta

Electronic Supplementary Material
The online version of this chapter (https://doi.org/10.1007/978-3-030-24326-5_21) contains supplementary material, which is available to authorized users.

A. Wulff-Abramsson (✉)
Department of Architecture, Design and Media Technology, Aalborg University Copenhagen, Copenhagen, Denmark
e-mail: awj@create.aau.dk

A. Lopez
School of Biological Sciences, University of Missouri-Kansas City, Kansas City, KS, USA
e-mail: aalhd7@mail.umkc.edu

L. Antonio Mercado Cerda
Universidad Autónoma de Nuevo León, FIME, San Nicolás de Los Garza, NL, Mexico
e-mail: luis.mercadocr@uanl.edu.mx

J. L. Contreras-Vidal et al. (eds.), *Mobile Brain–Body Imaging and the Neuroscience of Art, Innovation and Creativity*, Springer Series on Bio- and Neurosystems 10, https://doi.org/10.1007/978-3-030-24326-5_21

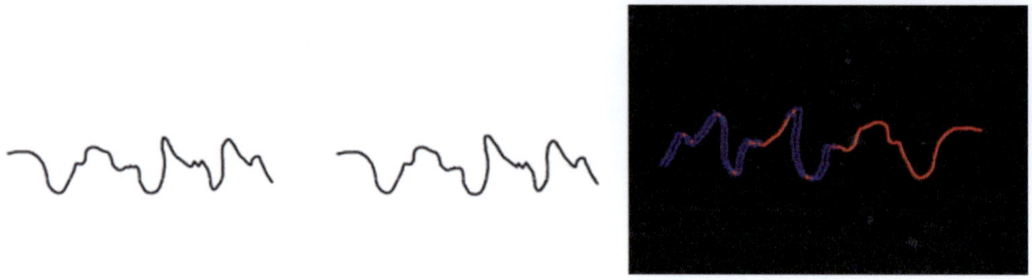

Fig. 1 Shows how colocalization theoretically works on any kind of waveforms. Located to the Left and middle: two individual waves. Right: two waves superimposed upon each other in order to see how similar they are

and delta band power at any electrode during color exchange. On the contrary Zhang and Tang [9] research on color and EEG signal, suggests the following relationship between colors, alpha and beta waves. Green provoking higher power of alpha than blue, and blue higher than red. While for the beta waves the opposite relationship is clear. The idiosyncratic nature of the relationship between colors and brain waves has not gone unnoticed by the BCI community.

Rasheed and Marini [5], presented an investigation to classify the EEG signals produced by a random visual exposure of the colors red, green, and blue. Using the event-related spectral perturbations (ERSP) as input features for a support vector machine (SVM), the ERSP showed significant power variations in the delta, theta, and alpha bands. In their study, the highest increase in power was seen in red, and the lowest in green during an interval from 100 to 400 ms within the delta and theta bands. These discriminative power changes allowed for the EEG signals to be successfully classified in Red, Green, and Blue (RGB) classes with a 98% of accuracy using SVM with a radial basis function kernel.

In the concept of classifying the EEG signals, Alharbi et al. [1] performed a single trial classification. Using various feature extraction methods like the ERSP, target mean with Fast Fourier Transform, Wavelet Packet Decomposition, Auto Regressive model, and Empirical Mode Decomposition residual. This last feature extraction method was found to be the best according to the accuracy of their results. Using a SVM for the classification, the evoked stimulus by the RGB colors required shorter time compared to other stimulus, such as imagery and spelling words. Their study focused on proving that classifying colors is more efficient and faster to give commands to BCIs.

3 The Prototype

As mentioned earlier this prototype takes inspiration from the active BCI systems, where the interactive loop is instantly without the need for a classifier to compare the signal up against a pre-learned template (see Fig. 2). Taken inspiration from the literature the power of different frequency bands will be associated with either red, green, or blue in order to control the color of the paint brush. To control the position of the brush similar associations will be made.

3.1 Setup

The EEG signal was captured through a wireless g.Unicorn EEG helmet (g.tec Neurotechnologies GmbH, Graz). The EEG configuration on said helmet follows the 10–20 system at the following eight positions (Fz, Cz, Pz, P3, P4, PO7, PO8, and Oz). The helmet was connected to Matlab's Simulink® interface through Bluetooth. In the Simulink® environment, the EEG signal was captured, pre-processed and communicated to Unity® through User Datagram Protocol (UDP) connection. In Unity® the signal is used to control the position and color of a paint brush, which paints a blank canvas. The subject could try to control the brush color by gazing at one of three virtual cubes, which were colored red, green, or blue respectively.

3.2 Simulink® Patch

When the EEG signal arrived to Simulink® a notch filter at 50 Hz was applied together with a band pass filter from 0.5 to 100 Hz to extract noise and power interference from the EEG signal. The signal was then divided into five frequency bands, Delta (0.5–3 Hz), Theta (4–7 Hz), Alpha (8–12 Hz), Beta (13–32 Hz), Gamma (33–50 Hz). Afterwards, all the delta signals from all the channels were added together, all the thetas were added together, alphas, betas and gammas respectively. The summed signal values were communicated further to Unity® through said UDP connection.

3.3 Unity® Program

Unity® receives accumulated Delta, Theta, Alpha, Beta, and Gamma input. Before implementing the painting experience,

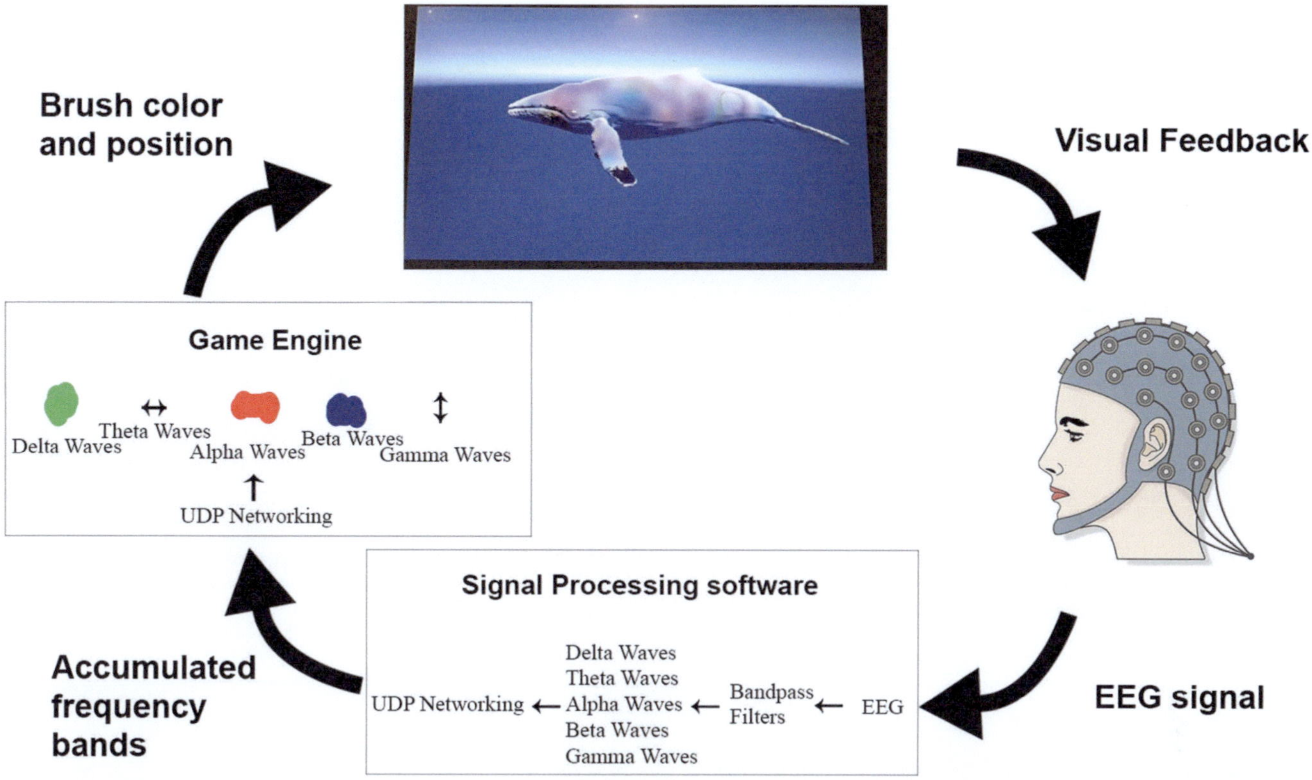

Fig. 2 Shows the whole feedback loop from EEG signal to visual painting application. It can be seen how the EEG signal is sent from the subject to the signal processing software, in this case Simulink®, from Simulink® the accumulated frequency bands' power is send further to a game engine, this time being Unity®, it interprets them as color and position of the paint brush, which is projected on the screen for the subject to watch. The subject then creates new EEG signal, which is interpret once again

we tested which frequency bands were most amplified when flashing red, green and blue 30 times each in front of 3 different subjects. Through the flash experiment we saw that Alpha was more amplified when red flashed in front of the user compared to green and blue, which is further confirmed by Yoto et al. study [7]. Beta seemed to be of higher amplitude when viewing blue compared to the two other colors and lastly Delta showed higher amplification when viewing green compared to the two others (see Fig. 3 for how the EEG looked like for one of the subjects). The results from that initial experiment gave us an informed choice to relate the sum of the recorded alpha amplitudes with the red color channel of the RGB brush, Delta with the green channel, and Beta for the blue channel. They were related in such way that the higher value of either Alpha, Beta, or Delta the brighter the color would become, e.g. (255.0.0) is pure bright red, (255.255.255) is white and (0.0.0) is black. It was seen that since the aggregated values from Simulink® were much higher than the accepted values between 0 and 255, all three color values were divided by 50 to reach a value between 0 and 255.

For a painting program to work the brush also needs to move. For horizontal movement the doubled Theta amplitude values were used, as the signal was not that strong for our preliminary test subjects. For the vertical axis, we used the Gamma amplitudes. These two relationships were arbitrarily chosen, and we did not preliminarily test whether or not there was a relationship between movement direction and amplitude of different brainwaves. However, one thing we ensured was to keep the paint brush within the virtual canvas by centering the brush after every single move in any direction.

3.4 Initial Tryout

To showcase the system at the Hackathon in which this prototype was developed, a new volunteer tried the system (see Fig. 4). She did not have any experience with the program, nor tried it when we developed it, thus making it an intriguing experience for us all. She successfully managed to make three different primary colors (red, green, blue) on the virtual canvas when verbally tasked to do so. The volunteer was also able to create shades of these colors when allowed to free paint. It resulted in the wished colors and a fair amount of random colors deviating away from the wished color. Thus deeming the prototype an initial good step towards a low effort active BCI painting application (Online Resource 14).

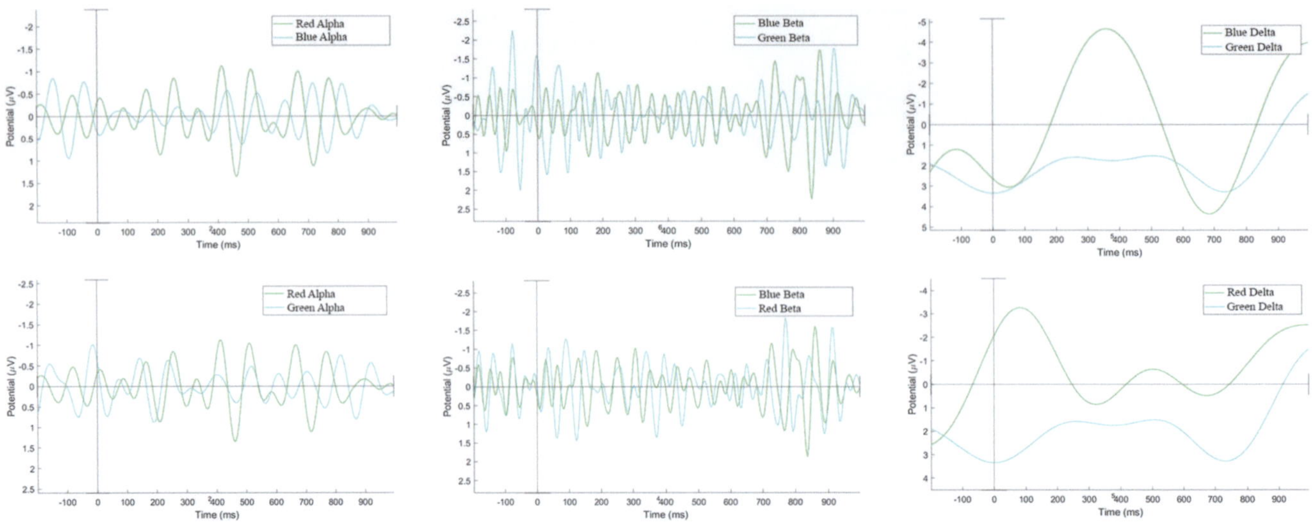

Fig. 3 EEG recordings of one subject's brain as a response to the different colors. To the left the alpha waves are shown in which red produces the highest peaks. In the middle the beta waves present showcasing blue producing the highest peak. To the right the delta waves are seen, where green color is the one producing most positive activity

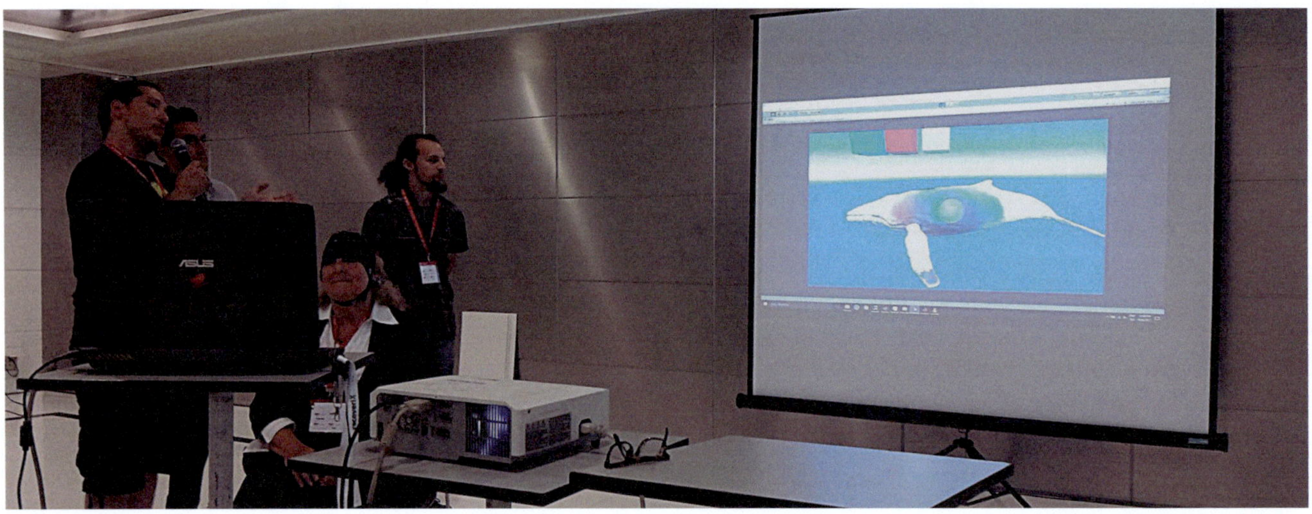

Fig. 4 Shows a demonstration from the Hackathon with a subject who has not tried the prototype before

4 Discussion

This painting interface is interesting in many ways. Compared to the P300 painting machine [10] it is faster and the user can quickly create art without waiting 15 s for each command to comply. However, despite the higher efficiency, the effectiveness is far lower as the P300 machine selects the commands with 80–100% accuracy, while this active BCI machine shows what can be interpreted as random colors, and it is up to the perception of the user to decide whether or not the machine is choosing the right colors. For example, the participant called out blue and recognized a slight blue tint indicating the right color selection. Despite the lower accuracy, the application could be of other use than the direct artistic endeavors of the subject. Seeing that the system portrays every single recorded brain wave as a distinct color and location the subject could be doing something completely different than art or real painting and get a distinct unique output from the system showcasing with abstract colors, what the subject has been doing both in terms of perception, affect and cognition.

Ultimately, when the relationship between the brainwaves have been fine-tuned and comply with the classification results found in Rasheed and Marini's and Alharbi et al. [1, 5] a more reliable active BCI painting system could be created. With it, a situation where the artistry will be more up to the conscious mind of the subjects who want to use this system, rather than the system picking everything up and converting it to colors. For the matter of fact such system is especially usable by paralyzed subjects, as they only need to think in order to create as opposed to the P300 painting machine, which needs the conscious attention from the gaze to create the paint commands [10]. Lastly, with reliable associations between the brain waves and the colors, this system can be of use for easier and quicker diagnosis of neural issues in patients, as distinct colors could light up in case there were missing waves or the composition of brain waves diverted from the norm with respect to certain cognitive or perceptual tasks. This system is close to the frequency analysis visualization tools usually used when looking at the amplitude and distribution of brain waves [6], but this concatenated visualization contains information from what used to be individual frequency visualization maps into one.

5 Conclusion

Based on the premise that an active BCI relates the processed EEG signal directly with what is in the head of the user, the work presented here further proved that the brain frequency bands can be associated with different colors allowing a volunteer to realize a quick sketch on a virtual canvas. However, the work presented here only remained a little above random chance. Further processing of the brain signals would be required to establish specific colors to different brain frequency amplitude thresholds. Furthermore, more colors could be found to be associated with these thresholds thus allowing a wider disposal of colors to the creativity of the user. However, considering the ambitious task and short time frame, the first steps taken by this project yielded remarkable and unignorable results given a larger lexicon of EEG signals mapped, a larger number of volunteers queried a root language for a finely tuned active BCI to use for greater functionality, and accuracy could be in the future. The invention of such a device could have countless applications for communicating with the unconscious to, allowing one to paint and print out a mental creation or even memory. Which may prove invaluable in the treatment of patients with impaired communication or mental illness.

6 Summary: Lessons Learned

In this chapter, an active BCI painting prototype was described. The process from choosing active BCI over reactive BCI was debated, as it is clear that active systems are more efficient than reactive systems. Choosing the relationship between EEG signals and application commands were researched based on literature and initial tests, which rendered red to be controlled by alpha waves, blue by beta waves, and green by delta waves. Additionally, theta and gamma waves were chosen to control the direction of the brush. Lastly, the chapter describes the findings from a live demonstration at the BOA'17 Hackathon together with a discussion related to the systems advantages and limitations, which revealed that the system is still immature; however, it has fine future prospects if the relationship between colors and EEG data is fine-tuned even more. Below is a list of learned lessons from said hackathon.

- The relationship between EEG and the output stimuli should be carefully crafted.
- EEG signals are individual from person to person, thus hardcoded relationships will not fit all.
- Isolating a brainwave and correlate it with a color is a dangerous task as all types of brainwaves occur at once.
- Complex BCI systems need more care than simple BCI systems.
- The EEG foot print from imagining a color is different than seeing a color.
- Imagining a clear color in your head requires training of the subject while seeing does not.

References

1. Alharbi, E., Rasheed, S., Buhari, S.: Single trial classification of evoked EEG signals due to RGB colors. Brain (Bacau) **7**(1), 29–41 (2016)
2. Andujar, M., Crawford, C.S., Nijholt, A., Jackson, F., Gilbert, J.E.: Artistic brain-computer interfaces: the expression and stimulation of the user's affective state. Brain Comput. Interfaces (Abingdon) **2**(2–3), 60–69 (2015)
3. Belitski, A., Farquhar, J., Desain, P.: P300 audio-visual speller. J. Neural Eng. **8**(2), 025022 (2011)
4. Hjelm, S.I., Browall, C.: Brainball-using brain activity for cool competition. In: Proceedings of NordiCHI, vol. 7, no. 9, October 2000
5. Rasheed, S., Marini, D.: Classification of EEG signals produced by RGB colour stimuli. J. Biomed. Eng. Med. Imaging **2**(5), 56 (2015)
6. Stiber, B.Z., Sato, S.: Visualization of EEG using time-frequency distributions. Methods Inf. Med. **36**(04/05), 298–301 (1997)
7. Yoto, A., Katsuura, T., Iwanaga, K., Shimomura, Y.: Effects of object color stimuli on human brain activities in perception and attention referred to EEG alpha band response. J. Physiol. Anthropol. **26**(3), 373–379 (2007)
8. Zander, T.O., Kothe, C.: Towards passive brain–computer interfaces: applying brain–computer interface technology to human–machine systems in general. J. Neural Eng. **8**(2), 025005 (2011)
9. Zhang, H., Tang, Z.: To judge what color the subject watched by color effect on brain activity. IJCSNS Int. J. Comput. Sci. Netw. Secur. **11**(2), 80–83 (2011)

10. Zickler, C., Halder, S., Kleih, S.C., Herbert, C., Kübler, A.: Brain painting: usability testing according to the user-centered design in end users with severe motor paralysis. Artif. Intell. Med. **59**(2), 99–110 (2013)

Andreas Wulff-Abramsson since his M.Sc. in medialogy at Aalborg University Copenhagen (2016), he has worked in the augmented cognition lab. Concerning human affective, cognitive, and perceptual processes in relation to different media, spanning from interactive immersive games, to art, film, and to light, by discerning psychophysiological footprints (EEG, eye-tracking, GSR, and HR).

Adam Lopez is a Ph.D. candidate UMKC research neuro-degenerative diseases of the brain and eye hopes to one day cure Alzheimer's disease. His awards include 2017 National Science Foundation funded travel award to participate in the Doctoral/Postdoctoral Consortium of the 2017 International Conference on Mobile Brain–Body Imaging (MoBI) and the Neuroscience of Art, 2017 UMKC SGS research grant award, and two time UMKC award for teaching excellence recipient.

Luis Antonio Mercado Cerda obtained his M.Sc. in the Universidad Autónoma de Nuevo León (UANL, 2014). Curently, he is finishing his doctorate in the same university. He is interested in research of BCI applications including: decodification of kinematics, and assistance for disabled people with exoskeletons and prosthesis.

Mediated Interdependence in Motion: A Co-op Augmented Reality (AR) and Brain–Computer Interface (BCI) Installation

Guillermo Herrera-Arcos and Daniel Pimentel

1 Introduction

Mutual trust, empathy, and cooperation are integral factors shaping the quality of human interactions across various contexts. During the developmental stages of our youth, educational and instructional experiences are designed to facilitate social cohesion, learning, and socialization. As adults, the importance of these factors become accentuated given the demands of increasingly diverse and collaborative work environments. For example, multinational corporations devote extensive resources towards inciting organizational change via management interventions that promote openness, trust, empathy, and collaboration among employees [11]. While such factors influence company performance in the business realm, these outcomes prove integral across disciplines.

Whether in educational, medical, or corporate environments, interventions designed to engender trust, empathy, and cooperation among in-group members have long relied on art-based activities. Psychiatric institutions regularly integrate art activities as a means by which to empower patients while simultaneously building acceptance and empathy [14]. Similarly, art-based programs for at-risk youth have shown promise in terms of generating similar benefits [4].

While art-based activities have proven effective in many respects, a major limitation has been the cost and limited accessibility associated with such initiatives. That is, such programs often require a trained individual to guide the activities, in addition to the costs of materials. One potential solution to this can be found in the development of scalable new media technologies, namely mobile-based video games. Such platforms have allowed for mass communities to engage in "social play," defined as any activity wherein the successive behavior of one individual in a group is contingent on the behavior of the other member(s). Social play itself is an effective means by which to teach trust, cooperation, and fairness [1]. Indeed, videogames consistently have merged principles of art, technology, and play to facilitate engaging experiences that improve group dynamics.

Acknowledging the evident benefits associated with game-based interventions, it should be noted that a commercially available interactive game for the purpose of team-building remains elusive. Given the absence of a standalone videogame experience dedicated to improving group dynamics, this paper proposes a game prototype addressing this gap. Combining principles of art, videogames, social play, and human–computer interaction (HCI), we present "*Art of War*" (AoW). AoW is a two-player cooperative augmented reality (AR) game leveraging a brain–computer interface (BCI) and a robotic agent to encourage trust, empathy, and cooperation among a dyad. In the following sections, we review pertinent literature detailing the justification behind choosing an AR/BCI-based solution. Afterward, we explain the development of the proof-of-concept, followed by a discussion on the social, organizational, and medical implications of such experiences.

2 Literature Review

2.1 Trust and Cooperation

Organizational psychologists define trust as a multidimensional variable characterized by the willingness to be vulnerable in a cooperative, interdependent relationship, and the belief that others will act in the best interest of those comprising that relationship. The absence of trust negatively

G. Herrera-Arcos (✉)
Tecnologico de Monterrey, School of Engineering and Science, Monterrey, Mexico
e-mail: guillermohra@gmail.com

D. Pimentel
Media Effects and Technology Lab, University of Florida, Gainsville, FL, USA

J. L. Contreras-Vidal et al. (eds.), *Mobile Brain–Body Imaging and the Neuroscience of Art, Innovation and Creativity*, Springer Series on Bio- and Neurosystems 10, https://doi.org/10.1007/978-3-030-24326-5_22

affects group dynamics in that individuals will cease acting for the collective betterment of the team, sacrificing potential gains resulting from cooperation. Media psychology research has demonstrated that online games facilitate trust and cooperative, an effect driven by an interdependence among players [2].

2.2 Empathy and Social Play

An important factor shaping trust and cooperation among individuals engaging in mediated experiences is empathy. Empathy, defined as one's ability to understand another human's feelings, has shown to be strongly related to interpersonal trust in virtual communities [5]. More pertinent to this research is that empathy can be strengthened through shared experiences. Considering the ability for videogames and similarly cooperative virtual experiences to facilitate shared experiences, social play represents an appropriate method by which to increase trust and empathy, and thereby improve group dynamics.

As previously mentioned, social play is a means by which to dynamically increase the three integral parts of group cohesion. It also allows for the integration of creative elements often associated with art-based interventions. For example, Nintendo's latest gaming console (Nintendo Switch) features a game *Deru—The Art of Cooperation*. The game leverages problem-solving elements with abstract art in a cooperative setting. According to the Libin Game Model, games present systems of engagement through which a player (or players) exchange actions with a gaming environment, one which is mediated by specific goals [12]. Considering the interactive and engaging nature of immersive media platforms like virtual reality (VR), it is evident that they may serve as a catalyst for users to engage in entertaining self-improvement with clear implications for education, well-being, and personal development.

Where Libin emphasized the co-relationship between user and system, we propose that such media platforms also strengthen, and benefit, co-op dyads using these systems in gaming contexts. In particular, we propose augmented reality (AR) as a favorable media platform with which to examine the effects of in-game interdependence among a player dyad on group dynamics, namely trust, cooperation, and empathy.

2.3 Augmented Reality (AR)

Where VR removes the user from the physical realm and immerses them in a computer-generated environment, AR integrates computer-generated elements into the user's physical environment, allowing the user to interact with a mixed-reality. AR thus allows interactions with virtual content specific to users' shared space, as shown in numerous science exhibits and museums where visitors can share interactions with extinct creatures in a shared room. This affordance has led HCI scholars to identify the collaborative potential of AR applications to facilitate peer-to-peer communication. Indeed, studies examining its impact in classrooms lend credence to its ability to foster engagement and cooperation during group activities [7]. Activities involving AR headsets are limited by bandwidth, such that only one user at a time has access to the shared virtual environment. However, this limitation is an inherent feature in various games, such as *Battleship*, where one player engages the other player's environment despite not being able to see it. This paradigm of play lends itself nicely to AR, as one player can have access points to the virtual realm, yet rely on another player to engage it through other inputs. One method of input which we investigate herein relies on the use of brain–computer interfaces and robotics.

2.4 EEG-Based Brain–Computer Interfaces (BCI)

Non-invasive EEG-based BCI's are capable of converting neural recordings into formats that computers communicate to the external world, a feature which complements AR-based cooperative play experiences. In order to decode the neural activity, processing techniques such as the P300 paradigm are required.

The P300 paradigm is based on Event-Related Potential (ERP), which is a stereotyped electrophysiological response to an external or internal stimulus. This response can be detected while the subject is classifying different types of events, with one of them occurring much more frequently than the others, this event elicits ERPs consisting of an enhanced positive-going signal component with a latency of 300 ms [3]. In order to classify events, several trials need to be averaged. The stronger the ERP signal, the fewer trials are needed.

A common format for classifying events is a matrix displayed on a screen that includes identifiable characters or symbols. While the rows and columns flash in rapid succession, the user should focus on the desired element by attending to it, the flash of the row and column that contains the desired element elicits a P300 response. By determining the row and column that elicits the P300 response, the BCI system is able to identify the element that the user wants to select.

Other than the classification of events, BCI systems allow quantification and visualization of human neurophysiological and behavioral elements otherwise only subjectively expressed, including human cognitive functions, emotions, motor intentions and human-human interaction dynamics in varied circumstances. Thanks to recent advances in mobile brain–body imaging, it is possible to record brain signals of freely behaving subjects [9] and characterize brain responses to particular stimuli [6]. Our installation leverages the use of EEG-based Brain–Computer Interfaces to provide a novel way to control a robotic agent with brain signals using the P-300 paradigm.

2.5 Robotic Agents

Robotic agents have been extended to promote playful interactions in different contexts and have demonstrated to be socially and emotionally engaging, making them good candidates to be incorporated into games. Some designs include robots for pediatric care contexts, where a friendly robot is used to interact with the children, with the objective of reducing any kind of stress and anxiety from clinical treatments [8]. Others have been used to facilitate therapies for children with disorders like autism.

Besides successful and promising implementations in therapeutic scenarios, some studies suggest that the presence of a physical, embodied robot enables more interaction, as well as more enjoyment of the interaction, especially when the robot uses gestures [10]. The use of robotic agents presents opportunities for providing positive stimulation to the users while playing, as well as exploiting embodiment to enhance gaming experiences. One such opportunity is its role as a conduit for a player engaging in a cooperative AR-based game wherein they cannot see the virtual elements embedded in their world, through a robotic agent, and cooperation with an AR-enabled user, they may exert influence over the outcome of game events. This ultimately forms the crux of our proposed game-based solution described in detail in the next section.

3 Methods

3.1 The Installation: Art of War

The "Art of War" project is a co-op, mixed-media interactive strategy game leveraging augmented reality (AR), brain–computer interface (BCI), and robotics to convey, experientially and artistically, the role of human interdependence during times of distress.

The play area is any flat space (e.g., table, floor). The system randomly generates enemy sentries on key locations on the map. The map is depicted via an augmented reality (AR) phone application (Augment), with already built-in 3D environments and models. The map is only accessible to one of the players (Player 1), which uses the integrated phone camera to project the sentries and map layout onto the designated surface.

Player 2 assumes the perspective of a prisoner-of-war, which is represented on the play map (surface) by a robotic ball, a Sphero SPRK + (Sphero). The player is tasked with escaping his captors by navigating through the map, avoiding collisions with the sentries. Should the player collide with a sentry, the game ends. Because the player cannot see the sentries, as he/she does not have access to the AR map, Player 1 must communicate to Player 2 (a) the position of the sentries and (b) the recommended navigation/movement via voice commands. Player 2 can then execute a move using the BCI, based on the command selected on the screen. The user has the option to select one command at a time over more than 25 different options, from turning left or right, to spinning, to executing S-like movements, following the P300 paradigm and sends the movement orders to the robotic ball. The time for the system to recognize the desired command can be set up in the configuration of the proprietary BCI software (g-tec), less time could derive in less accuracy, and more time could derive in greater accuracy. The average time is around 15 s. The robot is connected via Bluetooth to a port on a personal computer. Prior to the start of the game, Player 2 must undergo a calibration procedure with the BCI system and is recommended to undergo at least one training session to verify if the calibration was done correctly or if there is need of repeating the procedure. The BCI hardware used is a 16 gel-based electrode system (g-tec).

By using the BCI, it is conveying the message of deception and the need to rely on non-verbal and other overt communication cues to escape. In this case, brain waves function as a proxy for the user's intentions, unbeknownst to the captors. This is akin to the written word, or embedding hidden messages in songs (e.g., the Colombian Government's use of the song "Better Days" to rescue 19 hostages) to respond to the aid of others.

Furthermore, the use of AR is also done purposefully, as it represents the notion that those who assist the oppressed in their escape often see threats which the captor cannot. Ultimately, one player understands the threats but cannot execute the moves, whereas the other is blind to the threats but must trust the other player to execute the moves. See Fig. 1 for a general layout of the installation.

An optional feature is the use of paint to trace the player's movement generated by the robotic agent throughout the play area. This will serve as a visual representation of the journey individuals must take to evade oppression, as well as artistically representing trust, empathy, and cooperation performed by the players. See Fig. 2 for the initial play area and the art piece generated at the end of the game.

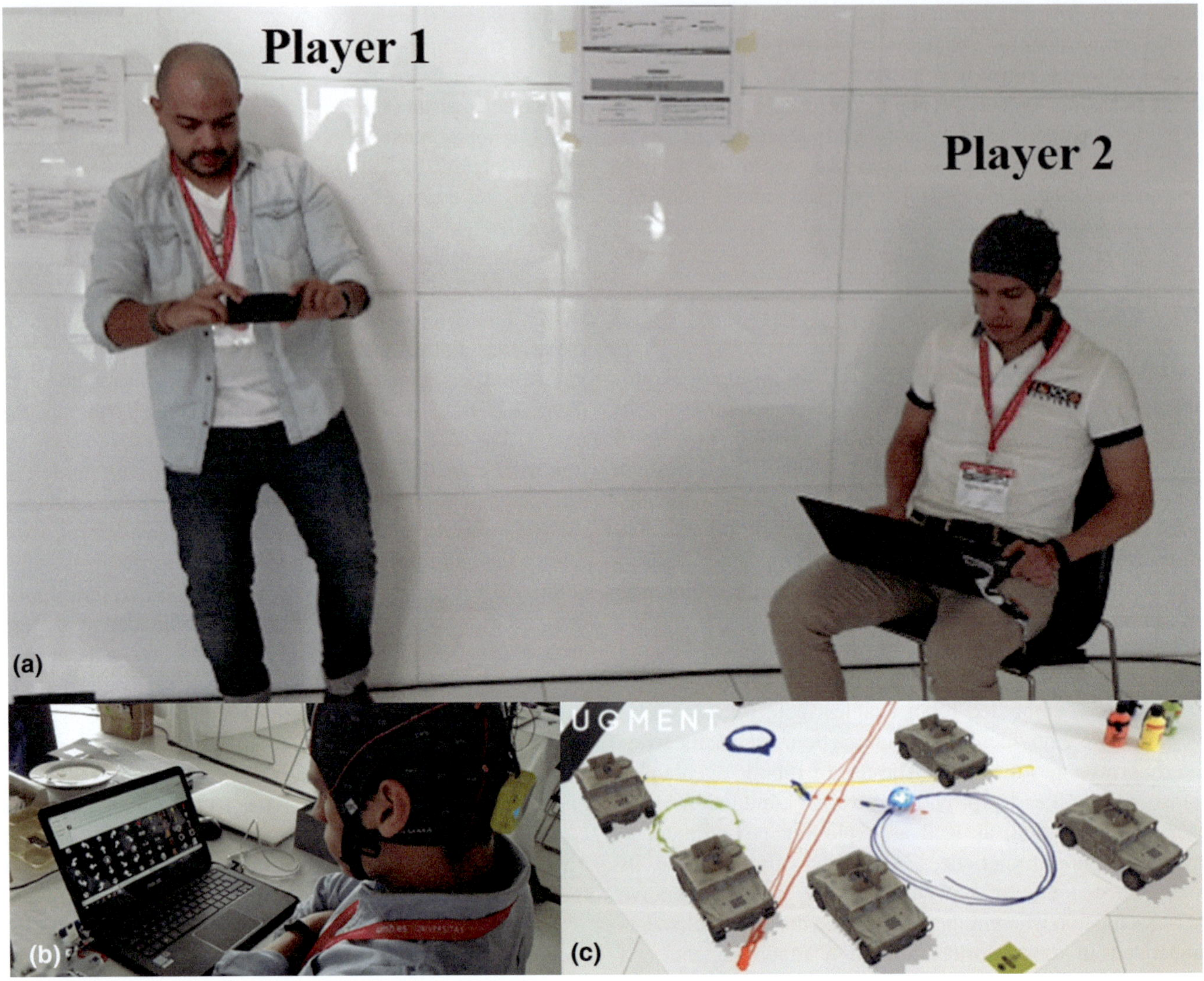

Fig. 1 General description of the installation. **a** Player 1 and player 2 playing the game. Player 1 contemplates the AR map on the mobile phone and sends voice commands to player 2, player 2 executes P300 responses to command the robot according to player 1 indications. **b** Player 2 uses a g.tec BCI system that decodes the P300 response elicited by looking at the command matrix at the screen. **c** AR map with sentries set and the robotic agent executing the movements

4 Discussion

Fostering human-to-human interactions to generate trust, empathy, and cooperation, using novel ways that combine BCI, AR, robots, and gaming, give us tremendous opportunities to enhance the way the brain process information, incorporates external cues (i.e. voice commands and visual targets), and develops new brain connections. The ability of these type of systems, like the one presented here, to promote human-to-human interaction through gaming, to interface with neural signals to control a robot, and to engage the users with AR scenarios and robotic agents, present potential benefits to the development of these desired social skills.

The proof-of-concept presented here requires high levels of communication, decision-making, and concentration from the users. Player 1 has to strategically develop a plan for the prisoner to escape the sentries and be able to communicate effectively the desired actions to player 1, thus promoting decision-making and communication skills on player 1. Player 2 has to be attentive to the instructions declared by Player 1, and then, immediately focus to generate the P300 responses, which will be elicited only if the user is actively engaged, thus, promoting concentration on player 2. This setup creates an environment where effective communication and execution must be met to succeed in the game.

Moreover, the use of a spherical robot to simulate a soldier, as the robot dynamically rolls in different ways, the

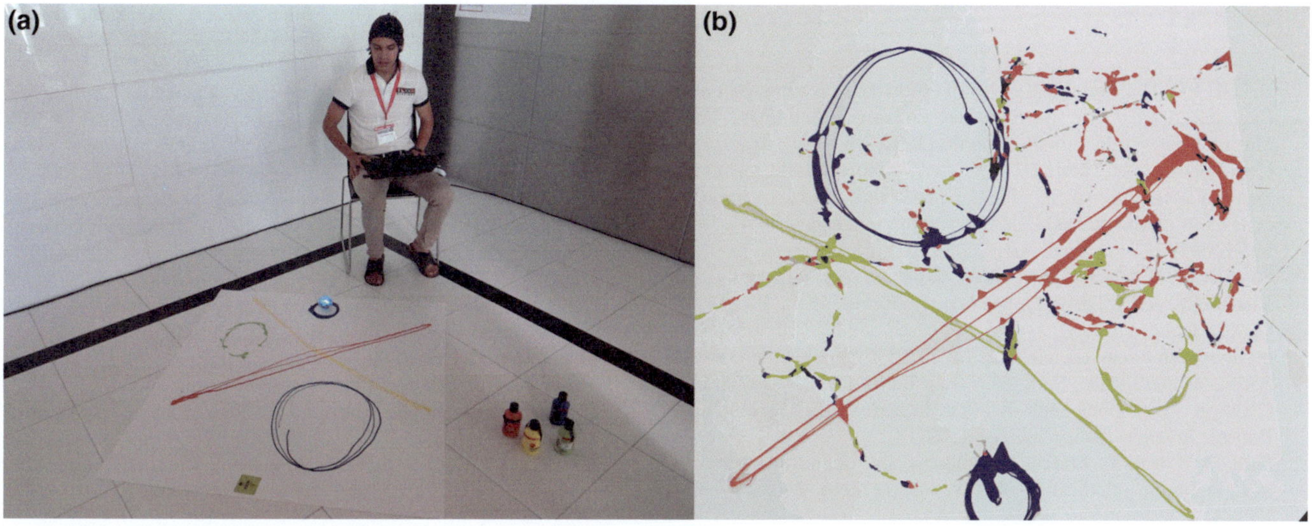

Fig. 2 Use of paint to trace the robot movements. **a** Initial setup of the play area with paint of four different colors and the robotic agent in place. **b** Art piece generated at the end of the game, visually representing the movements generated by the robotic agent. *Picture credit* Universitat Miguel Hernandez

movements generated present a more realistic way in which the user identifies with the soldier and thus, generates embodiment with the robotic agent and engagement with the experience.

Apart from mainstream pharmacological or behavior modification strategies, recent advancements in EEG signal processing have allowed the development of BCIs to offer original and therapeutic solutions by using neurofeedback, EEG-based imagery enhancement, and close sensorimotor loops [13]. Moreover, the use of robots has proven to reduce stress and anxiety [8] and to facilitate therapies for children with disorders. In this way, we envision this installation could be used as a therapeutic intervention. Thanks to the use of EEG signals, brain imaging could be performed to identify neural patterns related to abstract cognitive states like trust, cooperation, empathy and track them to evaluate the effectiveness of this installation when played by vulnerable groups. Additionally, the use of interactive technologies like robots and AR, may improve engagement and enjoyment of the game and make it suitable for children and patients with attention-related conditions.

It is evident that neuroscientists, game developers, artists, and clinicians must work together to share and discuss experiences. The ultimate goal as a scientific community is to integrate all disciplines without prejudice to enhance people's experiences. With this, the design of novel methods to promote social cooperation, raise empathy, and enhance cognition, would come naturally and with levels of innovation never seen before.

5 Summary: Lessons Learned

Here we present an installation that uses BCI, AR, and robotics technology to produce an intervention aimed to engender trust, empathy, and cooperation with a game-based and artistic experience among a dyad. During the game, the players execute a variety of cognitive tasks like decision-making, strategy planning, concentration, and communication, which need to be executed precisely and effectively to succeed in the game. Thanks to the use of novel technologies, the participants were deeply engaged during the whole game and were able to interact with the other participant using voice commands and visual targets while interfacing with the BCI. This work aims to enhance human-to-human interactions and develop social skills by extending the way the brain process information, incorporates external cues, and develops new connections.

During the hackathon, we learned:

1. To communicate ideas across disciplines.
2. To translate abstract thoughts into executable actions.
3. To take full advantage of the capacity of the technologies used (BCI, AR, robotics).
4. To harmoniously band together art, science, and technology into a project that could tell a story and solve a problematic.
5. To leverage interactive multi-modalities to build trust between users.

References

1. Bekoff, M.: The Emotional Lives of Animals: A leading Scientist Explores Animal Joy, Sorrow, and Empathy—And Why They Matter. New World Library, Novato (2007)
2. Choi, B., Lee, I., Choi, D., Kim, J.: Collaborate and share: an experimental study of the effects of task and reward interdependencies in online games. CyberPsychology & Behav. **10**(4), 591–595 (2007)
3. Chumerin, N., Manyakov, N.V., Combaz, A., Suykens, J.A., Yazicioglu, R.F., Torfs, T., Merken, P., Neves, H.P., Van Hoof, C., Van Hulle, M.M.: P300 detection based on feature extraction in on-line brain computer interface. In: 20199 Annual Conference on Artificial Intelligence. Springer-Verlag Berlin Heidelberg (2009)
4. Clawson H.J., Coolbaugh K.: The YouthARTS Development Project. Juvenile Justice Bulletin (2001)
5. Feng, J., Lazar, J., Preece, J.: Empathy and online interpersonal trust: a fragile relationship. Behav. Inf. Technol. **23**(2), 97–106 (2004)
6. Herrera-Arcos, G., Tamez-Duque, J., Acosta-De-Anda, E.Y., Kwan-Loo, K., de-Alba, M., Tamez-Duque, U., Contreras-Vidal, J.L., Soto, R.: Modulation of neural activity during guided viewing of visual art. Front. Hum. Neurosci. **11**, 581 (2017). https://doi.org/10.3389/fnhum.2017.00581
7. Ibáñez, M.B., de Castro, A.J., Kloos, C.D.: An empirical study of the use of an augmented reality simulator in a face-to-face Physics course. In: 2017 IEEE 17th International Conference on Advanced Learning Technologies (ICALT), pp. 469–471. IEEE (2017)
8. Jeong, S., et al: A social robot to mitigate stress, anxiety, and pain in hospital pediatric care. In: Proceedings of the Tenth Annual ACM/IEEE International Conference on Human-Robot Interaction Extended Abstracts, pp. 103–104 (2015)
9. Kontson, K.L., Megjhani, M., Brantley, J.A., Cruz-Garza, J.G., Nakagome, S., Robleto, D., White, M., Civillico, E., Contreras-Vidal, J.L.: Your brain on art: emergent cortical dynamics during aesthetic experiences. Front. Hum. Neurosci. **9**, 626 (2015). https://doi.org/10.3389/fnhum.2015.00626
10. Kose-Bagci, H., Ferrari, E., Dautenhahn, K., Syrdal, D., Nehaniv, C.: Effects of embodiment and gestures on social interaction in drumming games with a humanoid robot. Adv. Robot. **23**, 14 (2012). https://doi.org/10.1163/016918609X12518783330360
11. Lämsä, A.M., Pučėtaitė, R.: Development of organizational trust among employees from a contextual perspective. Bus. Ethics: A Eur. Rev. **15**(2), 130–141 (2006)
12. Libin, A.: Well-being focused gaming™: individualized engagement with plush toys, avatars, and personal robots. eLearning Papers (43), 8 (2015)
13. McFarland, D., Daly, J., Boulay, C., Parvaz, M.: Therapeutic applications of BCI technologies. Brain-Comput. Interfaces **47**(1–2), 37–52 (2017). https://doi.org/10.1080/2326263X.2017.1307625
14. Morris, F.J., Willis-Rauch, M.: Join the art club: exploring social empowerment in art therapy. Art Ther. **31**(1), 28–36 (2014)

Guillermo Herrera-Arcos is a Mexican technologist driven by how global health problems can be transformed with hard science and technology to make better people's lives. He is currently an undergraduate research assistant at the National Robotics Laboratory at Tecnologico de Monterrey, where he studies and designs hardware and software solutions in the biomechatronics and bionics fields.

Daniel Pimentel is a Ph.D. student and lab coordinator at the University of Florida's Media Effects and Technology Lab (MET) in the College of Journalism and Communications. Daniel conducts experimental research on the psychology of interactive media, namely, how immersive storytelling can contribute to pro-social outcomes.

Part VI

How Can the Arts and Neuroscience Describe and Promote the Processes of Learning and Creativity in K-12 and Higher Education?

Introduction

Chang S. Nam
Department of Industrial and Systems Engineering, North Carolina State University, Raleigh NC, USA

What kinds of skills should our students develop to survive in the twenty-first century? As part of an effort to answer this question, over the last decade a growing number of experts have proposed a new set of skills that are required for learners to acquire, such as complex problem-solving, critical thinking, collaboration, and creativity [3, 5]. Traditional Science, Technology, Engineering, and Mathematics (STEM) education programs have well served to train our students to learn such skills. However, it is also true that conventional STEM education faces the challenge of students losing interest in STEM-related subjects, leaving too many students behind [1, 4].

There are a growing group of advocates and educators who believe that STEM is currently missing a key component, the arts, and would like to add art to STEM to turn it into STEAM (Science, Technology, Engineering, Art, and Mathematics). Research has revealed a strong connection between the art and STEM education, because both inspire and demand creative thinking. In her Chapter "The Arts, Creativity, and Learning: From Research to Practice", Hardiman adds additional evidence to support that the arts should be an integral part of STEM education efforts—teaching with and through the arts, as the arts promote such twenty-first century skills, insights, and dispositions. Hardiman starts by demonstrating arts-integrated instruction that has the potential to improve retention of academic content and, compared to traditional ways of teaching, followed by an informative summary of previous studies on causal connections between arts-enhanced learning and better memory for content. In addition, several neuroscientific evidences are explained to show how arts-integrated education can promote creative thinking with anatomical changes in the brain. Finally, Hardiman wraps up her chapter by proposing changes and development of education practices and policies to increase access to the arts.

Part VI introduces one more chapter regarding effectiveness of arts in health. In Chapter "Intersectionality: The Confluence of Arts, Technology, and Wellbeing", Baefsky and Sonke present a summary of previous studies that arts-based approaches in healthcare and public health programs could enhance healthcare environments, reorient health services, and contribute to public health policy (e.g., National Organization for Arts in Health, NOAH, [2]). Importantly, Baefsky and Sonke continue by introducing an interesting concept, "social prescribing," where social, cultural, and artistic activities can be referred or prescribed by care providers. Next, Baefsky and Sonke share the same view with Hardiman that the arts and humanities need to be integrated into medical training, as well as STEM education. Finally, Baefsky and Sonke maintain that such the integration of the arts and humanities into medical or health education should be understood through the lens of neuroaesthetics that explores the neural processes underlying our appreciation and production of objects, artwork, and experiences including perception, interpretation, emotion, and action.

References

1. Bybee, R.: Advancing STEM education: A 2020 vision. Technol. Eng. Teach. **70**(1), 1 (2010)
2. National Organization for Arts in Health (NOAH): Arts, health and wellbeing in America. San Diego, CA (2017)
3. Partnership for 21st Century Skills: Learning for the 21st century. Partnership for 21st Century Skills, Washington, DC (2003)
4. Sanders, M.: STEM, STEM education, STEM mania. Technol. Teach. **68**(4), 20–26 (2009)
5. The World Bank Group: Lifelong learning in the global knowledge economy: Challenges for developing countries. World Bank, Washington, DC (2003)

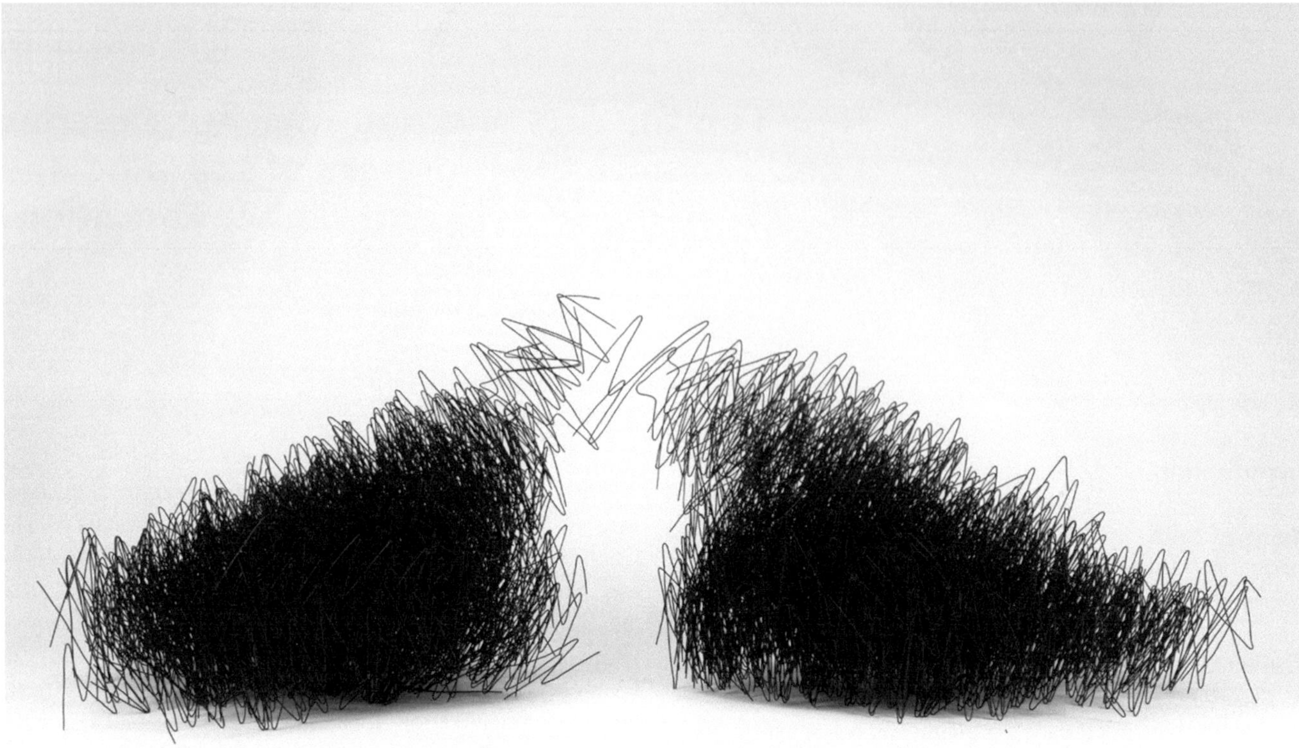

Julia Buntaine Hoel
Gamma Wave(s). 2' × 2 × 2.5 ft.
Rebar Wire
2013

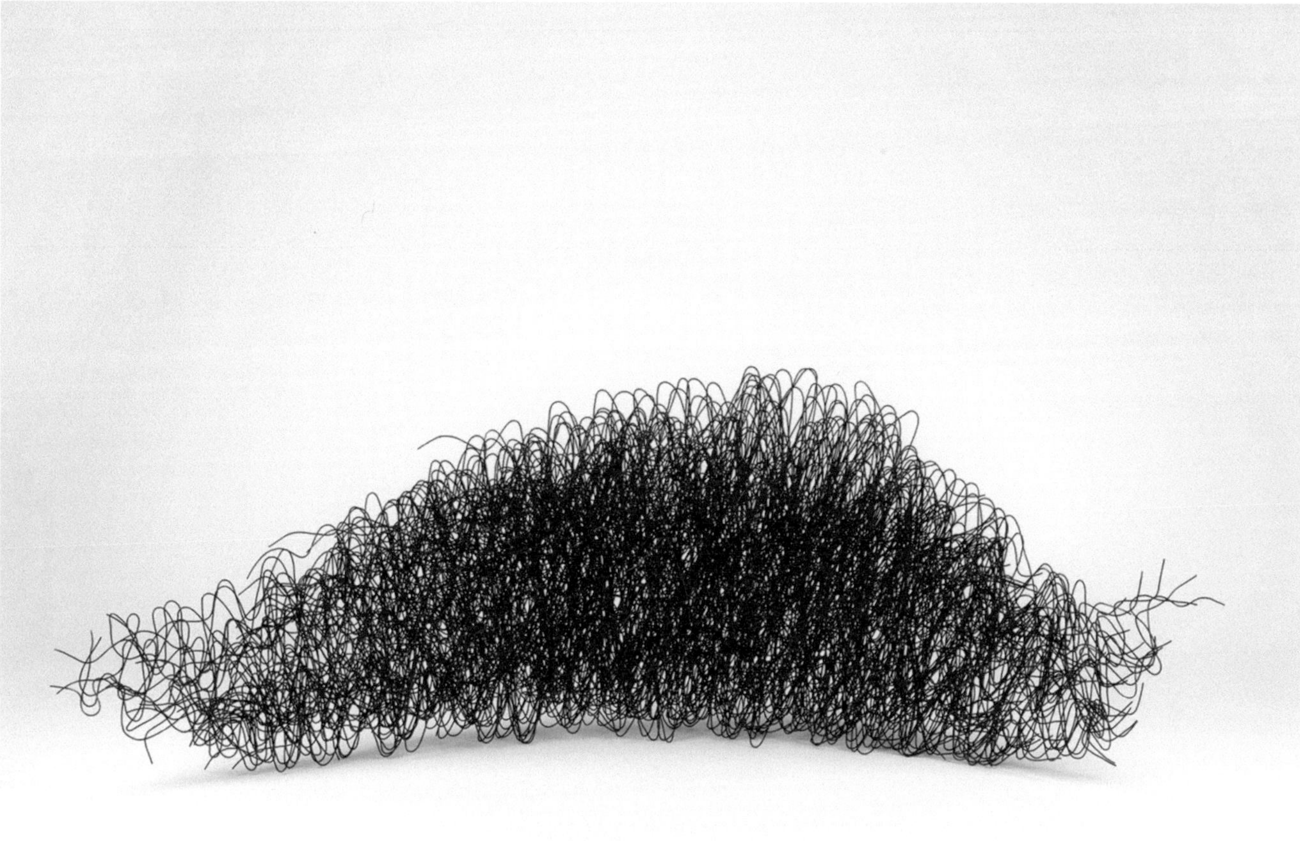

Julia Buntaine Hoel
Delta Wave(s)". 2 × 5 × 1 ft.
Rebar Wire
2016
The brain runs on electricity, and this electrical activity is one way in which we gain insight into the brain's mechanisms. Mathematically divided into five groups, our brain waves embody our conscious and unconscious activity. This piece is an exploration of the aesthetic potential of brain wave forms

The Arts, Creativity, and Learning: From Research to Practice

Mariale M. Hardiman

Art is not the possession of the few who are recognized writers, painters, musicians; it is the authentic expression of any and all individuality.

—John Dewey, Moral Principles in Education

The field of education is experiencing strong momentum for a seismic shift at every level of schooling, from preschool to higher education. In response to the demands of the world marketplace, educators are challenged to prepare students who display creative, innovative thinking and problem-solving skills. Across all sectors, especially in the areas of Science, Technology, Engineering, and Mathematics (STEM), industries are seeking workers capable of demonstrating skills well beyond content expertise. Often referred to as "Twenty-first Century Skills," professionals need to be adaptive, collaborative, flexible, and creative. In this chapter, I argue that the arts promote these twenty-first century skills, insights, and dispositions. Arts education and arts integration—teaching with and through the arts—may be the missing link in preparing students for the demands of the global workforce (Fig. 1).

1 Arts Integration and Learning

Many agree that the arts are important "for their own sake," yet strong evidence also suggests that arts can improve learning in other disciplines and help to engage students in subject matter more efficiently than traditional instructional techniques. When using the arts as a pedagogical approach in any area of instruction, educators are, in essence, teaching creative thinking. Common to all arts integration or arts-infused methods is the idea that non-arts content, such as language arts, social studies, math, and science, can be addressed through activities that incorporate the visual and performing arts. A growing body of research suggests that integrating the arts into non-arts subjects correlates with a host of positive outcomes for students, including engagement in learning, academic achievement, and deeper thinking dispositions. Workman [62] describes arts integration as a way of promoting the transfer of knowledge and skills from arts to non-arts domains by helping students draw connections between different disciplines within school curricula and engage in deeper learning skills. Using the arts as a pedagogical tool also encourages collaboration among learners. Moreover, a number of studies have linked arts integration to academic achievement in reading and mathematics. The A+ Schools Program whole-school reform initiative in North Carolina [18] and Oklahoma [4] reported achievement gains after schools instituted arts integration programs. They also found strong community and teacher support for the impact of arts integration on general learning outcomes. In a quasi-experimental four-year study, Scripp et al. [57] compared six arts-integrated Chicago Public Schools to six demographically matched control schools. Relative to the control schools, the arts-integrated schools produced higher scores on state assessments and narrowed the achievement gap between high- and low-performing students. Similarly, the Mississippi Arts Commission Whole Schools Initiative conducted a four-year study involving over 5000 students in public and independent elementary schools. Results showed that arts-integrated schools, particularly those with the highest level of implementation, increased the percentage of students scoring "proficient" in literacy [44]. A 10-year study sponsored by the Kennedy

Electronic Supplementary Material
The online version of this chapter (https://doi.org/10.1007/978-3-030-24326-5_23) contains supplementary material, which is available to authorized users.

M. M. Hardiman (✉)
John Hopkins University School of Education,
Baltimore, MD, USA
e-mail: mmhardiman@jhu.edu

J. L. Contreras-Vidal et al. (eds.), *Mobile Brain–Body Imaging and the Neuroscience of Art, Innovation and Creativity*,
Springer Series on Bio- and Neurosystems 10, https://doi.org/10.1007/978-3-030-24326-5_23

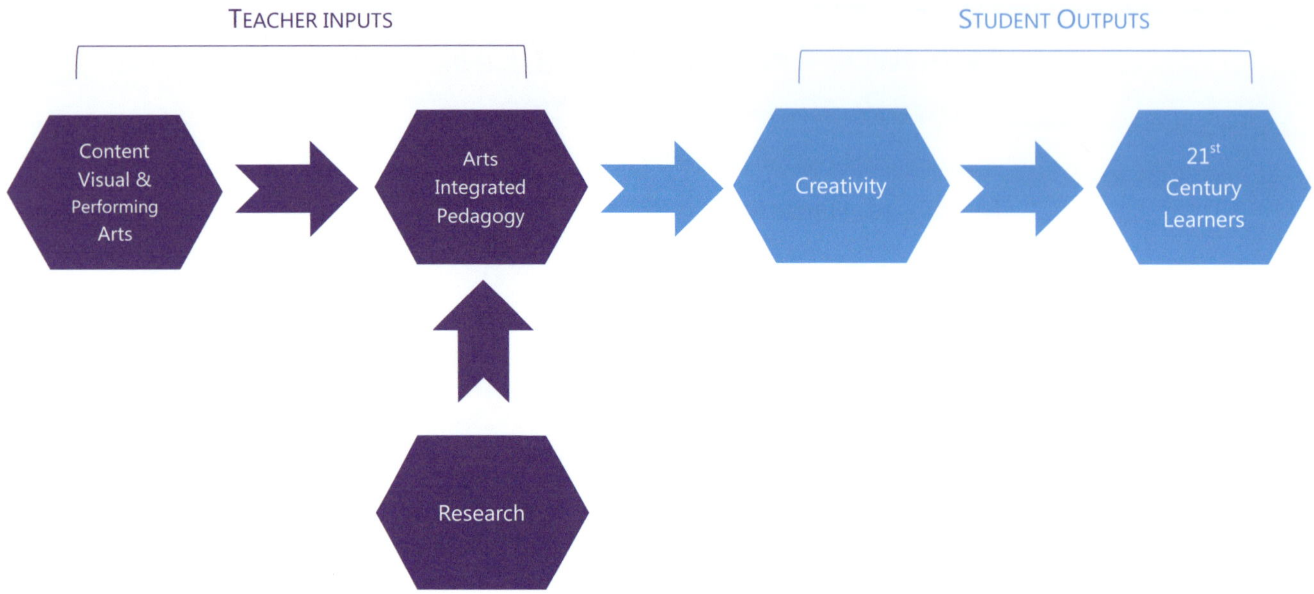

Fig. 1 Conceptual framework

Center found advantages for arts integration, especially for students at the lower levels of academic achievement [16]. Results also corroborated previous findings that teachers believe arts integration leads to deeper learning, including making connections between new learning and previously learned concepts, stronger analytical skills, and enhanced ability to synthesize information into global conceptual thinking. These findings support a recent comprehensive review of arts-integrated strategies conducted by the American Institutes for Research. Multiple studies show a positive effect on student learning, class engagement, pro-social skills, and problem-solving skills [35].

Integrating the arts into content instruction may also help students retain information. In a meta-analysis of studies that analyzed students' retention of content learned in school, Custers [12] reported that after two years, students retained about half of the academic content taught. Based on the idea that learning requires a certain number of repetitions for information to consolidate for long-term memory storage [27], arts-integrated instruction improves retention by prompting students to rehearse and elaborate on academic content through various visual and performing arts activities [20, 21]. In addition to rehearsal and elaboration, the arts commonly involve a variety of other ways of interacting with information that may improve retention of content. These "memory effects" are areas that have been well-researched in the learning sciences over the last several decades [21, 23, 52]. They include: (1) *elaboration* of content (e.g., [10, 28], which may include students writing dialogue or short stories to establish a stronger memory trace of the information learned; (2) *generation* of information (e.g., [58]), such as producing information from a cue to encourage creative and divergent thinking; (3) *enactment* of content (e.g., [40]), which might involve theater-based activities or role-playing; (4) *oral production* (e.g., [37]), such as generating content through the performing arts; (5) *effort-after-meaning* (e.g., [65]), which, for example, might include puzzling over the meaning of content as one might do in viewing a painting or reading a poem; (6) *emotional arousal*, central to all arts, which aids in information retrieval (e.g., [7]); and (7) *pictorial representation* of information (e.g., [38]), which has shown to produce greater memory than merely verbal input.

As these examples demonstrate, memory effects are naturally recruited through engagement with the arts. Thus, arts-integrated instruction has the potential to improve retention of academic content and, compared to traditional ways of teaching, offer rich and diverse ways to enhance instructional practice. To test this hypothesis, our research team at the Johns Hopkins University School of Education conducted a preliminary randomized control trial to explore whether or not students who learned science content through arts-integrated instruction would retain the content better than students who learned the same content through conventional teaching [23]. Our team developed arts-integrated and conventional versions of fifth-grade science units teaching the subject matters of ecology and astronomy. Four randomized groups of students received one body of content through an arts-integrated unit and a second body of content in a control unit that employed a traditional presentation. The units contained the same content but differed in the instructional delivery. For example, in the control condition, student displayed knowledge by completing a chart or presenting information orally; in the arts-integrated treatment

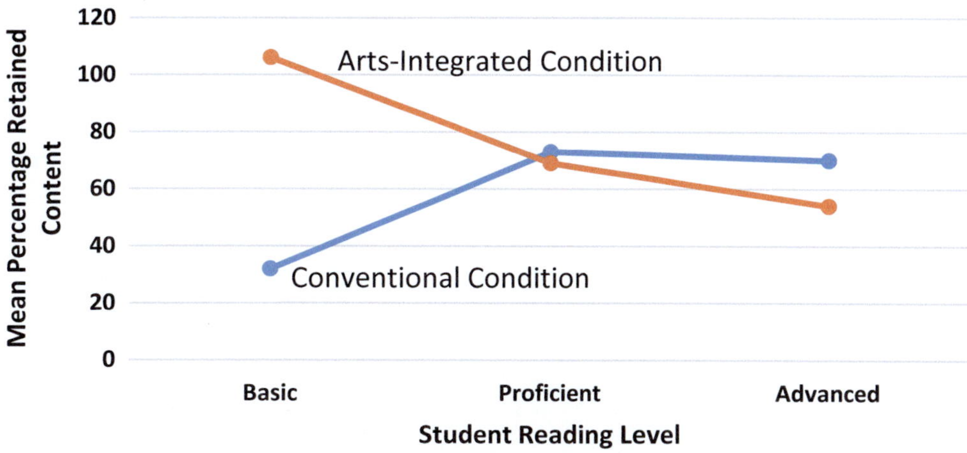

Fig. 2 Interaction of student reading level and study condition for the retained content

condition, students sketched, sang, chanted a rap, or used body movement such as tableau to demonstrate the content or concept. Curriculum-based assessments conducted at the conclusion of each of the units showed that from pre-testing to post-testing, students learned approximately the same amount of information regardless of the way they were taught. However, approximately 10 weeks later, delayed test scores were significantly better for the arts-integrated condition. The study found a differential benefit when comparing students according to levels of proficiency in reading: students at the lower levels of achievement were the most likely to retain significantly more science content when given arts-integrated lessons than when given traditional science instruction (see Online Resource 15, arts-integrated astronomy unit).

In an expanded pilot study with 16 randomly assigned groups, our team tested four sets of arts-integrated treatment units matched to control units using four science content areas. Similar to findings from the original study, students at the lower levels of achievement benefited the most [22]. These studies provided some preliminary causal connections between arts-enhanced learning and better memory for content for students who struggle with learning in conventional ways. Findings also raised some interesting questions about whether or not learning through the arts transfers residual benefits (Fig. 2).

Preliminary findings from the second study suggested the possibility that, once taught using arts-integrated instruction, students may later apply the strategies they learned, even during subsequent instruction through conventional methods [22]. Data shows that students who experienced the arts-integrated units first performed significantly better in subsequent non-arts-infused units than students who had never experienced the arts-integrated approach. While other factors such as familiarity with the unit structures may account for all students performing better the second time around, the findings nevertheless raise questions that bear further investigation. In particular, emerging research on creative thinking and problem-solving might connect learning with and through the arts as a fruitful alternative to conventional methods—as suggested by Dewey at the start of the twentieth century (Fig. 3).

2 Linking the Arts, Neuro-Cognitive Research, and Creativity

This study and others suggest the possibility that the arts can influence not only areas of academic attainment and engagement in learning but also creative thinking and problem-solving, which have become a signature focus in the call for teaching twenty-first century skills. Spearheaded by organizations such as the Partnership of Twenty-first Century Schools, those skills essential to a successful career include collaboration, effective communication, innovative thinking, and creative problem-solving. Yet, while educators are increasingly encouraged to design teaching activities that will promote creative thinking, there is little consensus on what creativity is or what it means in an educational context [15]. While definitions of creativity abound, neuroscientists point out the subjectivity involved in recognizing an endeavor or product as creative and the difficulty in realizing consensus within a group in the evaluation of creativity. Acknowledging that difficulty, Plucker [45] argues that researchers have begun to provide a better understanding of what creative thinking entails. Most creativity researchers agree that creative thinking requires producing original ideas that have value. More specifically, creativity most often includes general processes such as divergent thinking, originality of response, fluency and elaboration in generating ideas, and utility of ideas generated (see [31, 46]). Zhao [66], internationally known for his work in creativity, globalization, entrepreneurship, and technology, posits that creativity is multi-faceted and underlies all learning. He believes that

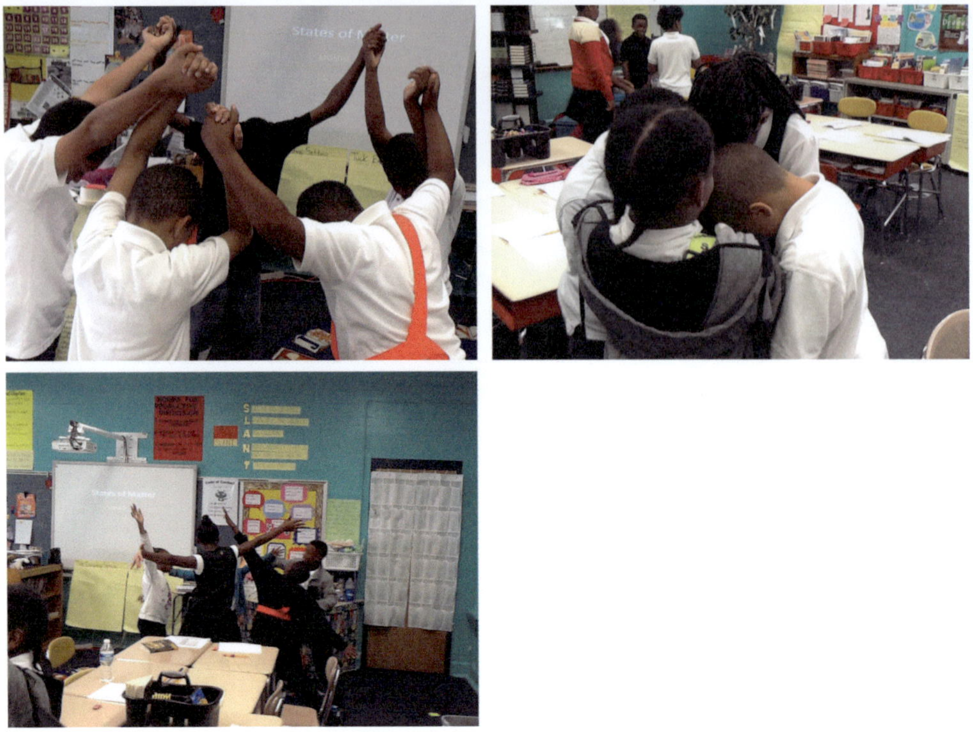

Fig. 3 Students learning about the states of matter by practicing acting out what happens in solids, liquids, and gas

creative thinking occurs within the cognitive ability of combining elements to produce a novel idea that has social value to others. Researchers studying creativity also note the importance of the social context within a specific domain. Many researchers believe that creativity occurs within a social context in which one has the emotional courage to take risks and embrace the learning that occurs from mistakes, setbacks, and failures [5, 51].

Given the breadth of these definitions, Gregory et al. [19] argue that within the educational context, creative thinking and problem-solving should be the focus of instruction for all students, not just those identified as gifted learners. This assumes that creativity is not a fixed attribute, a special gift bestowed on a lucky few. Rather, this idea advances the notion that all students can demonstrate creative thinking, especially when they experience instruction that encourages them to find connections among disparate concepts, varied solutions to problems, and application of content in novel contexts [47, 48, 54, 60]. Such strategies include collaboration on solving a problem having multipart tasks [14], support for scaffolding content such as the use of graphic organizers to help students make connections among concepts [56], and evaluating and revising ideas generated by others to aid in more original and fluent generation of one's own ideas [34].

While research cited earlier shows the power of the arts to improve student outcomes on domains of learning and memory, many believe that teaching with and through the arts is a powerful way to induce and empower students to think creatively and solve problems. To accomplish this, educators must be able to let go of traditional, structured approaches to learning by embracing and exploring ways to teach students to promote creative thinking. Rostan [53] argues that engaging in high-quality arts learning has been shown to develop creativity and provide an advantage for related forms of critical thinking. Csikszentmihalyi [11] describes the special role the arts play in cognition by highlighting the emotional responses that the arts can engender, creating novel ways of thinking that "…break through the gray affectless daily routines and expand the range of what it means to be alive" (p. 36). He describes how the arts can create a state of deep concentration and of "flow" that leads to the "aha" of creative thinking.

Welch et al. [61] argue that sketching plays a crucial role in generating and developing ideas, especially important in the language of design. They encourage the development of design drawing, which encourages students to construct and reconstruct different kinds of design ideas, freeing them from feeling that they have to produce a particular and expected kind of product. Hetland et al. [24] argue that the arts contribute to the development of more general thinking skills and dispositions that benefit school performance, such as envisioning, observing, reflecting, and engaging in multiple forms of expression. Perkins [43] and Arnheim [2] also emphasize that visual-thinking skills acquired through the arts can promote new and creative ways of viewing the world. Drawing on research in cognition, Perkins [43] presents arguments that endorse the use of the arts as a means for cultivating reflective thinking that motivates and engages students in all areas of learning. The benefits of looking at art include the development of dispositions of thinking, which he refers to as reflective intelligence—a set of skills, alertness to opportunities to utilize those skills, and the inclination to use them. Similarly, in the research compilation *Critical Links: Learning in the Arts and Student Academic and Social Development*, Deasy [13] reports on multiple studies that suggest the benefits of the arts for general learning in non-arts subjects, including self-motivation, social skills, tolerance, empathy, persistence and positive peer interaction.

Posner and Patoine [49] assert that the arts help to sustain attention, which they argue may improve learning by strengthening the brain's attentional networks. Others studying the arts through the lens of neuroscience have contributed to the understanding of how the arts may promote creativity and lead to anatomical changes in the brain. Dunbar [17], for example, studied differences between students who participated in performing arts experiences with those who did not. Results of fMRI studies showed that during tasks that required creative thinking, the performing arts group showed increased activity in the left frontal lobe, often associated with higher-order mental processing. Using standard measures of creative thinking, Dunbar also found that the students who had been engaged in the performing arts were more likely to generate creative ideas than peers who had no experiences in the performing arts. Others have found that musical training has shown neuroanatomical differences in brain regions associated motor and auditory processing [3] and the regulation of stress, arousal, and emotions by initiating reflexive brainstem responses [9]. Jung et al. [25] report anatomical changes in cortical thickness in the parietal lobe related to creative performance associated with divergent thinking tasks. Limb and Braun [33] found that spontaneous, creative improvisation activates different parts of the brain compared to memorized performance. They conducted fMRI studies of professional jazz musicians and found differences in brain activation while playing improvisational jazz compared to playing a memorized jazz music selection. Sawyer [55] also found that students participating in improvisational jazz and theater groups produced more novel ideas than non-arts peers. Supporting these studies in the performing arts, Kraus [32] found that playing an instrument may assist in processing speech and interpreting voice changes that influence language comprehension. The study of neuroanatomical changes related to artistic creative endeavors provides biological evidence of the power of the arts for learning and human development (for a thorough review see [1, 26]). These studies, in addition to a growing body of literature on the connection of arts and creative thinking, suggest that experiences in the visual and performing arts have the potential to help sustain attention, improve memory, create emotional connections to content, foster concentration leading to "aha" discoveries, and promote multiple, divergent solutions in problem-solving tasks.

3 Educational Practices and Policies

Given the compelling evidence of the power of the arts to generate creative thinking, it is reasonable that more creative types of schooling would be at the forefront of educational reform. Yet, the arts have generally been viewed as "fringe subjects" and have been victim of the curriculum narrowing that have plagued schools throughout the country, especially in school districts with fiscal constraints. Walker [59] found that 81% of elementary teachers reported that time devoted to math and language arts instruction resulted in less time for other subjects, especially the arts. The U.S. Department of Education reported that 40% of high schools did not require coursework in the arts for graduation [6]. Most alarming, however, is that children attending schools in low-income neighborhoods are the least likely to receive arts experiences [41]. O'Brien [42] reports that in high-poverty schools, just 59% of schools have dedicated space for visual arts instruction compared to 76% in low-poverty schools. Evidence shows that the children in low-income communities are paying a high price—in terms of academic and life achievement—for this disparity. Catterall et al. [8] found that students from low-income schools who had higher levels of arts experiences than peers without the same experiences were more likely to complete high school, attain higher grade point averages, enroll in college, become more involved in community activism, and express greater interest in current affairs. These studies also highlight a sobering fact: students in low-income communities who attend schools that do not offer sufficient arts courses are five times more likely to drop out of high school than their counterparts who had multiple courses in the arts [36]. Lack of access to the arts coupled with the high-stakes accountability movement in educational systems are working against the goals of a focus on creative thinking skills. Zhao [66] suggests that in many education contexts conformity is expected and rewarded, resulting in a diminishing ability for teachers to promote creativity within their curriculum and instruction.

To accomplish the goals of twenty-first century schooling, education systems must progress beyond the stringent accountability standards that have focused squarely on standardized testing in basic skills of reading and mathematics and consider broader domains of learning when assessing students' academic performance. The latest iteration of the Elementary and Secondary Education Act, the Every Student Succeeds Act (ESSA), is a step forward in broadening the view of educational success and providing all students with a "well-rounded education." Significantly, with the enactment of ESSA, the arts are designated as core subjects and art educators can access federal funds to expand programming. Many education advocates hope that this new approach will address some of the unintended consequences of No Child Left Behind—most importantly, the narrowed curriculum that led to diminishing arts programs in schools.

An important step in this direction can be seen in the role the arts have recently played in the teaching of STEM subjects. Adding the arts to the acronym, the STEAM movement has recently gained traction, likely in response to the often-espoused need for a more creatively productive workforce to increase U.S. global competitiveness. Additionally, Merten [39] argues that science and the arts are a natural combination as both scientists and artists seek to create something new, whether new knowledge or a new product. The processes of scientific discovery and arts creation similarly involve seeking novel ways of understanding, exploring multiple solutions to a problem, trying new approaches, synthesizing multiple elements to create a larger whole, and envisioning what is not yet seen or discovered. Many view the arts as a sort of springboard for imbuing traditionally taught STEM subjects with the creative application of knowledge that encourages innovation [64]. Studies have shown some compelling advantages in adding the arts to STEM subjects. For example, Kong et al. [29] and Kong and Huo [30] found that infusing STEAM activities into elementary schools resulted in statistically significant increases in positive attitudes toward science education, higher levels of self-efficacy for STEAM subjects, and increased interest in scientific learning. Yee-King et al. [63] investigated effects of STEAM approaches to teaching computing coding and programming. They reported that students who learned programming through an arts-integrated approach earned higher grades and developed more sophisticated programming skills compared to students who learned programming in conventional classes.

These findings support the argument for adding the arts to pedagogical approaches across all curricular areas. The challenge facing educators is how to design creative, arts-integrated instruction within the context of school accountability measures. Educators know that measures of high-stake assessments drive not only the content but also the approach to teaching. Broader and deeper school accountability measures that include a strong focus on the arts and creativity could have profound implications for educational practices and policies. This will require new metrics that measure performance and competencies that promote creative problem solving, allowing students to use multiple, authentic ways to demonstrate mastery of content, skills, and concepts. The reform of assessment systems will also embolden teachers to move beyond the silos of compartmentalized subjects to build bridges across different curricular offerings. To accomplish this, assessments should be informed by how creative products and processes are measured in other disciplines. Given the various perspective in which creativity researchers measure creative endeavors [1, 50], collaborations across the disciplines of education and the neuro-cognitive sciences could produce robust and authentic educational measures of creative products and processes scalable to all levels of schooling. For example, creativity assessments that require expert judges to measure creative products such as the Consensual Assessment Technique could be adapted for teams of educators to review student work in a particular subject matter (domain-specific creativity) or in general creativity measures such as assessing the number of novel ideas in a student product (domain-general creativity). This approach could be supplemented with student self-reports and reflections. Teachers could assess creative products using well-developed rubrics that clearly describe expectations and indictors for content-related knowledge, including art-related processes, and informed by the divergent thinking tests used by creativity researchers.

Changes in schooling must also be driven by robust programs of teacher preparation, continuing education, and professional development. Teachers should not be on their own to figure out what creativity looks like or how to measure it within the classroom setting. Moreover, no school should be denied arts educators who provide arts education to all students and collaborate with content teachers to develop and support arts-integrated pedagogy. Education researchers, policy makers, and funding organizations can address this need with a strong commitment to education research and the translation of research from the science of learning to relevant educational applications. This approach calls for a revolution in how student learning is assessed and significant changes in how teachers are prepared and evaluated.

Far from being a fringe subject, one might view the arts as the cornerstone of better schooling. From pre-school to adult learning, we know what arts education and arts integration can accomplish. The arts engage students by making them complicit in their own learning. Students follow structured, but never predefined, pathways to discover and weigh multiple solutions. Arts-integrated learning activities give students permission to be creative, but with a focus and a purpose. They remember more of what they learn because instead of memorizing content, they create and experience it. These are fundamental twenty-first century skills. Before we succeed in reforming the American education system, we must continue to pursue a goal that is within our grasp and facilitated through arts experiences: educating every child to be the innovative, inventive, and creative citizens of tomorrow.

Acknowledgements The author would like to acknowledge several colleagues from the Johns Hopkins University School of Education's Neuro-Education Initiative. The following individuals contributed valuable time and expertise to various components of this book chapter: Ranjini JohnBull, Assistant Professor; Clare Grizzard, Arts Integration Specialist; Joe Meredith, Executive Specialist; Kara Seidel, Research Assistant.

References

1. Abraham, A.: The Neuroscience of Creativity. Cambridge University Press, New York (2018)
2. Arnheim, R.: Visual Thinking. England, London (1969)
3. Barrett, K.C., Ashley, R., Strait, D.L., Kraus, N.: Art and science: how musical training shapes the brain. Front. Psychol. **4**, 713 (2013)
4. Barry, N.H.: Oklahoma A+ Schools: what the research tells us 2002–2007. In: Quantitative Measures, vol. 3. Oklahoma A+ Schools/University of Central Oklahoma. Retrieved from http://static1.1.sqspcdn.com/static/f/1313768/21019976/1353351929267/V3+final.pdf?token=qtH1HFHv6LqRbL9NQLH8vVPoty4%3D (2010)
5. Beghetto, R.A.: Taking beautiful risks in education. Educ. Leadersh. **76**(4), 18–24 (2018)
6. Bryant, S.: New NAMM foundation study shows parents and teachers in harmony about students learning music. Retrieved from https://www.nammfoundation.org/ (2015)
7. Cahill, L., McGaugh, J.L.: A novel demonstration of enhanced memory associated with emotional arousal. Conscious. Cogn. **4**, 410–421 (1995). https://doi.org/10.1006/ccog.1995.1048
8. Catterall, J.S., Dumais, S.A., Hampden-Thompson, G.: The arts and achievement in at-risk youth: findings from four longitudinal studies (Research Report 55). National Endowment for the Arts. Retrieved from: http://files.eric.ed.gov/fulltext/ED530822.pdf (2012)
9. Chanda, M.L., Levitin, D.J.: The neurochemistry of music. Trends Cogn. Sci. **17**(4), 179–193 (2013)
10. Craik, F.I., Tulving, E.: Depth of processing and the retention of words in episodic memory. J. Exp. Psychol. Gen. **104**, 268–294 (1975). https://doi.org/10.1037/0096-3445.104.3.268
11. Csikszentmihalyi, M.: Assessing aesthetic education: measuring the ability to "ward off chaos". Arts Educ. Policy Rev. **99**(1), 33–38 (1997). https://doi.org/10.1080/10632919709600763
12. Custers, E.: Long-term retention of basic science knowledge: a review study. Adv. Health Sci. Educ. **15**, 109–128 (2010). https://doi.org/10.1007/s10459-008-9101-y
13. Deasy, R.J. (ed.): The Arts and the Transfer of Learning. Critical Links: Learning in the Arts and Student Academic and Social Development. Arts Education Partnership, Washington, DC (2002)
14. Diehl, M., Stroebe, W.: Productivity loss in brainstorming groups: toward the solution of a riddle. J. Pers. Soc. Psychol. **53**(3), 497–509 (1987). https://doi.org/10.1037/0022-3514.53.3.497
15. Dietrich, A.: How Creativity Happens in the Brain. Palgrave Macmillan, NY (2015)
16. Duma, A., Silverstein, L.: A view into a decade of arts integration. J. Learn. Arts **10**(1) (2014)
17. Dunbar, K.N.: Arts education, the brain, and language. In: Asbury, C., Rich, B. (eds.) Learning Arts and the Brain: The Dana Consortium Report on Arts and Cognition. New York, pp. 81–92 (2008)
18. Fiske, E.B.: Champions of change: the impact of the arts on learning. Washington, DC: Arts Education Partnership and President's Committee on the Arts and Humanities. Retrieved from http://files.eric.ed.gov/fulltext/ED435581.pdf (1999)
19. Gregory, E., Hardiman, M., Yarmolinskaya, J., Rinne, L., Limb, C.: Building creative thinking in the classroom: from research to practice. Int. J. Educ. Res. **62**, 43–50 (2013). https://doi.org/10.1016/j.ijer.2013.06.003
20. Hardiman, M.: Connecting Brain Research with Effective Teaching: The Brain Targeted Teaching model. Lanham, Maryland (2003)
21. Hardiman, M.: The Brain-Targeted Teaching Model for 21st-Century schools. Thousand Oaks, California (2012)
22. Hardiman, M.: The Effects of Arts Integration on Retention of Content and Student Engagement (Grant No. R305A120451). Johns Hopkins University School of Education. Institute of Education Sciences, Washington, DC (2015)
23. Hardiman, M., Rinne, L., Yarmolinskaya, J.: The effects of arts integration on long-term retention of academic content. Mind Brain Educ. **8**, 144–148 (2014). https://doi.org/10.1111/mbe.12053
24. Hetland, L., Winner, E., Veenema, S., Sheridan, K.: Studio Thinking: The Real Benefits of Visual Arts Education. New York, NY (2007)
25. Jung, R.E., Segall, J.M., Jeremy Blockholt, H., Flores, R.A., Smith, S.M., Chavez, R.S., Haier, R.J.: Neuroanatomy of creativity. Hum. Brain Mapp. **31**(3), 398–409 (2010)
26. Jung, R.E., Vartanian, O. (eds.): The Cambridge handbook of the neuroscience of creativity. Cambridge University Press, New York (2018)
27. Kandel, E.R.: In Search of Memory: The Emergence of a New Science of Mind. New York (2006)
28. Klein, S.B., Kihlstrom, J.F.: Elaboration, organization, and the self-reference effect in memory. J. Exp. Psychol.: Gen. **115**(1), 26–38 (1986). https://doi.org/10.1037/0096-3445.115.1.26
29. Kong, Y.T., Huh, S.C., Hwang, H.J.: The effect of theme based STEAM activity programs on self efficacy, scientific attitude, and interest in scientific learning. Info **17**(10 (B)), 5153–5159 (2014)
30. Kong, Y.T., Huo, S.C.: An effect of STEAM activity programs on science learning interest. Adv. Sci. Tech. Lett. **59**, 41–45 (2014)
31. Kozbelt, A., Beghetto, R.A., Runco, M.A.: Theories of creativity. In: Kaufman, J.C., Sternberg, R.J. (eds.) The Cambridge Handbook of Creativity. New York, pp. 20–47 (2010)
32. Kraus, N.: Cognitive-sensory interaction in the neural encoding of music and speech. Paper presented at the American Association for the Advancement of Science Annual Meeting, San Diego, California (2010)
33. Limb, C.J., Braun, A.R.: Neural substrates of spontaneous musical performance: an fMRI study of jazz improvisation. PLoS ONE **3** (2), 1–9 (2008). https://doi.org/10.1371/journal.pone.0001679
34. Lonergan, D.C., Scott, G.M., Mumford, M.D.: Evaluative aspects of creative thought: effects of appraisal and revision standards. Creat. Res. J. **16**(2), 231–246 (2004). https://doi.org/10.1080/10400419.2004.9651455
35. Ludwig, M.J., Boyle, A., Lindsay, J.: Arts Integration Research Through the Lens of the Every Student Succeeds Act (ESSA). American Institutes for Research. http://www.wallacefoundation.org/knowledge-center/pages/essa-arts-evidence-review-report.aspx (2017)
36. Lynch, R.: Arts Education Transforms Societies. Huffington Post, Arts & Culture, New York City (2014)

37. MacLeod, C., Gopie, N., Hourihan, K., Neary, K., Ozubko, J.: The production effect: delineation of a phenomenon. J. Exp. Psychol.: Learn. Mem. Cogn. **36**, 671–685 (2010). https://doi.org/10.1037/a0018785
38. McBride, D.M., Dosher, B.A.: A comparison of conscious and automatic memory processes for picture and word stimuli: a process dissociation analysis. Conscious. Cogn. **11**(3), 423–460 (2002). https://doi.org/10.1016/S1053-8100(02)00007-7
39. Merten, S.: Enhancing science education through art. Sci. Scope **35**(2), 31–35. Retrieved from http://www.jstor.org/stable/43183128 (2011)
40. Mohr, G., Engelkamp, J., Zimmer, H.D.: Recall and recognition of self-performed acts. Psychol. Res. **51**(4), 181–187 (1989). https://doi.org/10.1007/BF00309146
41. Nathan, J.: Advocating for arts education. Educ. Week. Retrieved from http://blogs.edweek.org/ (2015)
42. O'Brien, A.: Can arts education help close the achievement gap? Learning First Alliance. http://www.learningfirst.org/can-arts-education-help-close-achievement-gap (2012)
43. Perkins, D.N.: The Intelligent Eye: Learning to Think by Looking at Art, vol. 4. California, Los Angeles (1994)
44. Phillips, J., Harper, J., Lee, K., Boone, E.: Arts Integration and the Mississippi Arts Commission's Whole Schools Initiative. Retrieved from http://www.mswholeschools.org (2013)
45. Plucker, J.A. (ed.): Creativity and Innovation: Theory, Research, and Practice. Prufock Press Inc., Texas (2017)
46. Plucker, J.A., Beghetto, R.A., Dow, G.T.: Why isn't creativity more important to educational psychologists? Potentials, pitfalls, and future directions in creativity research. Educ. Psychol. **39**(2), 83–96 (2004). https://doi.org/10.1207/s15326985ep3902_1
47. Plucker, J.A., Kaufman, J.C., Beghetto, R.A.: What We Know About Creativity. P21 Research Series. Partnership for 21st Century Learning, Washington, DC. Retrieved from http://www.p21.org/our-work/4cs-research-series/creativity (2015)
48. Plucker, J.A., Kennedy, C., Dilley, A.: What We Know About Collaboration. P21 Research Series. Partnership for 21st Century Skills, Washington, DC. Available at http://www.p21.org/our-work/4cs-research-series/collaboration (2015)
49. Posner, M.L., Patoine, B.: How arts training improves attention and cognition. Cerebrum, 2–4. Retrieved from https://www.researchgate.net/ (2009)
50. Qian, M., Plucker, J.A.: Creativity assessment. In: Plucker, J.A. (ed.) Creativity and Innovation Theory, Research, and Practice, pp. 223–234. Prufock Press Inc., Texas (2017)
51. Richardson, C., Henriksen, D., Mishra, P.: The courage to be creative: an interview with Dr. Yong Zhao. Tech Trends **61**, 515–519 (2017). https://doi.org/10.1007/s11528-017-0221-1
52. Rinne, L., Gregory, E., Yarmolinskaya, J., Hardiman, M.: Why arts integration improves long-term retention of content. Mind Brain Educ. **5**(2), 89–96 (2011). https://doi.org/10.1111/j.1751-228X.2011.01114.x
53. Rostan, S.M.: Studio learning: motivation, competence, and the development of young art students' talent and creativity. Creat. Res. J. **22**(3), 261–271 (2010). https://doi.org/10.1080/10400419.2010.503533
54. Runco, M.A.: Creativity: Theories and Themes: Research, Development, and Practice. San Diego, California (2014)
55. Sawyer, R.K.: Educating for innovation. Think. Skills Creat. **1**(1), 41–48 (2006). https://doi.org/10.1016/j.tsc.2005.08.001
56. Sawyer, R.K., Berson, S.: Study group discourse: how external representations affect collaborative conversation. Linguist. Educ. **15**(4), 387–412 (2004). https://doi.org/10.1016/j.linged.2005.03.002
57. Scripp, L., Burnaford, G., Vazquez, O., Paradis, L., Sienkiewicz, F.: Partnerships in Arts Integration Research final Reports. Arts Education Partnership, Washington, DC. Retrieved http://www.artsedsearch.org (2013)
58. Slamecka, N.J., Graf, P.: The generation effect: delineation of a phenomenon. J. Exp. Psychol.: Hum. Learn Mem. **4**, 592–604 (1978). https://doi.org/10.1037/0278-7393.4.6.592
59. Walker, T.: The testing obsession and the disappearing curriculum. The National Education Association Today. Retrieved from http://neatoday.org/ (2014)
60. Weisberg, R.W.: Expertise and reason in creative thinking: evidence from case studies and the laboratory. In: Baer, J. (ed.) Creativity and Reason in Cognitive Development. New York, pp. 7–42 (2006)
61. Welch, M., Barlex, D., Lim, H.S.: Sketching: friend or foe to the novice designer? Int. J. Technol. Des. Educ. **10**(2), 125–148 (2000). https://doi.org/10.1023/A:1008991319644
62. Workman, E.: Beyond the Core: Advancing Student Success Through the Arts. Education Commission of the States, Denver. https://www.ecs.org/beyond-the-core-advancing-student-success-through-the-arts/ (2017)
63. Yee-King, M., Grierson, M., d'Inverno, M.: STEAM WORKS: student coders experiment more and experimenters gain higher grades. In: Global Engineering Education Conference (EDUCON), 2017. IEEE, pp. 359–366 (2017)
64. Yusuf, S.: From Creativity to Innovation. The World Bank, Washington, D.C. (2007). https://doi.org/10.1016/j.techsoc.2008.10.007
65. Zaromb, F.M., Roediger, H.L.: The effects of "effort after meaning" on recall: differences in within-and between-subjects designs. Mem. Cogn. **37**, 447–463 (2009). https://doi.org/10.3758/MC.37.4.44
66. Zhao, Y.: Reach for Greatness: Personalizable Education for all Children. Thousand Oaks, CA (2018)

Mariale M. Hardiman EdD is professor at the Johns Hopkins University School of Education. She co-founded and directs the Neuro-Education Initiative, a cross-disciplinary program that brings to educators relevant research from the learning sciences. Her research includes randomized control trials investigating the effects of arts integration on long-term retention of academic content. A former school principal, Hardiman developed a teaching framework, the Brain-Targeted Teaching® Model, that promotes arts integration and creative problem-solving.

Intersectionality: The Confluence of Arts, Technology, and Wellbeing

Laurie Baefsky and Jill Sonke

1 Arts, Health and Wellbeing

Healthcare in America is a $3.5 trillion-dollar industry—$10,739 per person was spent in 2017 [1]. Despite this significant investment, the United States trails substantially behind other nations in health outcomes. With dramatic advances in medical devices and healthcare technologies, can a new wave of physicians be trained for understanding, prescribing, adapting, and maintaining emergent (neuro) technologies that empower individuals to gain control of their own health and wellbeing? How should higher education respond to "how we educate" in the medical, social services, and public health professions, and what roles do and can the arts and humanities play toward an equitable, resilient, healthy society? How can the arts transform the culture of care, wellbeing, and health to broaden access?

As our national healthcare system seeks solutions, arts and humanities, in tandem with the sciences and emergent technologies, have been partnering with each other, communities, higher education, and medical institutions to develop a growing suite of promising interventions. In May 2018, the National Academies of Sciences, Engineering, and Medicine issued a consensus report, *The Integration of the Humanities and Arts with Sciences, Engineering, and Medicine in Higher Education: Branches from the Same Tree.*[1] This report recommends integrative education, encompassing arts and humanities, science, technology, engineering, mathematics, and medicine at the undergraduate and graduate levels, to include training of health care professionals. The Association on American Medical Colleges, and the American Association of Colleges and Universities similarly endorse an integrated curriculum, with arts and humanities blended into traditional curricula as a high-impact practice.

Over the past several decades, evidence has mounted to demonstrate that the arts have positive and measurable impacts on individual and community health. A field of *Arts in Health* has developed from the rapidly expanding presence of arts programs in healthcare settings. Since 2004, arts programs have been documented at approximately half of accredited healthcare institutions in the United States [2]. This prevalence has led to the development of a recognized professional field, as well as to the development of an academic discipline that supports education and research in the field [3].

Arts in health broadly refers to the use of arts-based activities and interventions to promote or improve health in healthcare and community settings. Arts-based approaches in healthcare and public health programs have become a common means for promoting health, enhancing healthcare environments, reorienting health services, contributing to public health policy, strengthening community development and supporting personal/social development [3, 4].

Within arts in health, there is growing interest in how the arts and creative activity can affect health and wellbeing outside of clinical settings—in our daily lives, and where we live, learn, work, and play. In the past few decades, there has been mounting interest in examining and quantifying the impact of the arts, creativity, and cultural activities on **wellbeing**. It is widely accepted that the social and cultural environment has an impact on the quality of life, wellbeing, and health of the people who live, learn, work, and play in that environment. Many studies have measured impacts of arts and cultural participation on wellbeing, quality of life, and even on longevity. In a well-controlled seminal 9-year

[1]To download this report, see: https://www.nap.edu/catalog/24988/the-integration-of-the-humanities-and-arts-with-sciences-engineering-and-medicine-in-higher-education.

L. Baefsky (✉)
College of Arts & Media, University of Colorado Denver, Denver, CO, USA
e-mail: laurie.baefsky@ucdenver.edu

J. Sonke
Center for Arts in Medicine, University of Florida, Gainesville, FL, USA

J. L. Contreras-Vidal et al. (eds.), *Mobile Brain–Body Imaging and the Neuroscience of Art, Innovation and Creativity*, Springer Series on Bio- and Neurosystems 10, https://doi.org/10.1007/978-3-030-24326-5_24

study of arts participation with 15,198 participants, Bygren et al. [5] found that people who regularly attend cultural events live longer [5].

Similarly, Johansson et al. [6] conducted a study with 3,793 adults over an 8-year period designed to assess how changes in the habit of attending cultural events might predict self-reported health. It found a 65% increase in the risk of impaired perceived health among those who were not culturally active as compared with those who attended cultural activities. These findings are supported by more recent studies, including an 18-year study undertaken by Väänänen et al. [7], which associated increased cultural activity with decreased mortality, and work by Cuypers et al. [8] that documented associations between cultural participation and perceived health, anxiety, depression, and satisfaction with life, in a study of over 50,000 adults. Recently, Fancourt et al. [9] reported studies that found a lower incidence rate of dementia over a 10-year follow-up period, in adults aged 50 and older who visited museums every few months or more, compared with less-frequent visiting. In addition and equally promising, arts and cultural participation among older adults has been found to have protective effects on cognitive function and chronic pain [10, 11].

Data such as these are compelling the growing movement toward "social prescribing," wherein social, cultural, and recreational activities, namely, the arts, can be referred or prescribed by care providers. In some instances, health systems provide payment for these services [12, 13]. These programs are currently expanding in the UK, Denmark, Canada, Sweden, and Norway, and a few similar models are emerging in the United States as well [14]. The programs have demonstrated enhancements in self-management of health conditions and improvements in both physical and mental health among participants [15]. They also demonstrate what human beings have always seemed to know, that the arts and creative engagement make our lives—and us—better.

There is increasingly a heightened interest in the arts within the public health sector in the U.S. A national initiative, *Creating Healthy Communities: Arts + Public Health in America,* was launched in 2018 by ArtPlace America and the University of Florida Center for Arts in Medicine. This initiative was designed to accelerate the innovation taking root at the intersections of the arts, community building, and public health, to build healthier communities in alignment with national public health goals. Through a series of working group convenings, a national network, and a broad research agenda, the initiative is building strategic cross-sector collaborations and developing resources that translate evidence into practice and policy. Building on the widespread acceptance and utilization of the arts in the healthcare sector in recent decades, the arts in public health movement is scaling evidence-based applications of the arts to enhance health and wellbeing at the population level in the United States.

One example is the rise of prison arts initiatives designed to provide high-quality arts programs to incarcerated individuals (see the Prison Arts Coalition https://theprisonartscoalition.com/programs/). Often in partnership with universities, these programs address the lack of psychological and humanistic resources of the U.S. carceral system, which currently houses over 10 million individuals annually. There is also promising innovation underway pairing arts and arts therapy delivery with military and veteran populations. Creative Forces: NEA Military Healing Arts Network is a growing partnership of the National Endowment for the Arts, the Departments of Defense and Veterans Affairs, and state and local arts agencies. This program serves military patients and veterans with traumatic brain injury and psychological health conditions, as well as their families and caregivers. The program places creative arts therapies at the core of patient-centered care. The initiative also promotes access to community arts activities to promote health, wellness, and quality of life for military service members, veterans, and their families and caregivers (see: https://www.arts.gov/national-initiatives/creative-forces). Started at the Walter Reed National Military Medical Center in Bethesda, MD, there are now 13 Creative Forces clinical sites throughout the country. With the arts consistently one of the highest-ranking health interventions for returning soldiers (at very low relative cost and minimal negative side effects), the potential of positively impacting trauma patients is also significant.

Pedagogically, there is a similar shift occurring in education. The integration of the arts and humanities into medical training, as well as incorporated into the sciences and engineering are becoming increasingly accepted. Medical humanities is an interdisciplinary field dating back to the 1960s with an established pedagogy for including humanities, arts, and social sciences in medical education. The goals of a medical humanities curriculum are to: (1) ingrain aspects of professionalism, empathy, and altruism; (2) enhance clinical communication and observation skills; (3) increase interprofessionalism and collaboration; and (4) decrease burnout and compassion fatigue. Curricula includes bioethics, clinical ethics, and literature, and can include poetry, narrative, theater, or visual arts as part of a medical education. Designed around different ways of knowing, it can also foster increased tolerance for ambiguity, and increase interest in communication skills. It can be used as a way to help medical students develop diagnostic skills, as well as a pathway to create more humanistic physicians [16]. Over 30 medical schools and museums across the U.S. are known to collaborate to improve visual literacy, observational awareness, and visual perception (National Academies of Sciences and Medicine) [17].

Health humanities programs have likewise more than quadrupled from 2000 to 2016 [18]. In the 2015–2016 academic year, 94% of medical schools surveyed had required and/or provided elective courses in medical humanities [11, p. 155]. One study of the New Pathways program at Harvard University, which integrates social and behavioral sciences with the biological and clinical, found that students who came from the humanities-oriented curriculum were more prepared and inclined to pursue careers in humanistic medicine (such as primary care or psychiatry) compared to their peers who came from a more traditional medical curriculum. The students from the humanities-oriented curriculum were also more confident in managing patients' psychosocial issues [19].

Dr. Delphine Taylor, Associate Professor of Medicine at Columbia University Medical Center, emphasizes that arts-focused activities are important in training future doctors to be present and aware. This is increasingly difficult today given the pervasiveness of technology and media and a "digital attention crisis" facing society. "It's not just a nice idea to incorporate humanities into medical schools to make the education more interesting," Dr. Michael Flanagan at Penn State College of Medicine says of such programs. "It's protecting and maintaining students' empathy so that by the time they go off to practice medicine, they're still empathetic individuals." He notes that while medical students traditionally enter their third year with very high levels of empathy, after three years, this level decreases (M. Flannigan, Interviewee) [20].

2 Brain Science, Health, and Community

These trends, and the effectiveness of arts and humanities as a means for impacting learning, behaviour, and wellbeing, can be understood through the lens of neauroaesthetics. Neuroaesthetics is an emerging discipline focused on exploring the neural processes underlying our appreciation and production of objects, artwork and experiences including perception, interpretation, emotion, and action [21]. Neuroaesthetics frames aesthetic experiences as "emergent states that arise from the interaction between sensory–motor, emotion–valuation, and meaning–knowledge neural systems" [e.g., pp. 3]. From the neuroaesthetics perspective, aesthetic experience involves a blending of perceptual, emotional, and cognitive domains [22]. As holistic experiences, aesthetic experiences can provide an ideal basis for embodied insight, understanding and expression. Human beings have engaged in aesthetic expression and communication throughout time. Today, neuroaesthetics provides a lens through which we can define the more objective constructs involved, and learn to use the arts and aesthetic experience intentionally to craft enhanced engagement and learning outcomes.

Developments in neuroscience, along with the arts and art therapies, are progressively moving toward addressing human behavioral questions, such as mood response and pain management, as well as more intractable issues around traumatic brain injury, post-traumatic stress disorder, Alzheimers, and aging. Understanding the impact of aesthetic experiences on the brain has tremendous implications for improving the ways we live, heal and learn [23].

3 Social Impact

As we clamor today to translate the impact of the arts into biomedical terms and outcomes, we may lose sight of other important roles that the arts play in our lives and communities. The arts educate, they foster engagement and social change, and influence people's individual and collective behaviors. The arts illuminate and influence culture, and facilitate embodied consideration and understanding of abstract ideas that may be difficult to articulate in conventional language forms. Percy Shelly long ago referred to poets as "the unacknowledged legislators of our time," recognizing that art is a particularly engaging and persuasive language. Jane Hirschfield put that notion into similar terms more recently:

> Good art is a truing of vision, in the way a saw is trued in the saw shop, to cut more cleanly. It is also a changing of vision. Entering a good poem, a person feels, tastes, hears, thinks, and sees in altered ways. Why ask art into a life at all, if not to be transformed and enlarged by its presence and mysterious means? … And by changing selves, one by one, art changes also the outer world that selves create and share…. [24]

These notions suggest not only that we can learn *through* the arts, but that art can better enable learning by changing our ability to see, absorb and consider. Hirschfield goes on to note that artists "perceive the extraordinary within the ordinary by changing not the world but the eyes that look."

In October 2018, addressing ~200 fine arts deans in Seattle, WA, Jane Chu, former chairman of the National Endowment for the Arts, reminded the amassed cultural leaders, "We went into the arts because we loved the arts; not because we loved to measure them" (ICFAD National Conference, 2018). Likewise, while there has been ongoing academic and practical discussion surrounding the training of artists to be "artists" primarily (arts-for-arts sake) versus training toward broader practices that are use-inspired, we see the arts being increasingly applied in translational research and toward the "solving" of social problems. These applications build on the cultural assets of communities to solve some of our most intractable problems, including growing public health concerns such as trauma, chronic illness, mental health and addiction.

In this age of big data, rapidly-scaling technology, automation, digitization, and corporate global consolidations, the arts are critically important to the metaphysics of being human—which is perhaps where the true power of the arts lie. "Health is as much about caring as it is about curing." The systematic neglect of culture is the single greatest barrier to the advancement of the highest standards of health worldwide [25]. With *hope* being the leading predictor of wellbeing for children and adults, culturally-relevant artistic historical traditions are critical to our health and humanity.

4 Modernity, Technology, and Global Considerations: The Case for Realignment

The illiterate of the twenty-first century will not be those who cannot read and write, but those who cannot learn, unlearn, and relearn.

—Alvin Toffler, American Writer and Futurist

According to a 2016 report from the U.S. Department of Labor's Bureau of Labor Statistics, the median number of years that younger workers (ages 25–34 years) stayed in a single job was 4.2 years. These data suggest that graduates will be well-served by skills and competencies that are transferrable from one job to another, as well as by the ability to be adaptable, lifelong learners who can synthesize new knowledge they may need for each new job (National Academies of Sciences and Medicine [17]. A 2017 McKinsey Global Institute study estimates that by 2030, 70% of our jobs are yet to be identified, with up to one-third of the workforce needing to retool to retain or find new work [26]. The report also estimates that between 400 and 800 million jobs could be displaced by automation by 2030. As the very essence of work is re-envisioned, so must education retool.

In 2010, IBM issued a global study interviewing more than 1,500 Chief Executive Officers from 60 countries and 33 industries worldwide. Chief executives believed the number one attribute needed to successfully navigate an increasing complex world was creativity [27]. Frank Kern, then senior vice president, IBM Global Business Services stated, "… the biggest challenge facing enterprises from here on will be the accelerating complexity and the velocity of a world that is operating as a massively interconnected system." In survey-after-survey over the past several years, the top three skills companies want their employees to have are: complex problem solving, critical thinking, and creativity. And yet, in the areas of applying knowledge and skills in real-world settings, critical thinking, and written and oral communication, fewer than 30% of employers think that students are well prepared. More than 80% of employers feel that colleges and universities need to improve in helping graduates gain cross-cutting skills and knowledge (National Academies of Sciences and Medicine [17], p. 46). Employers are asking for a more hybrid workforce. Students are likewise demanding a different kind of education. Millennials and younger Gen Z students are digital natives with broad skills and interests—higher education must likewise retool to address a rapidly changing twenty-first century workforce.

As global urgencies accelerate and the need to make radical behavioral and societal changes become increasingly imminent, meaning-making and community cohesion remain primary domains of the arts. There are now an estimated 7.7 billion humans on planet earth. By the year 2050, that figure is estimated by the United Nations to rise to 10 billion. In 2002, Harvard University sociobiologist Edward O. Wilson estimated the earth could support 10 billion people if everyone became vegetarian and all arable land turned to food production. He underscored that "The constraints of the biosphere are fixed" [28]. At *current* consumption levels, the planet can support far less than 10 billion humans. Simultaneously, there has been an explosion of knowledge, with 130 million books recorded, including 2.5 million new science papers being published each year. From 1965 to 2009, 50 million papers were published. Information and the means to access that information is readily available, and yet human behavior and consumption habits do not easily change in response to scientific information—no amount of data seems to alter personal or societal behavior. Attitudes are dictated by habits of mind, cultural and community norms, and individual preferences.

Throughout human history, arts and humanities have played essential roles in human development, at both the individual and societal levels. Today, with increasing global and local pressures, consideration for development of the *human* can take a back-seat to the development of the *digital* in many settings. Additionally, massive global migration, cultural displacement and the erosion of cultures are results of civil war, rapid technological progress, and environmental climate change. These effects impose a high cost in regard to human and global health, and in turn may significantly impact our individual, collective, and planetary futures. To learn well, we must *be* well. Effective education is essential to the advancement of our societies. The arts have much to offer in bolstering conditions for optimal learning, collaboration and innovation. The arts also play an essential role in keeping hope, local culture, and imagination alive, as well as being instrumental in realigning technologies toward resiliency, ethical considerations, and the mutually-beneficial and sustainable interests of humanity. While the arts are being increasingly recognized and harnessed in medicine and public health, higher education and arts sectors must continue to reclaim and assert the native roles of the arts to create spaces for listening and more holistic communication, innovation and risk-taking, and for illuminating and driving culture in a time punctuated by massive and accelerating change.

References

1. U.S. Centers for Medicare & Medicaid Services: National Health Expenditure Data (2017)
2. State of the Field Committee: State of the field report: arts in healthcare 2009. Washington, DC Soc Arts Healthc (2009)
3. NOAH National Organization for Arts in Health: Arts, health and wellbeing in America (2017)
4. Cultural Ministers and Standing Council on Health: National Arts and Health Framework. Canberra (2013)
5. Bygren, L.O., Konlaan, B.B., Johansson, S.-E.: Attendance at cultural events, reading books or periodicals, and making music or singing in a choir as determinants for survival: Swedish interview survey of living conditions. BMJ **313**, 1577–1580 (1996)
6. Johansson, S.E., Konlaan, B.B., Bygren, L.O.: Sustaining habits of attending cultural events and maintenance of health: a longitudinal study. Health Promot. Int. **16**, 229–234 (2001)
7. Väänänen, A., Murray, M., Koskinen, A., et al.: Engagement in cultural activities and cause-specific mortality: prospective cohort study. Prev Med (Baltim) **49**, 142–147 (2009)
8. Cuypers, K.F., Knudtsen, M.S., Sandgren, M., et al.: Cultural activities and public health: research in Norway and Sweden: an overview. Arts Health **3**, 6–26 (2011)
9. Fancourt, D., Steptoe, A., Cadar, D.: Cultural engagement and cognitive reserve: museum attendance and dementia incidence over a 10-year period. Br. J. Psychiatry **213**, 661–663 (2018)
10. Fancourt, D., Steptoe, A.: Cultural engagement predicts changes in cognitive function in older adults over a 10 year period: findings from the English Longitudinal Study of Ageing. Sci. Rep. **8**, 10226 (2018)
11. Fancourt, D., Steptoe, A.: Physical and psychosocial factors in the prevention of chronic pain in older age. J. Pain **19**, 1385–1391 (2018)
12. Bickerdike, L., Booth, A., Wilson, P.M., et al.: Social prescribing: less rhetoric and more reality. A systematic review of the evidence. BMJ Open (2017)
13. Chatterjee, H.J., Camic, P.M., Lockyer, B., Thomson, L.J.M.: Non-clinical community interventions: a systematised review of social prescribing schemes. Arts Health **10**, 97–123 (2018)
14. Alderwick, H.A.J., Gottlieb, L.M., Fichtenberg, C.M., Adler, N.E.: Social prescribing in the US and England: emerging interventions to address patients' social needs. Am. J. Prev. Med. **54**, 715–718 (2018)
15. Redmond, M., Sumner, R.C., Crone, D.M., Hughes, S.: 'Light in dark places': exploring qualitative data from a longitudinal study using creative arts as a form of social prescribing. Arts Health 1–14 (2018)
16. Wachtler, C., Lundin, S., Troein, M.: Humanities for medical students? A qualitative study of a medical humanities curriculum in a medical school program. BMC Med. Educ. **6**, 16 (2006)
17. National Academies of Sciences, Engineering, and Medicine: The integration of the humanities and arts with sciences, engineering, and medicine in higher education: Branches from the same tree. National Academies Press (2018)
18. Berry, S.L., Lamb, E.G., Jones, T.: Health humanities baccalaureate programs in the United States. Cent Lit Med Hiram Coll Hiram, OH, USA (2016)
19. Peters, A.S., Greenberger-Rosovsky, R., Crowder, C., et al.: Long-term outcomes of the New Pathway Program at Harvard Medical School: a randomized controlled trial. Acad. Med. **75**, 470–479 (2000)
20. Flannigan, M.: Interviewee. Why Med Schools Are Requiring Art Classes
21. Chatterjee, A.: Neuroaesthetics—researchers unravel the biology of beauty and art. The Scientist **5**, 33–38 (2014)
22. Adajian, T.: The definition of art. Stanford Encycl. Philos. (2016)
23. Magsamen, S., Pitcock, S.: Impact thinking: a research approach to enhance human potential in health, wellbeing, and learning through the arts (2018)
24. Hirshfield, J.: Ten windows: how great poems transform the world. Knopf (2017)
25. Napier, A.D., Ancarno, C., Butler, B., et al.: Culture and health. Lancet **384**, 1607–1639 (2014)
26. Manyika, J., Lund, S., Chui, M., et al.: Jobs lost, jobs gained: Workforce transitions in a time of automation. McKinsey Glob Inst (2017)
27. IBM: The IBM 2010 Global CEO Study—capitalizing on complexity assessment. New York (2010)
28. Wilson, E.O.: The future of life. Vintage (2002)

Laurie Baefsky is Associate Dean for Research, Collaboration, and Innovation in the College of Arts & Media at University of Colorado Denver. Prior to moving to Denver, from 2014–2018, she served as executive director of ArtsEngine and the Alliance for the Arts in Research Universities (a2ru), based at the University of Michigan.

Jill Sonke is director of the University of Florida Center for the Arts in Medicine and Assistant Director of UF Health Shands Arts in Medicine. She serves on the faculty of the Center for Arts in Medicine, and is an affiliated faculty member in the School of Theatre & Dance, Center for African Studies, and Center for Movement Disorders and Neurorestoration.

Part VII
The Future of Neuroaesthetics

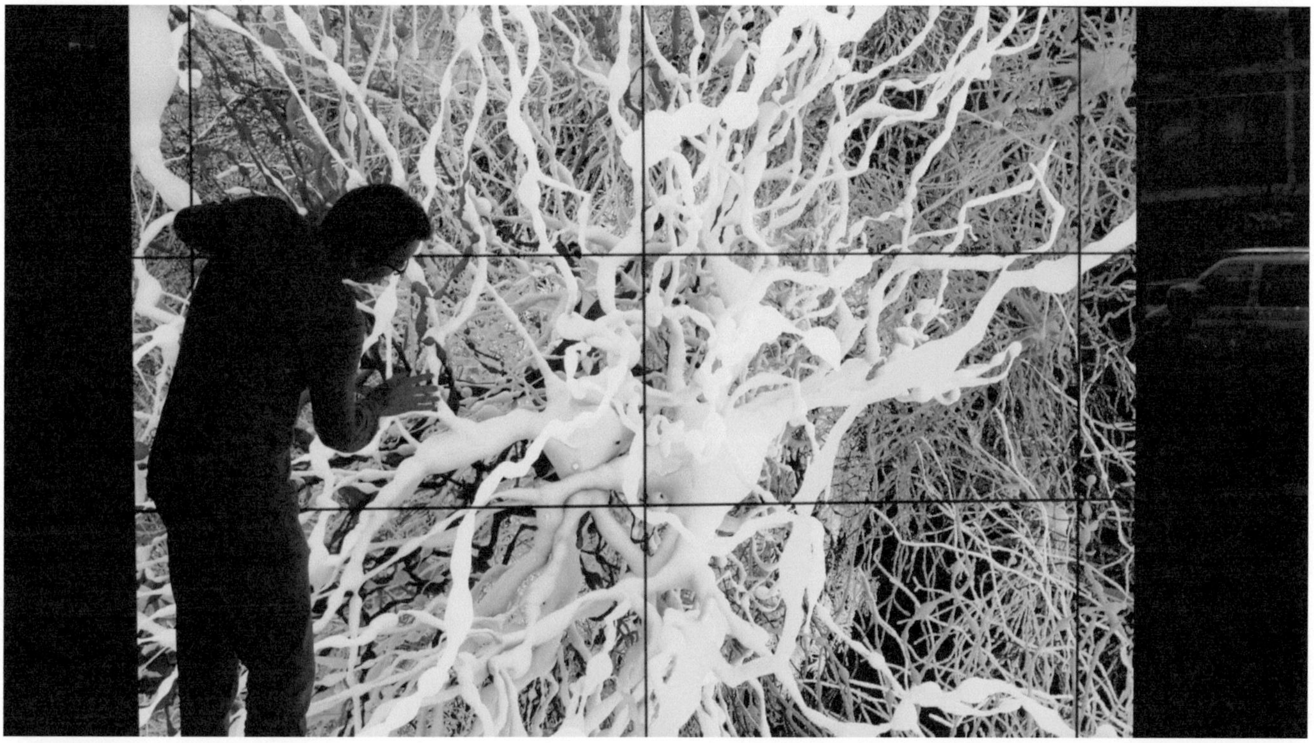

Noah Hutton
Touching the Brain
Digital Video Still. 2880 × 1620 pixels
December 2017

A scientist from the Blue Brain Project manipulates a digital image of a brain on a large array of displays in the project headquarters in Geneva, Switzerland

Towards a Roadmap for Neuroaesthetics

Jose L. Contreras-Vidal, Dario Robleto, and Jesus G. Cruz-Garza

1 On the Value of Art/Science Collaborations

Neuroaesthetics and the study of the human creative process are still emergent fields that have much work to do to establish its contributions to creativity and aesthetics research. While fascinating research has begun, and hopefully will continue to do so, we should not lose sight of another type of unique, unstated experiment occurring right in front of us: What does true collaboration look like between disciplines and people from disparate fields? Of the examples we can look at on this vital point, the challenges between the arts and sciences are significant (see Chapter "Brain Mechanisms of Creativity: What We Know, What We Don't" by Dietrich). Our respective fields are working against the perception that we do not have much in common in how we formulate questions and methodological approaches, how we determine "results" or what the broader impacts of such transdisciplinary research would entail. These assumed misunderstandings between art and neuroaesthetics research are perhaps most succinctly summed up in the criticism that neuroscience widely oversteps when it suggests that it will "explain" art through physiological processes alone (see Chapter "Unknown and Solitary Seas: Angelo Mosso's Nineteenth-Century Discovery of Imaging Dreams Through the Cerebral Pulse" by Robleto). For many critics, this solely physical approach overlooks the art itself and its role for the viewer and artist in the personal construction of meaning and emotional expression. Further, it removes the aesthetic experience from the environmental, social, and cultural context that is a crucial aspect in the contemplation and appreciation of art. In other words, the arts and humanities would argue that neuroscience is asking the entirely wrong questions in the pursuit of understanding the mysteries of aesthetic experience and the divide in approaches is so significant as to make further discussion pointless. While we should undoubtedly remain cautious from this valid criticism, it is too easy to assume we cannot overcome the difficulty of this problem and others. Instead, there is a principle and spirit to collaborate in the face of such suspicion *because* it is difficult. For both artist and scientist have an obligation to the pursuit of knowledge, following it no matter where it leads, while listening to other perspectives and be willing to adapt when challenged (see Chapter "Art-Science Collaborations: How to Break Boundaries Without Breaking Trust" by Biggs et al.).

In the growing field of neuroaesthetics, there is a need for expansion in the discussion of building relationships between scientists, artists, academia, and arts institutions. This has mostly been avoidable because much neuroaesthetic research has focused on the art of past centuries or more general physiological and evolutionary insight into perception and conceptions of beauty, not experimentally based studies of contemporary artists or museum patrons. But the promise of mobile brain–body imaging technology to creativity and aesthetics research is in its capabilities of investigating freely behaving artists and art audiences, including children, in the context of art settings (museums, studios, galleries, performance venues, etc.). This means that not only will access and building trust between these communities be essential (what are, for example, the logistical, privacy, and legal hurdles to performing a scientific experiment in a public museum, or recruiting artists as test subjects?), but new questions that challenge current methodologies and definitions will arise. This presents an exciting possibility, as it is extremely rare when a scientific field's advancement largely depends on developing long-term and meaningful relationships and collaborations with artists and arts institutions. As artists and scientists, we have to remain open to the idea that the expertise we have spent years developing and fine-tuning within our respective

J. L. Contreras-Vidal (✉) · J. G. Cruz-Garza
IUCRC BRAIN, University of Houston, Houston, TX, USA
e-mail: jlcontr2@central.uh.edu

D. Robleto
Artist in Residence in Neuroaesthetics, IUCRC BRAIN,
University of Houston, Houston, TX, USA

J. L. Contreras-Vidal et al. (eds.), *Mobile Brain–Body Imaging and the Neuroscience of Art, Innovation and Creativity*,
Springer Series on Bio- and Neurosystems 10, https://doi.org/10.1007/978-3-030-24326-5_25

traditions and methods may be useful and even revelatory outside our fields.

2 The Need for Convergent Research on Neuroaesthetics

It has been recognized by the scientific community that the solution of complex societal problems requires a deep integration of knowledge, methods, tools, approaches, data, and expertise from different disciplines to converge on innovative solutions to those problems (see, for example, the report on Convergent Research published by the National Academies [6]). This book highlights some examples of early convergence in research. For example, Cruz-Garza, Kopteva, Fleischhauer, and Contreras-Vidal (Chapter "Into the Mind of an Artist: Convergent Research at the Nexus of Art, Science, and Technology") describe the challenges, innovations, and potential impact of conducting longitudinal research at the nexus of art, science and technology during art making in natural settings. Robleto (Chapter "Unknown and Solitary Seas: Angelo Mosso's Nineteenth-Century Discovery of Imaging Dreams Through the Cerebral Pulse ") provides a beautiful example of convergent research integrating science history, humanities, and technology to envision and reimagine the critical role of scientists in researching some of the most complex capabilities of the human mind. Biggs, Eriksen, and Žiburkus (Chapter "Art-Science Collaborations: How to Break Boundaries Without Breaking Trust") tell us about the importance of maintaining trust when working across disciplines.

The National Academies have recognized the critical role of the arts in fostering innovation in science, engineering, and medicine and through the National Academies Keck Future Initiative (NAKFI) organized the Conference on Art and Science, Engineering, and Medicine Frontier Collaborations: Ideation, Translation & Realization (Arnold and Mabel Beckman Center in Irvine, California on November 12–14). The purpose of the meeting was to gather a variety of experts from the arts, design, science, engineering, medicine, physics, biology, economics, and behavioral science to address the challenge of communicating and working together from a diversity of expertise and perspectives to solve complicated interdisciplinary problems. At the core of this meeting was "the idea of toppling barriers and constructing nontraditional solutions (p. 3, [5])." One challenge identified by a working group at the meeting was that "art is used as a tool to improve science education but is not valued on its own merit. Instead, the group envisioned an educational (and research) system that valued collaboration and integration across all disciplines [5]."

Unfortunately, trans-disciplinary meetings that engage multiple stakeholders from the arts, K-12 schools, academia, medical institutions, industry, federal agencies and policy makers, and students are still rare. The series of *International Conferences of Mobile Brain–Body Imaging (MoBI) and the Neuroscience of Art, Innovation and Creativity* (or also known as the "Brain on Art" conferences) was created to fill this gap. The Brain on Art conferences, held in 2016 [10], 2017 [11], 2018 [12], and 2019 [13], bring together global thought leaders and innovators from around the world to share strategies and best practices, discuss the state of the art, challenges and opportunities for convergent research through invited lightening talks, roundtable discussions, special sessions, collaborative discussions, MoBI demonstrations, brain–computer interface hackathons, and a doctoral consortium for selected trainees. To promote discussion and collaborations, an "un-conference" format, where audience-driven discussions whose content is provided by the participants themselves is emphasized. The doctoral consortium provides an opportunity for graduate students and postdoctoral fellows to explore and develop their research interests in a trans-disciplinary conference, under the guidance of a distinguished group of international researchers and innovators.

The success of the Brain on Art conferences can be assessed in part by several success histories of art–science integration and convergent research, including this book and follow-up events triggered by the conference series. For example, the D.C. Art Science Evening Rendezvous (DASER), co-sponsored by Cultural Programs of the National Academy of Sciences and Leonardo, the International Society for the Arts, Sciences, and Technology, dedicated an event on August 31, 2018 to explore the topic of the art and the brain [3]. A team of conference participants (Contreras-Vidal, King, Robleto, Ruhle, Valls, and Witts) received an honorific mention by the Keck Futures Award Competition in 2018 for their trans-disciplinary and multi-institutional proposal entitled "Aesthetic Neurotherapeutics: Towards a Safe, Effective and Noninvasive Arts Prescription (ArtRx) Program to Treat Physical, Neurological and Mental Disability" [7].

3 Going Forward

Human neuroscience studies of creativity have shed some light (and generated extensive debate!) over the past decade. Improved neuroimaging methods and mobile technology have made it possible to explore the neural correlates of creative behavior across artistic domains, opening new possibilities for artists, scientists, engineers, physicians, and educators to address profound societal questions. In the next sections, several opportunities for convergent research are proposed that could constitute initial steps for a blueprint or roadmap for the field.

3.1 Individuality and Variance in Human Behavior Must Be Taken into Account

Most studies about the human creative process are based on psychometric tests in lab settings, making the inquiry on creativity highly artificial and therefore limited in scope. Moreover, these studies do not address the individuality and variance in human behavior. Going forward, the field must take into account the multi-faceted nature of creativity, which may manifest differently across individuals and artistic modalities. Additionally, most studies have been conducted in time and movement-restricted laboratory settings, far removed from an authentic, active, and dynamic experience. Recent efforts to study the human creative experience approach the concept in question—creativity—from a nuanced and task-specific perspective. Creativity is a complex multidimensional process, created by the interaction of distinct neural, cognitive, affective, and sensorimotor processes operating under environmental, cultural, and societal constraints. Thus, future experiments should study the human brain in action and in context to advance our understanding of neural individuality and variance in creative and aesthetic tasks. Indeed, the National Science Foundation and the National Institute of Health have both identified understanding the brain in action as a scientific priority with societal implications [8, 9]. These studies should also investigate the neural basis of social, environmental, cognitive, affective and reward aspects of collaborative art and teamwork.

3.2 Artificial Intelligence (AI) Approaches May Be Harnessed to Understand Individuality in Creativity and Aesthetic Experiences in Health and Disease

Innovative, multi-institutional collaborations, involving not only the creative arts, science, engineering, and the humanities but also K-12 schools, colleges, museums, and performing venues, will be critical for acquiring large volumes of diverse, multi-modality, longitudinal, context-aware, brain–body imaging data from freely behaving participants, including children, with rich demographics and contextual information. Museums and schools offer ideal real-world settings to record multimodal data from hundreds of participants with rich demographics while accounting for individual artistic histories, ethnicity, and gender.

These Big Data volumes can then be mined using advanced AI tools such as deep neural networks [4] and other machine learning techniques. The data volumes could be parsed to understand the effects of age, gender, health status, cultural background, education, and other factors on neural activity, movement and behavior in general. Assaying the brain response to the arts, in action and in context, in large numbers of people could lead to better understanding of how brain processes inform social, behavioral, cognitive, and emotional issues while characterizing the MoBI data at the individual and population levels.

Moreover, better understanding of aesthetic stimuli (e.g., aesthetic visual, music, or dance) and how they modulate brain activity and concomitant behavior may lead to precision medical treatment of physical, neurological, and mental disorders using aesthetic drivers used during creative art therapy. This knowledge could have implications for student training and assessment as well. In this scenario, MoBI data could provide a window to study the acquisition of advanced artistic motor skills required to achieve a proficient level of artistic production. Normative brain–body imaging trajectories could be used to detect "bottlenecks" in motor skill learning that could have diagnostic value and could lead to changes in how the art education (or therapy) is delivered.

3.3 Bridging Communication Between Artists and Scientists

It is essential to have events, opportunities, and spaces that promote trans-disciplinary collaborations between scientists, engineers, and technologists on one hand, and artists, humanists, designers, educators, and clinicians on the other hand. If neuroscience studies are to characterize the neural dynamics associated with the human creative process, in its various forms, then the contribution from the artists themselves about what constitutes a creative experience in a contextually valid experimental setting and what aspects of that experience (e.g., for data tagging and interpretation) are recognized as critical steps in the execution of the art pieces, are essential for sound experimental design and analysis. These collaborations can be long-term (e.g., Artist-in-residence programs) or short term (e.g., artistic brain–computer interfaces, see Part V of this book, art-science conferences and workshops; and trans-disciplinary undergraduate and graduate programs), which are described next.

Artist-in-residence (AIR) programs formalize and allow for the collaboration between artists and scientists to understand the human creative process in the artists' particular field of expertise. These AIR programs may provide to the artist laboratory space, institutional affiliations, access to university resources, seed funding, access to technical expertise, and equipment to support art installations, outreach activities, and others). Scientists in return are benefitted by input from highly skilled performers, high-level descriptions of planning, exploratory, and performance of art, which provide valuable data for training computers to recognize patterns of brain activity and volitional movement that may help elucidate the creative process.

Indeed, the adoption of art-making activities as experimental tasks that resemble authentic creative experiences are critical for understanding the brain in action and in context—a scientific priority identified by the National Science Foundation [9] and the National Institute of Health [8]. This is imperative to allow for the investigation of cognitive states (or processes) associated with creative tasks. Equal consideration for the artist's understanding of what constitutes a creative experience (see Chapter "Theme and Variations as a Window into the Creative Mind" by Brandt, and Chapter "Into the Mind of an Artist: Convergent Research at the Nexus of Art, Science, and Technology" by Cruz-Garza et al.), and ensuring that the creative experience during recording sessions is important for success.

With recent advances in mobile brain–body imaging technology, experiments can now be conducted in real-time from a diverse group of people, at the artist studio, museums, or venues deemed contextually relevant. However, the elusive question of authenticity remains. Even away from a laboratory setting, the context involves measurement and recording equipment, time constraints, and other artificial set of variables that will disrupt the authentic experience. Until neurotechnology becomes pervasive in daily life, like our smartphones and earphones, and we allow enough time to become familiarized with the recording devices, they are likely to exert an influence (albeit likely decreasing over time as we learn to wear and use them) on our creative efforts. These influences may also allow the opportunity for the artist to augment how she/he interacts with the audience or the environment (e.g., an artistic BCI device that allows affective and cognitive states of the artist to operate or control synthesizers, lights, and mechanical devices).

Trans-disciplinary graduate programs are an important mechanism for faculty and students to cross-fertilize and innovate across fields. These programs can bridge language from science, technology, engineering, arts, math, and medicine (STEAMM). From approaches, methods, and tools currently used in their particular fields, researchers are likely to benefit from complementary, often innovative solutions from other areas of knowledge (Chapter "Intersectionality: The Confluence of Arts, Technology, and Wellbeing" by Baefsky and Sonke). However, to be successful, trans-disciplinary graduate programs must be designed to be integrated, flexible, synergistic, and transformational rather than just the independent attainment of separate degrees. Clearly, differences in program length, financial cost, program requirements, facilities, credit hours, and other factors must be managed to support and enhance the student experience and training.

Transdisciplinary conferences are also an important dissemination mechanism that enables the exchange of ideas and diverse findings while promoting hands-on collaborations and experiences. The first four conferences in the series "Your Brain on Art, Innovation, and Creativity" were small (under 100 invitees) by design, in order to promote interaction between participants and lengthy, fruitful discussion (see [10–13]). With a larger community every year, and the challenges that such a multidisciplinary research field entails, it is important to find ways to scale-up these meetings in the US and abroad. At the national level, the Alliance for the Arts in Research Universities (a2ru) is a concerted effort of research universities in the United States to promote the role of the arts and design in research universities. The a2ru mission is to "advance the arts and design in research, teaching, scholarship, and creative practice [1]".

3.4 Developing Metrics for Neuroaesthetics Collaborations

While neuroaesthetic researchers and artists hopefully rise to the challenge of better communication and collaboration, there is the assumption, at least by the scientist, that the objectives of the partnership in a general sense are clear—using scientific standards to produce more knowledge with possible societal implications. On the other hand, artists may have an entirely different approach that focuses more on the process of discovery itself, remaining poetic and open-ended in interpretation. Doing this type of work across fields will likely produce provocative outcomes we are unprepared to properly value. If so, how do we create the metrics necessary to assess and appreciate these outcomes?

There is a question an artist should ask when working with a scientist, and which the scientist should ask back: Creatively speaking, what is the most uncomfortable position we can place ourselves in? How do we get somewhere truly unexpected and challenging, while not repeating the self-congratulatory back patting that art–science collaborations can sometimes entail? For example, instead of the expectation that the artist will visualize data sets after the actual science was done, would a scientist let an artist contribute to the experimental design itself, even analyzing, interpreting, or formulating high-level models of the data? Similarly, would an artist let a scientist into their studio and make major aesthetic or conceptual decisions in the production of a new piece? Further, are there suitable metrics that can be applied to both art and science? This is no trivial matter, as collaborations between artists and scientists will require a common language to move forward. By asking the artist to seriously analyze their process through the measurable and quantifiable standards of neuroscience, not only will they gain a nuanced perspective on the physical actions of their creativity, but also help to invent classification categories that will enable the science to progress.

From an artistic point of view, one must reflect on what these collaborations mean for the production of new art. In this regard, we have been challenged to ask some uncomfortable questions: Is this science or is this art? If it were neither, then how would we judge its value to either field? For a scientist to admit this may not be science, and an artist to accept it may not be art is an unfamiliar position to be in, but one that will offer new possibilities of what constitutes the "work" or "results." As we ponder if art and science, each with their differing needs and standards of "success," can be reconciled in a single object or scientific protocol, we must equally honor the science, art and a possible third outcome that is less defined but no less part of the work we do together. The burgeoning field of artistic BCIs presents an exciting path forward for the merger of artistic and scientific investigations.

3.5 Artistic Brain–Computer Interface (BCI) Hackathons

The "Brain on Art" conferences have featured a 3-day hackathon for the creation of artistic BCIs using off-the-shelf MoBI technology [2]. Undergraduate and graduate students from engineering, neuroscience, humanities, and the arts form teams on-site to work on an original or conference-suggested project ideas (for an overview, see Chapter "The Art, Science, and Engineering of BCI Hackathons" by Ortiz et al.) under the artistic and technical mentorship of experts in the field while gaining hands-on technical experience and mentoring from participating faculty.

4 Conclusion

It is a rare opportunity when a scientific field's advancement largely depends on developing long-term and meaningful relationships and collaborations with artists and arts institutions. One of the challenges moving forward will be how the neuroaesthetics field stays rigorous to its scientific roots while remaining open to new ways of thinking and working within the arts, design, and the humanities. Similarly, the arts, design, and the humanities will need to remain open to the possibility that a neuroscientific understanding of the creative process can enrich their practice.

What does true collaboration look like between disciplines and people from disparate fields when putting together a team of STEAMM researchers? Is there a consensus definition of creativity and aesthetics that applies to both the arts and neuroscience? What are the physical (neurological) underpinnings of creativity and aesthetics, and can they be recorded and quantified in action and in context? Can the physical, neurological, and anatomical understanding of creativity and the brain say anything revelatory about the lived, experiential relationship of creators, and viewers to art? Can artistic experiences advance our understanding of individuality in neural activity and how emotions shape our aesthetic beliefs? These and other questions present an opportunity for truly trans-disciplinary work across science, technology, engineering, arts, math, humanities, and medicine.

References

1. a2ru: The Alliance for the Arts in Research Universities (a2ru). https://www.a2ru.org (2019). Accessed 22 Mar 2019
2. BR4IN.OI: The Brain-Computer Interface Designers Hackathon. https://www.br41n.io/ (2019). Accessed 22 Mar 2019
3. CPNAS and Leonardo: Cultural Programs of the National Academy of Sciences (CPNAS) and Leonardo, the International Society for the Arts, Sciences, and Technology. DC Art Science Evening Rendezvous (DASER). http://www.cpnas.org/events/daser-092018.html (2018)
4. Craik, A., He, Y., Contreras-Vidal, J.L.: Deep learning for Electroencephalogram (EEG) classification tasks: a review. J. Neural Eng. **16**(3), 031001. https://doi.org/10.1088/1741-2552/ab0ab5 (2019)
5. National Academies Keck Futures Initiative: Art, Design and Science, Engineering and Medicine Frontier Collaborations: Ideation, Translation, Realization: Seed Idea Group Summaries. National Academies Press (2016)
6. National Academies of Sciences, Engineering, and Medicine: A New Vision for Center-Based Engineering Research. National Academies Press (2017)
7. National Academies of Sciences, Engineering, and Medicine: NAFKI Challenge. https://www.nakfichallenge.org/#winners (2017). Accessed 22 Mar 2019
8. National Institutes of Health: The Brain Initiative. https://www.braininitiative.nih.gov/ (2013). Accessed 25 Mar 2019
9. NSF IUCRC: National Science Foundation; BRAIN: Brain Research Through Advancing Innovative Neurotechnologies. https://www.nsf.gov/news/special_reports/brain/initiative/ (2017). Accessed 25 Mar 2019
10. The 2016 International Conference on Mobile Brain-Body Imaging (MoBI) and the Neuroscience of Art, Innovation and Creativity, Cancun, Mexico, July 24–27. http://yourbrainonart.egr.uh.edu/ (2016). Accessed 16 Mar 2019
11. The 2017 International Conference on Mobile Brain-Body Imaging (MoBI) and the Neuroscience of Art, Innovation and Creativity, Valencia, Spain, September 10–13. http://yourbrainonart2017.egr.uh.edu/ (2017). Accessed 16 Mar 2019
12. The 2018 International Conference on Mobile Brain-Body Imaging (MoBI), Berlin, Germany, July 11–14, 2018; mobi2018.bpn.tu-berlin.de/. (2018)
13. The 2019 International Graphonomics Conference: Graphonomics and your Brain on Art, Creativity and Innovation, Cancun, Mexico, June 9–12. http://igs2019-boa.egr.uh.edu/ (2019). Accessed 3 Jul 2019

Jose Luis Contreras-Vidal Ph.D., IEEE Fellow, is Cullen Distinguished Professor of Electrical & Computer Engineering, and Director of the NSF Industry-University Collaborative Research Center on Building Reliable Advances and Innovations in Neurotechnologies (IUCRC BRAIN) at the University of Houston. He has pioneered the development of noninvasive brain–machine interfaces to control robots, exoskeletons, virtual avatars, and prosthetic limbs to restore motor function in individuals with spinal cord injury, stroke, and limb amputation. He has also pioneered the concept of "The Museum and the Classroom as Laboratories" to understand the brain in action and in context of free-behaving individuals, with applications to medical devices, art therapy, aesthetics, neuroscience, and regulatory science. His work has been supported by the NIH, NSF, VA, DARPA, ONR, Foundations, donors, and industry. Dr. Contreras-Vidal's research has been highlighted by The Economist, Nature, Science, Science News, Der Spiegel, NSF, Wall Street Journal, SFN, O&P, Scientific American, NPR's Science Friday, and Neurology Today among others. His career development in biomedical engineering has been highlighted in the magazine Science.

Webpage: http://www.ece.uh.edu/faculty/contreras-vidal.

Webpage: https://www.houstonmethodist.org/faculty/joseluis-contreras-vidal/.

IUCRC BRAIN: http://brain.egr.uh.edu/.

Dario Robleto is a transdisciplinary artist based in Houston, Texas. He currently serves as Artist-in-Residence in Neuroaesthetics at the University of Houston's Cullen College of Engineering and as Artist-at-Large at Northwestern University's McCormick School of Engineering. In 2016, he was appointed as the Texas State Artist Laureate.

Jesus G. Cruz-Garza is a Ph.D. candidate in Electrical Engineering at the University of Houston and Graduate Fellow on Translational Research with Houston Methodist Research Institute and University of Houston. He graduated from BS Engineering Physics at Tecnologico de Monterrey, Mexico.

Appendix

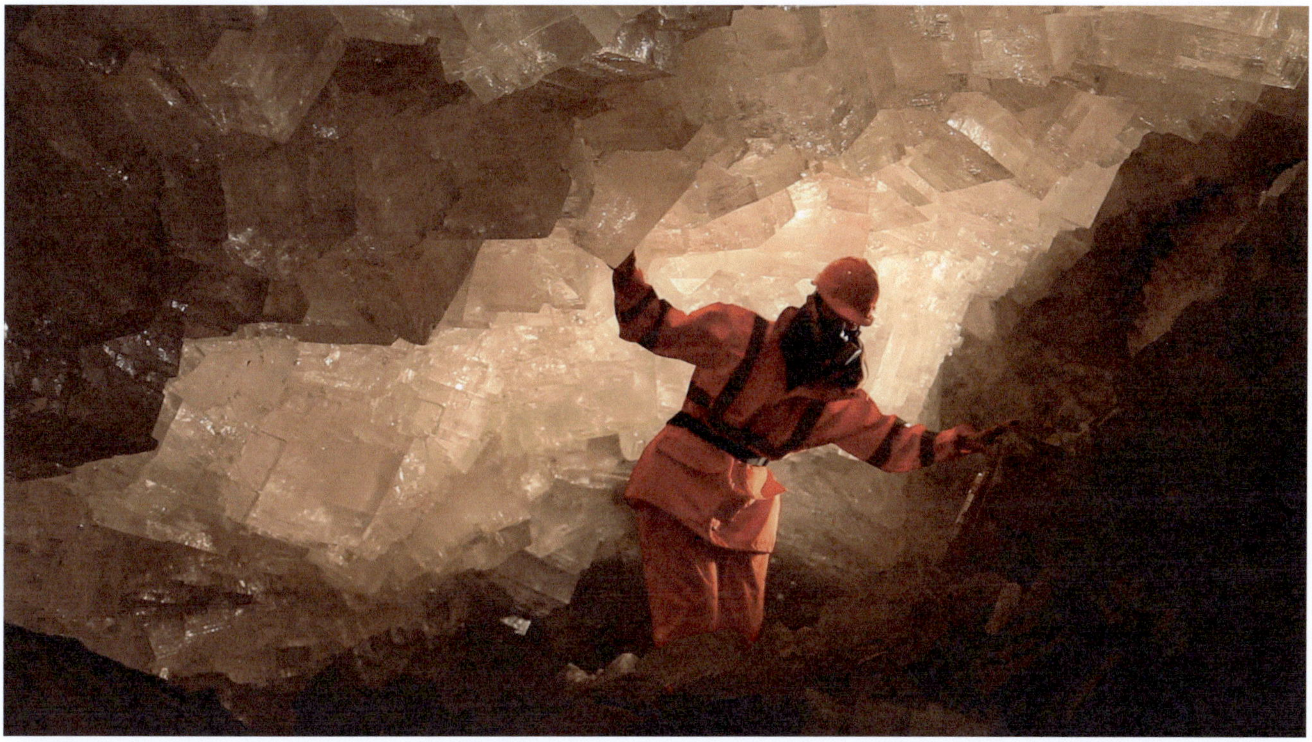

Janet Biggs

Can't Find My Way Home (detail).
Four-channel HD video installation with sound, size variable 2015.
Courtesy of the artist, Cristin Tierney Gallery (New York, NY), Analix Forever (Geneva, Switzerland) and CONNERSMITH (Washington, DC).

"Can't Find my Way Home" juxtaposes footage shot in the crystal caverns below the German Merkers salt mine with documentation of neurological research conducted in laboratories in New York and Houston. In doing so, Biggs draws visual connections between the structure of these crystals and the proteins that determine the biochemical conditions of a hyper-excited brain, such as one afflicted with Alzheimer's. By physically exploring the Merkers crystal cavern, Biggs figuratively sets out to investigate the diseased brain of her grandfather, tracing fading memories and making astonishing discoveries as she herself experiences disorientation and confusion, some of the same symptoms endured by Alzheimer's patients.

J. L. Contreras-Vidal et al. (eds.), *Mobile Brain-Body Imaging and the Neuroscience of Art, Innovation and Creativity*, Springer Series on Bio- and Neurosystems 10, https://doi.org/10.1007/978-3-030-24326-5

MIX
Papier aus verantwortungsvollen Quellen
Paper from responsible sources
FSC® C105338

If you have any concerns about our products,
you can contact us on
ProductSafety@springernature.com

In case Publisher is established outside the EU,
the EU authorized representative is:
Springer Nature Customer Service Center GmbH
Europaplatz 3, 69115 Heidelberg, Germany

Printed by Libri Plureos GmbH
in Hamburg, Germany